Archimedes

New Studies in the History and Philosophy of Science and Technology

Volume 67

Archimedes has three fundamental goals: to further the integration of the histories of science and technology with one another; to investigate the technical, social and practical histories of specific developments in science and technology; and finally, where possible and desirable, to bring the histories of science and technology into closer contact with the philosophy of science.

The series is interested in receiving book proposals that treat the history of any of the sciences, ranging from biology through physics, all aspects of the history of technology, broadly construed, as well as historically-engaged philosophy of science or technology. Taken as a whole, Archimedes will be of interest to historians, philosophers, and scientists, as well as to those in business and industry who seek to understand how science and industry have come to be so strongly linked.

Submission / Instructions for Authors and Editors: The series editors aim to make a first decision within one month of submission. In case of a positive first decision the work will be provisionally contracted: the final decision about publication will depend upon the result of the anonymous peer-review of the complete manuscript. The series editors aim to have the work peer-reviewed within 3 months after submission of the complete manuscript.

The series editors discourage the submission of manuscripts that contain reprints of previously published material and of manuscripts that are below 150 printed pages (75,000 words). For inquiries and submission of proposals prospective authors can contact one of the editors:

Please find on the top right side of our webpage a link to our *Book Proposal Form.*

W. Bernard Carlson • Erik M. Conway
Editors

Electrical Conquest

New Approaches to the History of Electrification

Springer

Editors
W. Bernard Carlson
TechInnovation
University of Galway
Galway, Ireland

Erik M. Conway
California Institute of Technology
Pasdena, CA, USA

ISSN 1385-0180 ISSN 2215-0064 (electronic)
Archimedes
ISBN 978-3-031-44590-3 ISBN 978-3-031-44591-0 (eBook)
https://doi.org/10.1007/978-3-031-44591-0

This work was supported by Maastricht University, Faculty of Arts & Social Sciences

This Springer imprint is published by the registered company Springer Nature Switzerland AG
The registered company address is: Gewerbestrasse 11, 6330 Cham, Switzerland

Paper in this product is recyclable.

Acknowledgements

This book is the product of the Research Institute for the History of Science and Technology, a joint enterprise between the California Institute of Technology and the Huntington Library in San Marino. To develop this volume, our original plan was to bring together the contributors for a two-week meeting at the Huntington in June 2020 so that the contributors could utilize the Huntington's significant research collections related to electrical history. In March 2020, as California began its first "Safer at Home" order at the onset of the COVID-19 pandemic, we postponed the workshop hoping to hold it later in the year. As it gradually became clear that the Huntington would not be re-opening for research during 2020, we reconfigured the workshop as virtual-only, using web-based videoconferencing for twice-weekly sessions during September 2020. We wish to thank our contributors for their patience with the changing plans, and for sticking with us through this difficult and chaotic year. We regret never having met in person, though, and we regret losing our opportunities for research in the Huntington's collections as well.

We wish to thank Jed Z. Buchwald, director of the Research Institute, and Dan Lewis of the Huntington, the Institute's associate director, for the opportunity to pursue this project. We are grateful to Caltech's Fran Tise, and Steve Hindle, Catherine Wey-Miller, Juan Gomez, and Natalie Serrano of the Huntington for their assistance in planning the in-person workshop prior to the pandemic. Amy Fisher actively participated in the virtual workshop, but obligations at her university related to the pandemic prevented Amy from contributing a paper to this volume. In Amy's place, Will Hausman and colleagues shared with us their paper on electrification in Cuba. Ruth Sandwell also participated in our virtual workshop but, due to the lengthy delay in preparing this volume, chose to publish her essay elsewhere. Thanks also to Jonathan Coopersmith and an anonymous referee who provided valuable feedback on this volume. Melissa Ferrell at the University of Virginia provided assistance in the preparation of the manuscript for this volume.

Contents

Contributors

W. Bernard Carlson manages the M.Sc programs in TechInnovation and AgInnovation at the University of Galway in Ireland. He is also the Joseph L. Vaughan Emeritus Professor of Humanities at the University of Virginia. He is also a lecturer in the Tech Innovate program at the National University of Ireland Galway. He has written widely on inventors and electrical history, and his books include *Innovation as a Social Process: Elihu Thomson and the Rise of General Electric, 1870–1900* (Cambridge University Press, 1991) and *Tesla: Inventor of the Electrical Age* (Princeton University Press, 2013).

Erik M. Conway is an independent scholar and an author on seven books, including *Merchants of Doubt* and *The Big Myth*, both with Naomi Oreskes, and *Exploration and Engineering: The Jet Propulsion Laboratory and the Quest for Mars.*

Chapter 1
Introduction

W. Bernard Carlson and Erik M. Conway

Abstract Over the last 150 years, electrification—the process of inventing, building, and operating systems for generating and distributing electric power—has profoundly changed the human and natural worlds. Most of us in the developed world take for granted the electricity that turns the wheels of industry, lights our homes, and powers our communication and information networks. But how have historians made sense of the evolution of electrical systems? We suggest that historians have been investigating how people have developed power systems by focusing on four themes: the social and cultural impact of electricity, the political economy of electrification, electricity and finance, and the envirotechnical view. We further argue that the future studies of electrification should use two conceptual tools, energy transitions and technological imaginaries. In building and enlarging their networks, electrical entrepreneurs often have to interact with more customers, maneuver in more political jurisdictions, and raise more capital; to capture this complexity and conflict, we advocate that scholars view electrification not as a linear process but rather as a series of transitions. Moreover, to mobilize the resources needed to build new and larger electrical networks, historical actors have to articulate a vision—what Sheila Jasonoff calls an imaginary—that captures the imagination of financiers, politicians, and the public. In concluding, we use these two concepts, transitions and imaginaries, to summarize the volume's chapters and suggest how the history of electrification provides a robust and vibrant perspective that informs both policy and our general understanding of how electricity has—and will continue to—transform our lives.

W. B. Carlson (✉)
University of Galway, Galway, Ireland
e-mail: wc4p@virginia.edu

E. M. Conway
California Institute of Technology, Pasadena, CA, USA
e-mail: erik@erikmconway.com

W. B. Carlson, E. M. Conway (eds.), *Electrical Conquest*, Archimedes 67,
https://doi.org/10.1007/978-3-031-44591-0_1

Keywords History of electrification · Social impact of electricity · Political economy of electrification · Electricity and finance · Envirotechnical approach · Energy transitions · Technological imaginary

In critiquing traditional history in the 1920s, the French poet Paul Valèry challenged historians to investigate the "conquest of the earth by electricity" since electricity has had "more meaning and greater possibilities of shaping our immediate future than all political events combined."[1] This conquest is clearly illustrated by a NASA satellite composite image of the earth at night showing millions of twinkling lights scattered across the continents (Fig. 1.1). Behind all of these lights is the process of electrification, of inventing, building, and operating systems for generating and distributing electric power. Even though we often take electricity for granted, how is it that some societies have been able to electrify? What is the process by which people have used electricity to conquer so much of the earth in just 150 years?

Over the past 40 years, historians have responded to Valèry's challenge by taking a variety of approaches. In this introduction, we will summarize these multiple approaches in order to set the stage for how the papers in this volume enlarge and enrich our understanding of the enormous electrical networks that invisibly deliver power to tens of millions of people worldwide.

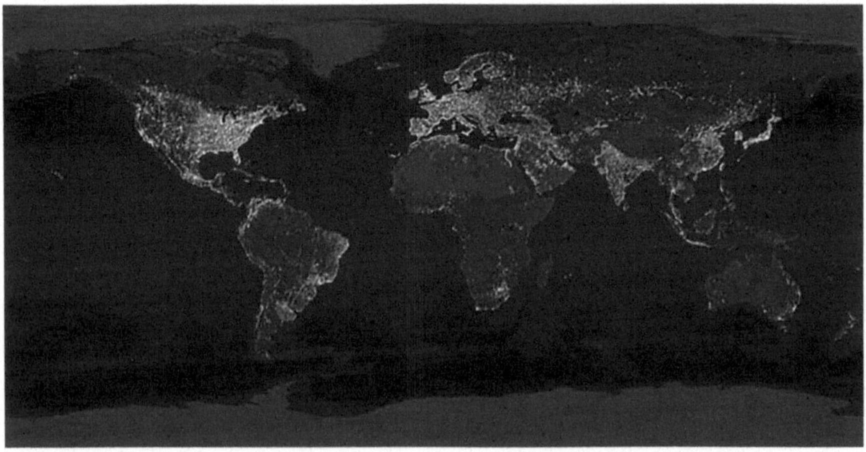

Fig. 1.1 "Earth at Night," C. Mayhew & R. Simmon (NASA/GSFC), NOAA/NGDC, DMSP Digital Archive. (Source: https://apod.nasa.gov/apod/ap001127.html, 23 Feb. 21)

[1]The quotes from Valèry come from a quote in Marc Bloch's *The Historian's Craft* and served as the opening epigram in David Nye's *Electrifying America: Social Meanings of a New Technology* (Cambridge: MIT Press, 1990), ix.

1.1 The Systems Approach of Thomas P. Hughes

Before the 1980s, if one wanted to learn about the history of electrification, one had to turn to a limited literature that focused on heroic inventors such as Thomas Edison, the origins of electrical engineering as a discipline, accounts of the rise of the great electrical manufacturing firms such as General Electric, or anniversary histories of individual utility companies.[2] There was scant emphasis on the most important way that people have used electricity to conquer the earth, namely the creation of networks for distributing power.

All of this changed dramatically in the early eighties starting with the work of Thomas P. Hughes.[3] Studying first engineering and then history at the University of Virginia, Hughes became fascinated about electric power systems, thanks to his mechanical engineering professor, Frederick T. Morse. In teaching students how to design electric power plants in the 1940s, Morse emphasized that electric power was at the heart of the modern industrial world. In a manner paralleling Lewis Mumford's argument in *Technics and Civilization,* Morse observed that while mechanical power in the nineteenth century had created new industries, it had deskilled workers and forced them into crowded cities.[4] Electricity in the twentieth century, promised Morse, would eliminate these social ills while allowing the economy to grow and prosper. But for this to occur, engineers had to design entire systems; as Morse stated, "The power plant must function as a unit, not as a collection of individual pieces of equipment."[5] Moreover, electric power systems, insisted Morse, could not be viewed in isolation but should be designed with an awareness of the social, financial, and political environment in which they would operate. Morse showed Hughes that it was possible to use engineering to create orderly systems; Morse, Hughes later recalled, "had the intellectual strength to use elegant electrical science in solving problems within a context of economic, political, and geographical factors. He solved not by excluding variables but by bringing to bear powerful and complex analysis and order."[6]

Inspired by Morse, Hughes devoted much of his scholarly career to investigating the evolution of electric power systems, culminating in his magnum opus, *Networks of Power.* In *Networks,* Hughes highlighted two grand themes: first that we needed to

---

[2]See Matthew Josephson, *Edison: A Biography* (New York: McGraw-Hill, 1959) and Harold C. Passer, *The Electrical Manufacturers. 1875–1900: A Study in Competition, Entrepreneurship, Technical Change, and Economic Growth* (Cambridge: Harvard University Press, 1953). A typical utility history is Nicholas B. Wainwright, *History of the Philadelphia Electric Company, 1881–1961* (Philadelphia: Philadelphia Electric Company, 1961).

[3]W. Bernard Carlson, "From Order to Messy Complexity: Thoughts on the Intellectual Journey of Thomas Parke Hughes," *Technology and Culture* 55:945–52 (October 2014).

[4]Lewis Mumford, *Technics and Civilization* (New York: Harcourt, Brace, 1934).

[5]Frederick T. Morse, *Power Plant Engineering and Design: A Text for Engineers and Students of Engineering, Covering the Theory and Practice of Stationery Electric Generating Plants* (New York, D. Van Nostrand, 1932; subsequent editions, 1953 and 1964), 3.

[6]Hughes, *Networks of Power,* ix.

understand how inventors and engineers brought together individual devices—generators, transmission lines, transformers, lights, motors—to create coherent systems, and second, how these systems mirrored the politics and culture in which they were embedded. By comparing power systems in the United States, Britain, and Germany, Hughes argued that there was a general pattern by which engineers scaled up electrical systems from serving neighborhoods in the early 1880s to providing power across nationwide grids by the 1930s. For Hughes, the process was one in which engineers identified gaps or bottlenecks [what he called "reverse salients"] and then devised technical, political, and economic solutions that allowed their systems to expand. For Hughes, the conquest of the earth by electricity was benign, in which more energy was delivered to more people, thus raising their standard of living.

Networks of Power was published in 1983, and the decades following saw a significant rise in historical studies of electric power. Students and disciples of Hughes filled in various parts of the systems paradigm by writing about other key inventors, the development of hydropower in Ontario, the electrification of the Ruhr industrial region, the PJM regional interconnection, and the politics of nuclear energy in France.[7] At the same time, Hughes' work prompted other historians to look at electrification in ways that both complemented and challenged his synthesis.

1.2 Electrical History Since Hughes

1.2.1 The Social History of Electrification

Broadly speaking, electrical history since Hughes can be seen as following four strands. One thread traces the social and cultural histories of electrification. David Nye's *Electrifying America* presented a social history of electrification: "The central subject becomes not genius, not profits, not machines, not scientific discovery, but the human experience of making electricity part of city, factory, home, and farm."[8] In his *Consumers in the Country,* Ronald Kline explored resistance to electrification in the American countryside, finding that far from the new technology being uncritically accepted, rural people were able to shape how electricity reached them

[7]W. Bernard Carlson, *Innovation as a Social Process: Elihu Thomson and the Rise of General Electric, 1870–1900* (New York: Cambridge University Press, 1991); Robert B. Belfield, "The Niagara Frontier: The Evolution of Electric Power Systems in New York and Ontario, 1880–1935" (Ph.D. dissertation, University of Pennsylvania, 1981); Bayla S. Singer, "Power to the People: The Pennsylvania-New Jersey-Maryland Interconnection, 1925–1970" (Ph.D. dissertation, University of Pennsylvania, 1983); Edmund N. Todd, "Industry, State, and Electrical Technology in the Ruhr Circa 1900," *Osiris* 5(1) (1989) https://www.journals.uchicago.edu/doi/10.1086/368689; Gabrielle Hecht, *The Radiance of France: Nuclear Power and National Identity after World War II* (Cambridge: MIT Press, 1998).

[8]Nye, *Electrifying America*, xi.

(or didn't), and how it was used (or wasn't). They had agency. Rural people were not passive observers of the march of electrical progress. Taking a similar look at the adoption of electricity in English households prior to 1914 over then-standard gas lighting and appliances, Graeme Gooday contended that domestication required overcoming fear of electricity's threat to household residents—especially women.[9] Electricity was deliberately gendered female via industry advertising. But Gooday's study was also a story of agency—women had influence and made choices in English households, and (male) utility executives understood they had to convince women as well as men that invisible electricity was safe and effective.

Recently, historians have started looking at places that electrification has not yet reached. In *Power Lines*, a history of electrification of the region surrounding Phoenix, Arizona, Andrew Needham examined the role of another population typically left out of electricity stories, Native Americans. Spanning the intersection of the states of Utah, Colorado, Arizona and New Mexico, the Navajo Reservation had vast quantities of coal; in the 1950s, as economy of the American West began to grow explosively, businessmen and Navajo leaders signed contracts to convert Navajo coal into electricity to support that growth. Yet the Navajo gained few jobs from these deals and little of the revenue generated but hosted the destroyed land and air and water pollution. It was an energy transition of injustice.[10]

But it is not just groups in the United States who have not experienced the benefits of electrification. In their critiques of the "Earth at Night" satellite image (Fig. 1.1), Sara Pritchard and Ute Hasenhohrl call our attention to the dark spaces in South America, Africa, and central Asia, challenging us to think about what this image reveals about how poverty and the colonial and post-colonial legacies have shaped electrification.[11] If Hughes' perspective was one of inevitability and universality, the grand human experiment in electrification has yet to achieve either. Electrification remains an unfinished energy transition, despite the passing of more than a century and a half.

[9]Graeme Gooday, *Domesticating Electricity: Technology, Uncertainty and Gender, 1880—1914, Domesticating Electricity: Technology, Uncertainty and Gender, 1880—1914*, Science and Culture in the Nineteenth Century, 7 (London: Pickering & Chatto, 2008).

[10]Andrew Needham, *Power Lines: Phoenix and the Making of the Modern Southwest Phoenix and the Making of the Modern Southwest* (Princeton, NJ: Princeton University Press, 2015). For justice considerations in energy transitions more generally see Gwen Ottinger, "The Winds of Change: Environmental Justice in Energy Transitions," *Science as Culture* 22, no. 2 (June 2013): 222–29. https://doi.org/10.1080/09505431.2013.786996

[11]Sara B. Pritchard, "The Trouble with Darkness: NASA's Suomi Satellite Images of Earth at Night," *Environmental History* 22, no. 2 (April 2017): 312–30, https://doi.org/10.1093/envhis/emw102; Ute Hasenöhrl, "Rural Electrification in the British Empire," *History of Retailing and Consumption* 4, no. 1 (January 2, 2018): 10–27. https://doi.org/10.1080/2373518X.2018.1436220

1.2.2 Electricity and Political Economy

If Hughes focused on systems builders, there were already indications that alternative approaches existed. An early example of a different approach was Jonathan Coopersmith's study of electrification in Tsarist Russia and the Soviet Union. Employing a social construction of technology approach, Coopersmith traced the international flows of engineers and money that advanced electrification and considered the constituencies inside Russia who favored electricity. Advocates of electrification during the Tsarist regime had little access to political power and electrification proceeded slowly. Under the Soviet regime, however, electrification became a demonstration of modernization and of state power. Politically connected engineers were able to gain the new leadership's support. People and politics mattered. Nevertheless, Coopersmith concluded that Russia never reached the level of electrification prevalent in the Western social democracies. Soviet electrification was not inevitable, and it hadn't passed through the common set of stages.[12]

While Coopersmith revealed how political ideology mattered in Soviet electrification, most historians have concentrated on the political economy of electrification in the United States. A classic in this area is Thomas McCraw's *TVA and the Power Fight*, which traced the conflict between Franklin Roosevelt's 'New Deal' government and private utilities over bringing hydroelectricity to the Tennessee Valley.[13] Prior to the 1930s, electrification in the US had largely stopped at city boundaries and the Roosevelt administration intended to use government power to electrify rural America. As a government chartered and financed corporation, the Tennessee Valley Authority was seen by the private utility industry as both a huge threat and a highly controversial approach. Along with the TVA, the Roosevelt administration also sought to promote electrification via low-interest loans provided to private cooperatives. As Abby Spinak in this volume will argue, this approach, executed by the Rural Electrification Administration, was used by the US to spread electrification overseas after World War II to decidedly mixed results.

In addition to directly providing electricity to new regions during the 1930s, the Roosevelt administration established a federal regulatory framework for utilities. Utility holding companies ("Power Trusts," they were sometimes called) often owned utilities in several states, making them largely immune to state regulation. After the sudden collapse of Samuel Insull's Middle West Company in 1932, Roosevelt was able to pass a sweeping utility reform law. This Public Utility Holding Company Act of 1935 primarily regulated utility *finance*, and in so doing, it fundamentally altered the structure of American utilities. One consequence was to separate electric transport (or "traction") from electrical utilities, making electric streetcars and interurban railways and the generation and transmission of power into

[12] Jonathan Coopersmith, *Electrification of Russia, 1880-1926*. (Ithaca, NY: Cornell University Press, 1992).

[13] Thomas K. McCraw, *TVA and the Power Fight, 1933-1939*, Critical Periods of History (Philadelphia: Lippincott, 1971).

two different businesses (and different markets). The law also explicitly empowered state regulation of the shrunken utilities for in-state generation and transmission, while granting authority to the Federal Power Commission (now the Federal Energy Regulatory Commission) to regulate interstate transmission.

Historian Richard Hirsh contends that, for decades after passage of the Public Utility Holding Company Act, a "utility consensus" emerged based around regulated vertical monopolies whose prices were set not by managers, but by state regulators.[14] This consensus, however, didn't survive the tumultuous 1970s. Late in the 1960s, the formerly reliable "march of technological progress" that had enabled ever-larger generators to reliably produce more electricity at lower per-kilowatt costs suddenly ended. The newest generation of turbines were less reliable than their forerunners. Bigger was no longer better. The 1973 oil embargo sent fuel costs, and therefore electricity rates, soaring, and consumer anger followed. Higher fuel costs encouraged businesses—both large and small–to start accounting for, and addressing, energy costs as an input they could control by improving efficiency of their operations and systems. On account of both consumer anger and changes in business practices, the electricity market in the US suddenly stopped growing. This destroyed one potential energy transition, to large-scale nuclear generation, because large nuclear plants needed growth to justify their high capital costs.

But the energy crises of the decade enabled a different transition, away from the "utility consensus" and towards a new economic model. As part of its response to the intertwined economic challenges of high inflation and high energy costs, the Carter administration drafted an Energy Act that sought to reintroduce competition to the regulated electricity market. While the administration's original bill was broken up into numerous pieces and heavily revised in Congress, the resulting Public Utilities Regulatory Procedures Act (PURPA) of 1978 highlighted the idea of market competition. It required the still-regulated utilities to buy power from third-party generators, introducing competition at the producer level. This enabled, somewhat unexpectedly, small-scale renewables generation to begin entering the new generation market.

This action has often been called "deregulation," but that term is misleading. Regulation remained but had been transformed in a more market-friendly direction—the Carter administration *marketized* electricity (as well as trucking, air travel, and railroads). Marketization did not originate with utility managers, who were generally hostile to the effort and lobbied to stop it via a "states rights" argument: states should have the right to keep their own regulated monopolies intact. This worked, and marketization was voluntary but encouraged under PURPA. Nor did marketization come from technological change. It emerged from the field of economics, and within the Carter administration, from Cornell University's Alfred Kahn. Kahn's vision of a regulated marketplace for power replaced that of utility managers.

---

[14]Richard F. Hirsh, *Power Loss: The Origins of Deregulation and Restructuring in the American Electric Utility System* (Cambridge: MIT Press, 2002).

Utility marketization highlights the role of political economy in energy transitions. The dominant political ideology after 1970 in the United States, especially after the sudden end of the Cold War, came to be "neoliberalism." The root concept of this ideology is the creation of markets and their encasement by law, insulating them as much as possible from democratic processes.[15] In the US, though, divided government (aka "federalism") ensured this revolution could never be complete. As Julie Cohn shows in this volume, efforts to develop large-scale wind generation in Texas were politically contested, with state and local governments sometimes cooperating and sometimes not, with various organized interest groups taking roles that further complicated the process.

A number of scholars have written about the politics of electricity from a very different perspective: the political project of unifying Europe. The development and deployment of shared infrastructure, including electrical networks, has been a central, if sometimes hidden, goal of that effort.[16] "Infrastructural Europeanism" is one term for the process of building a European identity upon infrastructure and infrastructure networks.[17] Not all of the impetus for this program of infrastructure-based unification came from explicitly political actors. Electrical engineers and their professional organizations, for example, saw themselves as "apolitical" actors whose involvement allowed the idea of transnational networks to be presented as merely "technical" accomplishments fostering efficiency, load balancing, and reliability.[18] Efforts along these lines long preceded the post-war effort to create a "United States of Europe," that ultimately produced the European Union.[19]

[15] Quinn Slobodian, *Globalists: The End of Empire and the Birth of Neoliberalism* (Cambridge: Harvard University Press, 2018).

[16] Thomas J. Misa and Johan Schot, "Introduction. Inventing Europe: Technology and the Hidden Integration of Europe," ed. Arne Kaijser et al., *History and Technology* 21 (2005): 1–19; and Vleuten, Erik van der, and Arne Kaijser. "Networking Europe." *History and Technology* 21, no. 1 (March 2005): 21–48. https://doi.org/10.1080/07341510500037495

[17] Frank Schipper and Johan Schot, "Infrastructural Europeanism, or the Project of Building Europe on Infrastructures: An Introduction," ed. Anastasiadou et al., *History and Technology* 27 (2011): 245–64; Erik van der Vleuten and Arne Kaijser, "Networking Europe," *History and Technology* 21, no. 1 (March 2005): 21–48; Vincent Lagendijk, "To Consolidate Peace? The International Electro-Technical Community and the Grid for the United States of Europe," *Journal of Contemporary History* 47 (2012): 402–26. https://doi.org/10.1080/07341510500037495

[18] Schot and Lagendijk refer to this as technocratic nationalism. Johan Schot and Vincent Lagendijk, "Technocratic Internationalism in the Interwar Years: Building Europe on Motorways and Electricity Networks," *Journal of Modern European History* 6, no. 2 (September 1, 2008): 196–217. https://doi.org/10.17104/1611-8944_2008_2_196

[19] Vincent Lagendijk. *Electrifying Europe: The Power of Europe in the Construction of Electricity Networks*. Technische Universiteit Eindhoven, 2008. https://research.tue.nl/en/publications/electrifying-europe—the-power-of-europe-in-the-construction-of-electricity-networks(6f7ef2a6-660a-49cd-9bd7-019a38591bea).html; Vincent Lagendijk, "'To Consolidate Peace'? The International Electro-Technical Community and the Grid for the United States of Europe," *Journal of Contemporary History* 47, no. 2 (April 1, 2012): 402–26. https://doi.org/10.1177/0022009411431722. Ronan Bolton, Vincent Lagendijk, and Antti Silvast, "Grand Visions and Pragmatic Integration: Exploring the Evolution of Europe's Electricity Regime," *Environmental Innovation and Societal Transitions*, How History Matters for the Governance of Sociotechnical Transitions, 32 (September 1, 2019): 55–68. https://doi.org/10.1016/j.eist.2018.04.001

1.2.3 Electricity and Finance

While much has been written about the political economy of electrification, scholars have paid less attention to the financing of electrification. This is surprising, given that electric light and power systems are expensive to build and maintain, especially as one seeks to increase the scale of the network and reach more customers. Generally speaking, customers are only willing to pay the cost of generating the electricity they consume, meaning that utilities had to find innovative ways to finance long-term capital costs. As several historians have noted, these innovations included electrical manufacturers devising ways to provide credit to their utility customers, the creation of engineering and financial intermediaries such as Stone & Webster, and the formation of holding companies that reduced risk by bringing together individual power companies from different markets [urban, suburban, and rural] and different regions.[20]

But one of the most interesting aspects of the financing of electrification is to recognize that it took place on a global scale. In the late nineteenth century, European financiers invested heavily in electrical utilities in the US and Canada whereas in the twentieth century American investors helped build electrical systems in Latin America, China, and India. This is most clearly revealed in a thorough study undertaken by a group of economic historians led by William J. Hausman, Peter Hertner, and Mira Wilkins. In their 2008 volume, *Global Electrification,* this group examined the flow of capital, technology, and knowledge around the world, revealing how financiers and engineers found ways to profitably invest in electrical networks in dozens of countries.[21] In particular, Hausman and his colleagues traced how these global flows rose and ebbed. Prior to 1930, there was significant investment by European and American financiers, leading to ownership of utilities by multinational enterprises, but starting with the Great Depression and continuing into the 1970s, the trend went toward domestic regulation, investment, and ownership of electrical networks. In the 1970s, with changing views and policies about energy markets, financiers once again began providing capital to utilities around the world and again acquiring equity stakes in power companies. Overall, these economic historians remind us that electrification was the conquest of the whole earth, tying nations together not just electrically but also financially. In their contribution to this volume, Hausman, Neufeld and Pereira trace the ebb and flow of investment in Cuban electrical infrastructure during the first half of the twentieth century.

[20] Hughes, *Networks of Power,* 386–401; Sidney Alexander Mitchell, *Sidney Z. Mitchell and the Electrical Industry* (New York: Farrar, Straus & Cudahy, 1960); and Forest McDonald, *Insull* (Chicago: University of Chicago Press, 1962).

[21] William J. Hausman, Peter Hertner, and Mira Wilkins, *Global Electrification: Multinational Enterprise in the History of Light and Power, 1878–2007* (New York: Cambridge University Press, 2008).

1.2.4 The Envirotechnical View

Another strand of post-Hughesian electrical history descends from the American Western historian Richard White.[22] Writing about development along the Columbia River, White sought to integrate natural and human history via the lens of work. He fused the utopian fantasies of electricity boosters, contests between public and private power advocates, the transformation of the river itself into an "organic machine" and the livelihoods of the people living along it. The promise of vast quantities of cheap hydroelectricity to supply industrial development along with subsidized water for agriculture had long-standing impacts. One key insight in White's work was that our reconstruction of "nature" into machinery alters it, and us, but still doesn't result in our control of that machine. Another is our inability to escape the history of our refashioning of the everchanging river. It's a relationship requiring everlasting work.

In her history of the post-war redevelopment of the Rhone river in France as a site for the production of hydroelectric and nuclear power, Sara Pritchard built on White's work as she developed an "envirotechnical analysis" of these integrated, large-scale systems that are at once part "natural," part "technological," and part "political." The developers of the Rhone system defined "nature" and "technology" in ways that aided their efforts, exploiting the malleability of the concepts. They constructed what Pritchard called an envirotechnical system, borrowing Hughes' idea of "system" but expanding it to include the (re-engineered and transformed) non-human environments that serve as inputs to, and components of, such systems. These envirotechnical systems were, in turn, embedded within "envirotechnical regimes," which were "the institutions, people, ideologies, technologies, and landscapes that together define, justify, build, and maintain a particular envirotechnical system as normative."[23] The concept of a regime allows examination of the roles or politics and power relations within the development of envirotechnical systems like the Rhone, or, for that matter, any other river redesigned for industrial and/or energy production, like the Columbia or Snake rivers in the United States.

A very different envirotechnical approach to the history of electricity has been the study of the resources developed and consumed to make the hardware of electric infrastructure, and the human and environmental cost of that resource extraction.[24] Hanna Vikstrom explored the resource constraints on lightbulb manufacturing at the

---

[22] Richard White, *The Organic Machine* (NY: Hill and Wang, 2001).

[23] Sara B. Pritchard, *Confluence: The Nature of Technology and the Remaking of the Rhône*, *Confluence: The Nature of Technology and the Remaking of the Rhône*, 2011, p. 23. She draws the regime concept in part from Gabrielle Hecht's *The Radiance of France*.

[24] Hanna Vikström, "The Rush for Greenlandic Metals," *Technology's Stories* 5, no. 2 (2017); Hanna Vikström, "Producing Electric Light: How Resource Scarcity Affected Light Bulbs, 1880–1914," *Technology and Culture* 61, no. 3 (2020): 901–22. https://doi.org/10.1353/tech.2020.0078. Hanna Vikström, "Risk or Opportunity? The Extractive Industries' Response to Critical Metals in Renewable Energy Technologies, 1980–2014," *The Extractive Industries and Society* 7, no. 1 (January 1, 2020): 20–28. https://doi.org/10.1016/j.exis.2020.01.004

beginning of the electrical age. Nathan Ensmenger is exploring the history of computing from a similar perspective.[25]

In *Power on the Hudson*, Robert Lifset examined the intersections of electrical development and the transformation of the American environmental movement. In 1962, Storm King Mountain on the Hudson River was proposed as the site of a large pumped storage reservoir and subsequently became a signature battle against a particular kind of development. The reservoir would have allowed the utility to run its largely coal-fired power plants more cost-effectively—and simply more. It would increase air pollution. While Lifset's aim was at understanding the transformation of environmentalism from a conservation-based set of ideals to one organized around preservationist ideals, his book is also a cautionary tale for a future renewables transition. Pumped storage is one technology available to help solve the intermittency problem that is posed by wind and solar power. Batteries are the other relatively well-established technology available.[26] Both have substantial environmental implications that have made them subject of protest in the past.[27]

1.3 Energy Transitions and Imaginaries

While scholars working within these four approaches both enriched and challenged the Hughesian paradigm, they have not fully grappled with a fundamental issue embedded in the paradigm. Perhaps reflecting the optimism of Morse, his old engineering professor, Hughes saw the development of electrical networks as an inevitable process—that once one builds a small power system, it's easy to scale up to larger and larger networks, first regionally and then nationally. Indeed, Hughes was confident that economies of scale were feasible, and that once established, technologies acquire a certain momentum. Overall, Hughes saw the development of electrical power as a smooth, linear process with few bumps in the road.

Yet, there have been lots of bumps, both technically and politically. It is not an easy matter to build larger systems and to sell larger amounts of electricity. To do so, you have to interact with more customers, maneuver in more political jurisdictions, raise more capital, and none of these activities are easy. Indeed, scaling up is often fraught with give-and-take, with conflict. Electrification has not been a smooth, linear process but rather a story of transitions.

[25] Nathan Ensmenger, "The Environmental History of Computing," *Technology and Culture* 59, no. 4 (2018): S7–33. https://doi.org/10.1353/tech.2018.0148.ens

[26] Others may yet emerge. Two more possibilities are thermal storage and conversion of renewables to hydrogen for either use in fixed power stations or in vehicles.

[27] Robert Lifset, *Power on the Hudson: Storm King Mountain and the Emergence of Modern American Environmentalism* (Pittsburgh: University of Pittsburgh Press, 2014).

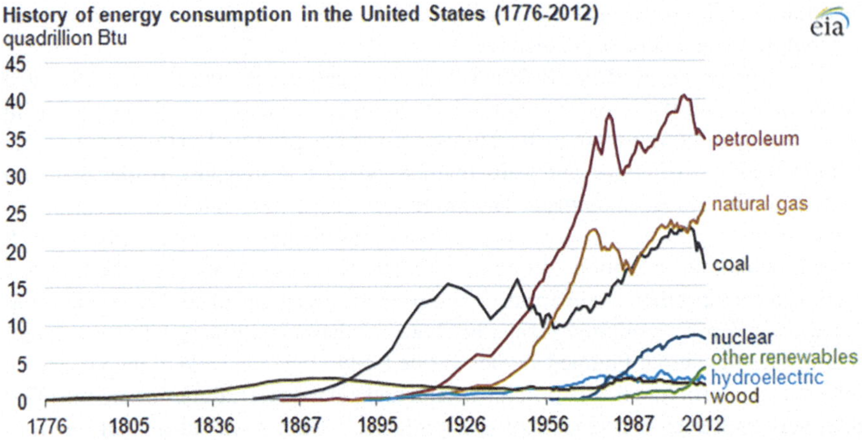

Fig. 1.2 History of Energy Consumption in the United States (1776–2012). (Source: U.S. Energy Information Administration)

In his *Routes of Power*, Christopher F. Jones contended that the history of energy in the United States has been one of transitions. His study began with coal and the redevelopment of rivers in the mid-Atlantic states to transport it to cities more efficiently. Initially, coal was seen as an industrial fuel, but even then demand for coal had to be created—it didn't suddenly appear. Discovery of oil brought new competition for coal, as did the development of hydropower. Ultimately, of course, all three of these sources of primary energy were harnessed to make electricity, but again, demand had to be created. This was done through both advertising and political activity by the would-be energy producers. These intertwined energy transitions were not inevitable.

As Jones' story suggests, the story of American energy transitions isn't a story of one fuel replacing another. Oil didn't replace coal. Instead, oil and its boosters expanded overall use of energy, as did hydropower. Figure 1.2 reveals that the combined use of these resources resulted in a new economy of energy abundance.[28] The lesson here is that fossil fuels are unlikely to just disappear during a renewables transition; indeed, policy will almost certainly have to drive them out of the marketplace.

Jones also remarks on the historical reality that these energy transitions "reconfigure[d] social power as well as mechanical power." Each new energy technology delivered benefits and costs unequally. They also brought power shifts—both political and economic—granting power and wealth to some and taking it from

[28] See also Martin V. Melosi, *Coping with Abundance: Energy and Environment in Industrial America* (Philadelphia, Pa.: Temple University Press, 1985); David E. Nye: *Consuming Power: A Social History of American Energies* (Cambridge, MA: MIT Press, 1998).

others.[29] The same will happen in future transitions, too. One might hope for a more egalitarian transition than those of the past two centuries, but whether that happens depends on policy and politics.

Building on Jones' work, the papers in this volume move away from a linear model of electrification and instead argue that the story of electricity is one of *transitions*—transitions that are not inevitable but rather profoundly shaped by contingent social, political, and environmental factors. As the authors in this volume reveal, these transitions come about when various historical actors—engineers, managers, bureaucrats—perceive new opportunities, in response sometimes to technical innovations but more often to social and political winds blowing through their cultures.

To take advantage of these winds and mobilize the resources needed to build new and larger electrical networks, historical actors have to articulate a vision, argument, or story that captures the imagination of key stakeholders such as financiers, politicians, and the general public. Like Sheila Jasonoff and other scholars, we call these visions and stories *imaginaries*. Jasanoff and Sang-Hyun Kim define these as "collectively imagined forms of social life and social order reflected in the design and fulfillment of nation-specific scientific and/or technological projects."[30] As our authors argue, each transition is not simply the installation of some new piece of electrical technology; rather, the real work in a transition is the articulation of a new imaginary, its contestation, resolution, and its instantiation in hardware. Hughes would claim that the origin of transition is the new technology and its affordances; in contrast, we argue that the origin of transition is in the articulation of an imaginary.

If the articulation of an imaginary is the origin of an energy transition, part of that imaginary's instantiation is within an envirotechnical regime. All energy systems draw resources from the "natural" world, and most return "waste" to that environment. Ideologies of consumption and conservation, of environmental management (or the lack thereof), and of political economy, undergird and guide the transformation of energy landscapes, of energy technologies, and of human (and even animal) relations. (Much of the world still uses significant amounts of animal power, even if most developed countries no longer do). Envirotechnical regimes are the obdurate form of the imaginary. Their political, economic, and ideological power make envirotechnical regimes resistant to efforts to change them, even if the imaginary that guided their founding is challenged and begins to fail.

---

[29] Christopher F. Jones, *Routes of Power: Energy and Modern America* (Cambridge, MA: Harvard University Press, 2014).

[30] Sheila Jasanoff, and Sang-Hyun Kim, "Containing the Atom: Sociotechnical Imaginaries and Nuclear Power in the United States and South Korea." *Minerva* 47, no. 2 (June 2009): 120. https://doi.org/10.1007/s11024-009-9124-4

1.4 The Organization of this Volume

In applying the concepts of transitions and imaginaries to the history of electrification, we have organized this volume in the following manner. The first essay by David Nye lays out the general idea of electrical transitions in history and argues for a heterogeneous model of energy transitions. Contrary to the Multi-Level Perspective [MLP] framework and the path dependence model from economics and Hughes' conception of technological momentum, Nye uses the reality that every nation and region has had a different experience with electrification to remind us that contingency, resource endowment, and choices matter to electrical transitions.[31] In the second essay, Cyrus Mody challenges the MLP framework from a different standpoint: the oil, gas, and electricity industries were deeply entangled in the United States during the "long 1970s," and separating them into neat analytical categories ignores that history. Oil companies were flush with money in that decade and invested in all sorts of alternative energy ventures, including solar, nuclear, and even geothermal.

The following four papers examine energy transitions in different times and different cultures, looking at the imaginaries and tactics used by historical actors as they sought to expand power networks. Nathan Kapoor takes us to New Zealand where we see that, unlike other societies which only gradually moved from local to national networks, New Zealanders sought to develop a national power grid from the outset. They did so by imagining the potential for large-scale hydropower as means for establishing their identity as an independent commonwealth and for securing the self-sufficient energy to develop industry and cities.

In the next paper, William Hausman, John Neufeld, and Rui Pereira explore electrification in the developing world, focusing on the case of Cuba from the 1920s to the 1970s. They show how American business invested in Cuban electrification via the American & Foreign Power Company, imagining that it would be relatively a straightforward and profitable process to build local utilities across the island. However, in the 1930s, American & Foreign Power discovered that a constellation of labor organizations, consumers, and government regimes objected to high prices for electric service and profits being taken out of the country by a "Yankee" holding company. While managers at American & Foreign Power were able to broker a deal with the Cuban government in the '40s and '50s, the challenges of the '30s set the stage for the eventual nationalization of the Cuban grid by Fidel Castro in 1960.

[31] Frank W. Geels, "Technological Transitions as Evolutionary Reconfiguration Processes: A Multi-Level Perspective and a Case-Study," *Research Policy*, 31, no. 8 (December 1, 2002): 1257–74, https://doi.org/10.1016/S0048-7333(02)00062-8; Frank W. Geels, "A Socio-Technical Analysis of Low-Carbon Transitions: Introducing the Multi-Level Perspective into Transport Studies," *Journal of Transport Geography* 24 (September 2012): 471–82, https://doi.org/10.1016/j.jtrangeo.2012.01.021; Frank W Geels, "Regime Resistance against Low-Carbon Transitions: Introducing Politics and Power into the Multi-Level Perspective," *Theory, Culture & Society* 31, no. 5 (September 2014): 21–40. https://doi.org/10.1177/0263276414531627

Stathis Arapostathis and Yannis Fotopoulos expand our understanding of the relationship between electrification and nation-building by narrating technical and political developments in Greece from the 1940s to the present. They trace how foreign engineering advisors in the late '40s and early '50s imagined and built regional grids based on imported coal and oil, only to be replaced 20 years later when Greek politicians and engineers put forward an imaginary of energy independence by promoting hydropower and the burning of locally produced lignite. Although lignite allowed several Greek regimes to build strong relationships with the unions and select business groups, it also resulted in Greece having significantly more air pollution. Consequently, in the 1980s, a new imaginary took shape that envisioned Greek energy independence based on wind power and other renewables, but the leaders promoting this imaginary had to content with the geopolitical challenges stemming from the vision of the European Union seeking to knit Europe more closely together through new transnational natural-gas pipelines.

The fourth paper in the volume's middle portion continues to examine the transition from fossil fuels to renewable energy sources, using the introduction of wind power in Texas as a case study. In her chapter, Julie Cohn introduces us to a fresh group of stakeholders, each of which evolved their own imaginary about the advantages and disadvantages of wind power. While engineering and political leaders saw wind power as means of reducing dependency on fossil fuels and increasing the diversity of energy inputs into the Texas grid, local groups had mixed views about the new transmission lines needed to carry the additional power developed on wind farms. While residents in several counties in the Texas panhandle imagined that wind power would provide a boost to the local economy, by contrast farmers in the Texas Hill Country anticipated that new power lines would be a blight on the landscape. Cohn reveals how these citizens articulated these imaginaries and how their actions made the introduction of wind power in Texas a far more complicated technical and political process. Cohn reminds us that energy transitions are frequently contentious and involve a wide range of actors.

While the middle papers in this volume examine the articulation of imaginaries in specific contexts, authors of the final two papers reveal how engineers and bureaucrats tie their imaginaries to broader cultural and political themes such as modernity, global economic development, and neoliberalism. Abby Spinak examines USAID's use of the co-operative model of electrification developed for the rural United States to foster electrification and development in Ecuador and Costa Rica. Marketed as a "middle way" between Communism and capitalism that could foster local democracy, co-ops often instead were co-opted into national government agendas. These electrification projects tended not to empower production, either. Most consumption went to lighting, entertainment and irons, not to appliances and machinery. Yet different outcomes from the same model make clear that energy transitions happen differently in different national contexts.

In his essay, Matt Eisler ties the nascent idea of using private vehicle batteries to stabilize California's failing electrical infrastructure to the reigning American political ideology of neoliberalism. This "vehicle to grid" [V2G] imaginary involved seeing consumers as entrepreneurs willing to 'sell' part of their vehicular vehicle battery capacity and longevity to the state's regulated, but semi-privatized, electricity market. This imaginary collapsed when the state found other means to stabilize its electricity market and never provided the incentives necessary to render the vehicle-to-grid imaginary real. V2G represents another 'energy transition' that is, as yet, unrealized.

1.5 Conclusion

Taken as a whole, the papers in this volume build on the previous work on the history of electrification to enlarge our view of the social and political processes by which societies have come to have electric power. By looking at electrification as a series of transitions shaped by the imaginaries constructed by multiple groups involved with this technology, we now have a more nuanced picture that moves us away from the linear vision of Thomas Hughes. We can now see how electrification affects all sorts of people, the economic and political welfare of nations, and it should be no surprise that electrification has been both celebrated and contested. Touching our lives in so many ways, people have indeed used electricity to conquer the earth.

But in revising the story of electrification, we are not merely concerned with updating one tiny corner of the history of technology. The papers in this volume provide context for thinking about what lies ahead as humans continue their conquest of the earth through electricity. As we are increasingly dependent on electricity to power our lights, heat and cool our homes, turn the wheels of industry, and keep our telecommunications, health care, and information systems humming, so we are increasingly vulnerable when the grid runs into trouble. On November 4, 2006, for example, millions of residents of Germany, France, Belgium, Italy and Spain lost power for 2 hours, leading to an investigation aimed at understanding system reliability.[32]

More recently, it has become clear that climate change is revealing how fragile our electrical systems have become. In California, years of neglecting maintenance on the transmission network triggered numerous wildfires over the past decade in forests where forest-management practices allowed excess underbrush to accumulate. The worst of these was the Camp Fire in 2018 which killed 86 people and destroyed over 18,000 structures. The utility responsible for the fire, Pacific Gas and Electric, filed for bankruptcy, facing on the order of $30 billion in liability for that and earlier fires.[33] In response, the state of California authorized "Public Safety

[32] Erik van der Vleuten and Vincent Lagendijk, "Transnational Infrastructure Vulnerability: The Historical Shaping of the 2006 European 'Blackout,'" *Energy Policy* 38, no. 4 (April 1, 2010): 2042–52. https://doi.org/10.1016/j.enpol.2009.11.047

[33] Kurtis J. Alexander, J. D. Morris, and Peter Fimrite. "PG&E Caused Camp Fire, Cal Fire Says." San Francisco Chronicle, May 16, 2019. https://www.sfchronicle.com/business/article/PG-E-power-lines-have-long-been-the-leading-13848463.php; NIST, "New Timeline of Deadliest California Wildfire Could Guide Lifesaving Research and Action," February 8, 2021. https://www.nist.gov/news-events/news/2021/02/new-timeline-deadliest-california-wildfire-could-guide-lifesaving-research

Power Shut-offs," allowing utilities to cut power to areas expected to experience high winds. These shut-offs, of course, impose their own economic and human costs.

And as we were editing this volume in February 2021, a blast of Arctic air made its way south to Texas, causing power generators that had not been winterized (to save costs) to fail, resulting in a multi-day blackout that affected millions. The crisis was further compounded when parts of the state's natural gas and water treatment infrastructure also failed in the frigid weather.[34]

All of these examples reveal that the electrical systems put in place over the last 150 years are far more fragile—indeed, brittle—than their inventors, promoters, or managers ever imagined. To redesign these systems so that they are resilient—capable of responding to climate change and keeping pace with our growing demand for energy—engineers will need to take these systems through another transition, balancing the technical, social, political, and economic considerations, much as their predecessors did in previous transitions. The conquest of the earth by electricity is unfinished, with more transitions to come, especially transitions that ensure electricity is available to all people in all corners of the planet.

Bibliography

Alexander, Kurtis, J. D. Morris, and Peter Fimrite. 2019. *PG&E Caused Camp Fire, Cal Fire Says*. San Francisco Chronicle, May 16. https://www.sfchronicle.com/business/article/PG-E-power-lines-have-long-been-the-leading-13848463.php.

Belfield, Robert B. 1981. *The Niagara Frontier: The Evolution of Electric Power Systems in New York and Ontario, 1880–1935*. Ph.D. dissertation: University of Pennsylvania.

Bolton, Ronan, Vincent Lagendijk, and Antti Silvast. 2019. Grand Visions and Pragmatic Integration: Exploring the Evolution of Europe's Electricity Regime. *Environmental Innovation and Societal Transitions, How History Matters for the Governance of Sociotechnical Transitions* 32: 55–68. https://doi.org/10.1016/j.eist.2018.04.001.

Carlson, W. Bernard. 1991. *Innovation as a Social Process: Elihu Thomson and the Rise of General Electric, 1870–1900*, Studies in Economic History and Policy. New York: Cambridge University Press.

———. 2014. From Order to Messy Complexity: Thoughts on the Intellectual Journey of Thomas Parke Hughes. *Technology and Culture* 55 (4): 945–952.

Coopersmith, Jonathan. 1992. *Electrification of Russia, 1880–1926*. Ithaca, NY: Cornell University Press.

Douglas, Erin and Ross Ramsey. 2021. *No, Frozen Wind Turbines Aren't the Main Culprit for Texas' Power Outages*. The Texas Tribune, February 17. https://www.texastribune.org/2021/02/16/texas-wind-turbines-frozen/

[34]Conservative media and politicians in the US attempted to shift blame from failed gas infrastructure to wind turbines, part of a two-decades long effort to forestall a transition to renewable energy. Erin Douglas and Ross Ramsey. "No, Frozen Wind Turbines Aren't the Main Culprit for Texas' Power Outages." *The Texas Tribune*, February 17, 2021. https://www.texastribune.org/2021/02/16/texas-wind-turbines-frozen/

Ensmenger, Nathan. 2018. The Environmental History of Computing. *Technology and Culture* 59 (4): S7–S33. https://doi.org/10.1353/tech.2018.0148.

Lagendijk, Vincent. 2008. *Electrifying Europe: The Power of Europe in the Construction of Electricity Networks*. Technische Universiteit Eindhoven. https://research.tue.nl/en/publications/electrifying-europe%2D%2Dthe-power-of-europe-in-the-construction-of-electricity-networks(6f7ef2a6-660a-49cd-9bd7-019a38591bea).html.

———. 2012. 'To Consolidate Peace'? The International Electro-Technical Community and the Grid for the United States of Europe. *Journal of Contemporary History* 47 (2): 402–426. https://doi.org/10.1177/0022009411431722.

Lifset, Robert. 2014. *Power on the Hudson: Storm King Mountain and the Emergence of Modern American Environmentalism*. Pittsburgh: University of Pittsburgh Press.

Geels, Frank W. 2002. Technological Transitions as Evolutionary Reconfiguration Processes: A Multi-Level Perspective and a Case-Study. *Research Policy* 31 (8): 1257–1274. https://doi.org/10.1016/S0048-7333(02)00062-8.

———. 2012. A Socio-Technical Analysis of Low-Carbon Transitions: Introducing the Multi-Level Perspective into Transport Studies. *Journal of Transport Geography* 24: 471–482. https://doi.org/10.1016/j.jtrangeo.2012.01.021.

———. 2014. Regime Resistance against Low-Carbon Transitions: Introducing Politics and Power into the Multi-Level Perspective. *Theory, Culture & Society* 31 (5): 21–40. https://doi.org/10.1177/0263276414531627.

Gooday, Graeme. 2008. *Domesticating Electricity: Technology, Uncertainty and Gender, 1880–1914*, Science and Culture in the Nineteenth Century. London: Pickeering & Chatto.

Hasenöhrl, Ute. 2018. Rural Electrification in the British Empire. *History of Retailing and Consumption* 4 (1): 10–27. https://doi.org/10.1080/2373518X.2018.1436220.

Hausman, William J., Peter Hertner, and Mira Wilkins. 2008. *Global Electrification: Multinational Enterprise and International Finance in the History of Electric Power, 1878–2007*. New York: Cambridge University Press.

Hecht, Gabrielle. 1998. *The Radiance of France: Nuclear Power and National Identity after World War II*. Cambridge, Mass: MIT Press.

Hirsh, Richard F. 1999. *Power Loss: The Origins of Deregulation and Restructuring in the American Electric Utility System*. Cambridge, Mass: MIT Press.

Hughes, Thomas Parke. 1993. *Networks of Power: Electrification in Western Society, 1880–1930*. Baltimore, Md: John Hopkins University Press.

Jasanoff, Sheila, and Sang-Hyun Kim. 2009. Containing the Atom: Sociotechnical Imaginaries and Nuclear Power in the United States and South Korea. *Minerva* 47 (2): 119–146. https://doi.org/10.1007/s11024-009-9124-4.

Jones, Christopher F. 2014. *Routes of Power: Energy and Modern America*. Cambridge, Mass: Harvard University Press. https://www.hup.harvard.edu/catalog.php?isbn=9780674970922.

Josephson, Matthew. 1959. *Edison: A Biography*. New York: McGraw-Hill.

McDonald, Forrest. 1962. *Insull*. Chicago: University of Chicago Press.

McCraw, Thomas K. 1971. *TVA and the Power Fight, 1933–1939*. Critical Periods of History. Philadelphia: Lippincott.

Melosi, Martin V. 1985. *Coping with Abundance: Energy and Environment in Industrial America*. Philadelphia, PA: Temple University Press.

Misa, Thomas J., and Johan Schot. 2005. Introduction. Inventing Europe: Technology and the Hidden Integration of Europe. Edited by Arne Kaijser, Erik van der Vleuten, Helmuth Trischler, Hans Weinberger, David Arnold, Onno de Wit, Adri Albert de la Bruhe`ze, and Ruth Oldenziel. *History and Technology* 21: 1–19.

Mitchell, Sidney Alexander. 1960. *S.Z. Mitchell and the Electrical Industry*. New York: Farrar, Straus & Cudahy.

Morse, Frederick Tracy. 1932. *Power Plant Engineering and Design: a Text for Engineers and Students of Engineering Covering the Theory and Practice of Stationary Electric Generating Plants*. New York: D. Van Nostrand.

Mumford, Lewis. 1934. *Technics and Civilization*. New York: Harcourt, Brace and Company.

Needham, Andrew. 2015. *Power Lines: Phoenix and the Making of the Modern Southwest Phoenix and the Making of the Modern Southwest*. Princeton, NJ: Princeton University Press.

NIST, New Timeline of Deadliest California Wildfire Could Guide Lifesaving Research and Action, February 8, 2021. https://www.nist.gov/news-events/news/2021/02/new-timeline-deadliest-california-wildfire-could-guide-lifesaving-research

Nye, David. 1998. Consuming Power: A Social History of American Energies. In *Cambridge*. London: The MIT Press.

Nye, David E. 1990. *Electrifying America: Social Meanings of a New Technology*. Cambridge, MA: MIT Press.

Ottinger, Gwen. 2013. The Winds of Change: Environmental Justice in Energy Transitions. *Science as Culture* 22 (2): 222–229. https://doi.org/10.1080/09505431.2013.786996.

Passer, Harold C. 1953. *The Electrical Manufacturers, 1875–1900, a Study in Competition, Entrepreneurship, Technical Change and Economic Growth*. Cambridge, MA: Harvard University Press.

Pritchard, Sara B. 2011. *Confluence: The Nature of Technology and the Remaking of the Rhône*. Cambridge, MA: Harvard University Press.

———. 2017. The Trouble with Darkness: NASA's Suomi Satellite Images of Earth at Night. *Environmental History* 22 (2): 312–330. https://doi.org/10.1093/envhis/emw102.

Schipper, Frank, and Johan Schot. 2011. Infrastructural Europeanism, or the Project of Building Europe on Infrastructures: An Introduction. Edited by Anastasiadou, Hans Buiter, Johan Schot, Vincent Lagendijk, Le'onard Laborie, Christian Henrich-Franke, Isabel To'lle, Pascal Griset, and Vale'rie Schafer. *History and Technology* 27: 245–264.

Schot, Johan, and Vincent Lagendijk. 2008. Technocratic Internationalism in the Interwar Years: Building Europe on Motorways and Electricity Networks. *Journal of Modern European History* 6 (2): 196–217. https://doi.org/10.17104/1611-8944_2008_2_196.

Singer, Bayla S. 1983. *Power to the People: The Pennsylvania-New Jersey-Maryland Interconnection, 1925–1970*. Ph.D. dissertation, University of Pennsylvania

Slobodian, Quinn. 2018. *Globalists: The End of Empire and the Birth of Neoliberalism*. Cambridge: MA: Harvard University Press.

Stokes, Leah Cardamore. 2020. *Short Circuiting Policy: Interest Groups and the Battle over Clean Energy and Climate Policy in the American States*. New York: Oxford University Press.

Todd, Edmund N. 1989. Industry, State, and Electrical Technology in the Ruhr Circa 1900. *Osiris* 5 (1): 242–259. https://doi.org/10.1086/368689.

van der Vleuten, Erik, and Arne Kaijser. 2005. Networking Europe. *History and Technology* 21 (1): 21–48. https://doi.org/10.1080/07341510500037495.

Vikström, Hanna. 2020a. Producing Electric Light: How Resource Scarcity Affected Light Bulbs, 1880–1914. *Technology and Culture* 61 (3): 901–922. https://doi.org/10.1353/tech.2020.0078.

———. 2020b. Risk or Opportunity? The Extractive Industries' Response to Critical Metals in Renewable Energy Technologies, 1980–2014. *The Extractive Industries and Society* 7 (1): 20–28. https://doi.org/10.1016/j.exis.2020.01.004.

———. 2017. The Rush for Greenlandic Metals. *Technology's Stories* 5 (2): 1–15.

Wainwright, Nicholas B. 1961. *History of the Philadelphia Electric Company, 1881–1961*. Philadelphia, PA: Philadelphia Electric Company.

White, Richard. 2001. *The Organic Machine*. New York: Hill and Wang.

W. Bernard Carlson manages the M.Sc programs in TechInnovation and AgInnovation at the University of Galway in Ireland. He is also the Joseph L. Vaughan Emeritus Professor of Humanities at the University of Virginia. He is also a lecturer in the Tech Innovate program at the National University of Ireland Galway. He has written widely on inventors and electrical history, and his books include *Innovation as a Social Process: Elihu Thomson and the Rise of General Electric, 1870–1900* (Cambridge University Press, 1991) and *Tesla: Inventor of the Electrical Age* (Princeton University Press, 2013).

Erik M. Conway is an independent scholar and an author on seven books, including *Merchants of Doubt* and *The Big Myth*, both with Naomi Oreskes, and *Exploration and Engineering: The Jet Propulsion Laboratory and the Quest for Mars*.

Chapter 2
A Model for Heterogeneous Energy Transitions

David E. Nye

Abstract The theory that economic forces and political decisions shape energy transitions fails because it does not recognize that the process of invention constrains choices. Alternately, the theory of "deep transition" contends that technologies drive toward convergence in homogeneous systems. This determinism ignores the heterogeneous outcomes of previous transitions. The proposed alternative theory combines the work of Thomas Hughes with the path dependency school of economic historians. It distinguishes between transitions in production and in consumption and between pioneering and colonial transitions. It identifies six unavoidable stages in energy transitions and explains why they result not in convergence but heterogeneity.

Keywords Energy transitions · Decarbonization · Fossil fuels · Consumption · Heterogeneity · Theory · Technological momentum · Path dependency

2.1 Introduction

Many energy transitions have occurred in the last thousand years, notably from muscle power to water and wind power; from hydraulic to steam power; from coal to oil and gas; from gaslight to electric light; from horse-drawn vehicles to motorcars; and so forth. Each new form of energy was woven into the socio-economic system, and it never entirely replaced earlier forms. Water power persisted long after steam power emerged, and it eventually became a source of electricity. Muscle-powered bicycles remain an important form of transportation in The Netherlands and Denmark. Wood is still burned to heat houses. Horses still provide transportation to some remote places and for groups like the Amish who refuse to use automobiles. The multiple energy systems in each society interact with one another, and the particular mix of energies used in each is unique. A large volume would be required to evaluate

D. E. Nye (✉)
University of Southern Denmark, Odense, Denmark
e-mail: nye@sdu.dk

W. B. Carlson, E. M. Conway (eds.), *Electrical Conquest*, Archimedes 67,
https://doi.org/10.1007/978-3-031-44591-0_2

theories of energy transitions against the historical record, and the present chapter will not attempt such a comprehensive task.[1] Instead, it focuses on one transition, electrification, and evaluates three sorts of arguments about it:

1. Energy transitions result from economic and political choices in response to historical events;
2. Energy transitions are driven by an underlying "deep transition" that unfolds over centuries;
3. Energy transitions involve similar processes of invention, adoption, and development, but this leads not to uniformity but to heterogeneous regional systems.

This paper rejects the first two arguments in favor of the third.

2.2 General Characteristics of Energy Transitions

Regardless of theoretical approach, one can establish some facts about the energy transitions. Vaclav Smil noted in 2016 that renewable energies have been growing at an annual rate of 3% per year for a quarter century, which is slower than the growth of oil, coal, or natural gas when each of them was gaining global market share.[2] Overall, in 2015 fossil fuels, primarily coal and natural gas, supplied 79% of electrical energy, with most of the rest coming from nuclear plants and hydroelectric dams. Smil found that fossil fuels were losing market share at the rate of only 0.3% per year.[3] Smil notes hopeful signs of change, because "some of the long-established, gradually progressing energy transitions—declining energy intensities, gradual decarbonization of global energy supply, rising share of electricity in the final energy use—will continue regardless of successes or failures of specific energy sources and conversions."[4] Energy transitions are incremental, requiring more than a generation, and they concern not just the source of energy but how it is converted, stored, and used.

American energy transitions, for example from waterpower to steam power or from coal to oil and gas, have usually required about half a century.[5] This does not mean that waterpower or coal ceased to be used, as both were repurposed to generate electricity. Smil's international comparisons led to a similar conclusion: "all of the past shifts to new sources of primary energy have been gradual, prolonged affairs,

[1]Vaclaw Smil, *Energy Transitions: Global and National Perspectives*, 2nd edition (Praeger, 2017); Benjamin K. Sovacool, "How long will it take? Conceptualizing the temporal dynamics of energy transitions," *Energy Research and Social Science* 13 (2016): 202–207.

[2]Vaclav Smil, "Examining energy transitions: A dozen insights based on performance," *Energy Research and Social Science* 22 (2016): 195.

[3]Ibid., 195.

[4]Smil, *Energy Transitions*, 223.

[5]David E. Nye, *Consuming Power: A Social History of American Energies* (Cambridge: The MIT Press, 1998), 251–254.

with new sources taking decades from the beginning of production to become more than insignificant contributors, and then another two to three decades before capturing a quarter or a third of their respective markets."[6] Benjamin Sovacool's survey of social science literature underscores this agreement that energy transitions are slow, often taking half a century or more. They accelerate after a slow beginning, so that the graphic representation of their growth usually takes the form of an "S curve."[7]

The current transition is not only a change from fossil fuels to wind turbines and solar panels. In addition to decarbonization of production, it requires higher energy efficiency in consumption. When these two goals are pursued simultaneously, the cost to consumers can remain stable. The European Union has mandated higher efficiency in appliances and electric motors, increased lighting efficiency, pushed up the fuel economy of new automobiles, and promoted better home insulation. Well-insulated houses cost more to build, but they cut energy use in half, as do the most efficient appliances. Hybrid cars also cost more but they use half the energy of the average gasoline car. More efficient lightbulbs may cost several times as much, but they last much longer and cost less to use. A viable theory of electrical energy transitions must as a minimum explain why such transitions take half a century or longer, and it ought to take account of rising energy efficiency as well as changes in electrical generation.

Reduction in overall consumption is not merely a theoretical goal for it has been realized on a large scale. California made a decisive change in its energy policy during the 1970s, and its per capita electricity use leveled off during the next three decades, while it doubled for the United States as a whole.[8] By 2002 Californians were using just over 6000 kilowatt hours of electricity per person each year, compared to more than 14,000 kwh used by Texans, and almost 13,000 kwh by Floridians. California's result was achieved through state programs that focused on improved building design, higher energy efficiency, and elimination of wasteful practices. California also was an early leader in developing the alternative energies of windmills, solar power, and burning biomass. By 2010, compared to Texans and Floridians, Californian households were saving $1000 per family every year on their energy bills.[9] Had the entire nation changed course in the 1970s, as did California, Massachusetts and New York, US per capita energy use in 2020 would have been close to the European average. Instead, it is twice as high.

These examples demonstrate a number of vital points. First, decarbonization and greater energy efficiency can be achieved while lowering the energy bill for the consumer. It need not entail economic sacrifices for the economy as a whole or for individual consumers. Second, it is possible to slow the growth of an energy regime, even when it remains dominant, as was the case with fossil fuels for decades after c. 1975. Third, even inside the United States, change is not a monolithic process,

[6] Smil, *Energy Transitions*, 224.

[7] Sovacool, "How long will it take?" 202–207.

[8] William Calvin, *Global Fever* (Chicago: University of Chicago Press, 2008): 207.

[9] Ibid., 207.

moving toward a similar endpoint. Just as the four Scandinavian nations have taken different energy paths, different parts of the United States have different energy trajectories. North and South Dakota have moved aggressively to develop wind turbines; Arizona has less wind, but its solar systems receive nearly constant daily sunshine. Moreover, as Julie Cohn explains in this volume, the electrical system in Texas is literally disconnected from the rest of the United States, and it has evolved somewhat differently than the rest of the country. She found that when it came to building renewable energy systems "local voices have become loud, legitimate, and highly influential in shaping electricity infrastructure."[10] Localities differ. Some states have increased their reliance on alternative energies and increased their energy intensity and efficiency far more than others. Likewise, as the former mayor of New York, Michael Bloomberg, and former Sierra Club Executive Director Carl Pope noted in *Climate of Hope*, some cities and businesses are decarbonizing without waiting for Washington.[11]

Local initiatives were also the driver of the electrical transition in the United States between 1875 and 1920, when cities and corporations electrified without much direction or inspiration from the federal government. The city of Washington DC was never a leader in electrification, and long remained poorly lighted. In 1891 its public lighting was inferior to that in Minneapolis or Jersey City. Between 1880 and 1915 expositions were the most important showcases for electrification, and the world's most spectacular lighting effects were continually improved and displayed at world's fairs in Chicago (1893), Omaha (1898), Buffalo (1901) and San Francisco (1915).[12] The national government did not take the lead. Congress never was the driving force behind an international exposition, and Washington was never the site for one, in contrast to European capitals. If the national government did not lead the early transition to electrification, why should one expect the divided Congress of 2020 to lead a decarbonization transition?

In the 1870s and 1880s, when American cities and states oversaw development of the electric generating and transmission systems, it was by no means a straightforward process. There were many systems on offer. In the 1880s the now forgotten "moonlight towers" rose 200 or more feet above Detroit, New Orleans, Kansas City, Denver, and many west coast cities.[13] In the East, different arc light systems built closer to the ground battled against one another and against gas lighting. As late as 1900, when Cincinnati decided to install a new system of public lighting, it sent a committee to visit ten other cities to compare the available gas and electrical systems.[14] British policy in the 1880s was more centralized, but on the whole it

---

[10] Julie Cohn, "Large-scale Renewables and Local Gatekeepers: Moving Wind and Solar Power Across the Landscape," this volume.

[11] Michael Bloomberg and Carl Pope, *Climate of Hope* (New York: St. Martin's, 2017), 97–104.

[12] David E. Nye, *American Illuminations* (Cambridge: MIT Press, 2018), 70–71, 116–131.

[13] Ibid., 85–108.

[14] Ibid., 56–59.

impeded rather than assisted the transition.[15] So great were the fears that electricity would be monopolized that the Electric Lighting Act of 1882 was designed to make it difficult. Among other things, it decreed that British municipal authorities would have the right to take over private lighting companies 15 years after their charters went into effect. This and other provisions made investment so uncertain that few private parties chose to invest. In 1884 *The Electrician* complained that in Britain: "The fanatical dread of monopoly has resulted in there being no business to monopolize."[16] In Germany, the state played a more helpful role, and electrification was more rapid than in Britain.[17] But whatever the role of politicians, the transition in all three nations required decades and was based on the interplay between technical improvements, business competition, and political interventions. In the US electricity slowly triumphed over gas in the marketplace for public lighting, but in the 1880s only some city centers adopted arc lighting. On most streets in most cities, arc lights were the exception rather than the norm until after 1900. Even in a fast-growing city like Chicago that replaced much of its infrastructure after the fire of 1871, half the public lighting in 1913 was supplied either by gas or gasoline lamps.[18] In Britain, gaslight remained dominant for a generation longer, while Berlin surpassed London in this regard by c. 1900.

The energy transition to electricity was not similar from one nation to another, as each relied on a diversity of energy sources, developed divergent patterns of consumption, and passed quite different national, state, and local laws. An energy transition is a slow socio-technical process that requires about half a century. It is a many-sided transformation that varies from one context to another.

2.2.1 Three Arguments

The first argument presents political decisions and economic forces as the motors of change, while technology is a dependent variable. Historians of technology call these externalist arguments, because they typically pay little attention to how new energy systems are invented and developed. Instead, external demands are the "mothers of invention."[19] The central actors are entrepreneurs and politicians. Journalists often adopt this argument, describing the decarbonizing shift away from fossil fuels as movement controlled by politics and guided by a roadmap. The main question is how fast each society is able to move down the road to minimal

[15] Ibid., 64, 78.

[16] R. H. Parsons, *The Early Days of the Power Station Industry* (Cambridge: Cambridge University Press, 1940), 189–190.

[17] Thomas Parke Hughes, *Networks of Power* (Johns Hopkins University Press, 1983), 175–200.

[18] Nye, *American Illuminations*, 53.

[19] David E. Nye, *Technology Matters: Questions to Live With* (Cambridge: MIT Press, 2006), 17–28.

CO_2 emissions. The destination is clear, and the problem is one of mustering the necessary political will. This is the perspective of former Vice President Albert Gore, who challenged Americans to make a decarbonizing transition in a single decade. Just before the election of Donald Trump the policy analyst Peter Sircom Bromley argued "that a rapid transition away from fossil fuels is possible" and that it only required "a universal acknowledgement of the climate crisis."[20] However, during the last quarter century few nations have acted decisively in the face of that crisis. Summits have often agreed that there is an international emergency, yet few nations reduced their carbon footprint much between 1995 and 2015, and many actually increased their CO_2 emissions.[21]

Why so slow? Michaël Aklin concluded in *Renewables: The Politics of a Global Energy Transition* that the fossil fuel system is so entrenched in national political and economic structures that only an external shock such as an abrupt rise in oil prices can force change.[22] From this perspective, the failure of nuclear power to take hold was due to the shock of the accidents at Chernobyl and Three Mile Island. However, in part that failure also was due to the anti-nuclear publicity of coal producers and to popular resistance to nuclear energy that was present before the famous accidents. On the other hand, the shock of energy shortages during the 1970s did prompt innovations in wind power in a few countries, notably Denmark, which now produces more than half its electricity with wind turbines.

Nevertheless, in historical perspective, external shocks seldom play a central a role in energy transitions. Americans did not adopt coal because wood was in short supply but because coal was a readily available form of concentrated energy that was easy to transport and to store. If there had been no US coal fields, then external shocks to the supply might have been a factor in the decline of coal in favor of other fossil fuels. But this was not the case, and none of the US energy transitions before 1970 were driven by shortages or external shocks. There were price shocks in the oil market during the 1970s, but this did not fundamentally alter the US energy system, which for decades afterwards remained committed to fossil fuels for transportation, power stations, and home heating. Indeed, despite the oil shortages of the 1970s Americans increased their electricity consumption by 50 percent, drove 20 percent more miles, continued to move out of cities into suburbs, and shifted the majority of their purchases to new shopping malls. Cities abandoned mass transit and railroads continued to decline.[23] Even that external shock did not cause an energy transition.

The processes of invention and the forces of competition are more important than economic forces or political decisions. In the 1870s there was a strong interest in

---

[20] Peter Sircom Bromley, "Extraordinary Interventions: Toward a framework for rapid transition and emission reductions in the energy space," *Energy Research & Social Science* 22 (2016): 165–171.

[21] See Robert S. Emmett and David E. Nye, *The Environmental Humanities* (Cambridge: MIT Press, 2017), 50–55.

[22] Michaël Aklin and Johannes Urpelainen, *Renewables: The Politics of a Global Energy Transition* (Cambridge, MA: MIT Press, 2018).

[23] Nye, *Consuming Power*, 221–223.

adopting electric streetcars, which would eventually prove to be faster, cheaper, and cleaner than horsecars, but the technologies did not yet exist in a reliable form. In the 1880s cities that adopted early systems found that they worked poorly, and horse drawn streetcars persisted until the technical obstacles had been overcome at the end of that decade.[24] Often, as in this case, technical problems cannot be quickly solved. In the 1970s the demand for an alternative to fossil fuels was strong, but the technologies of wind and solar power still required decades of research and development. Even as late as the first Obama administration (2009–2013) they had not won much market share in the US. Obama supported alternative energy through a loan program, but Solyndra and several other solar power companies that received support nevertheless went bankrupt.[25] Most loans were paid back, and the default rate overall was only 2.28%, but fossil fuels remained overwhelmingly dominant. The US tried to hasten this transition, but it did not move as rapidly into wind and solar power as Germany, Spain, Portugal, or Denmark, each of which had different geographies and political structures, but they had in common more expensive energy markets. The US only moved more decisively toward alternative energies after 2015, when new solar panels and wind turbines could deliver electricity more cheaply than coal-fired power plants. This price advantage combined with widespread concern about climate change promoted more rapid change. The Trump administration tried to prop-up coal-fired power plants, in part by weakening environmental regulations to allow them to pollute more than before. Nevertheless, in 2019 and 2020 alone, more than 75 US coal-burning power stations shut down.[26] As these examples suggest, technical problems prevented a rapid energy transition that the Obama administration promoted, while technical solutions to those problems hastened that transition even though the Trump Administration worked against it. In short, neither external shocks nor political will by themselves can explain the process of energy transitions. The process of invention must be understood in detail in order to assess the timing and progress of a transition. Politics and economics together are important but insufficient by themselves to explain energy transitions. The first kind of argument is therefore inadequate.

The second kind of argument takes the diametrically opposed view, positing that each energy transition inexorably occurs, regardless of voters, businessmen, consumers or politicians. Technology itself drives change, and the results are rather homogeneous. Societies adopt energy forms in a particular sequence, moving from muscle power to water to steam to electricity and so on. This argument comes in many variants, but it leads toward uniformity. Change is understood to be inevitable. Political and economic decisions may speed or retard the process, but the transition

[24]David E. Nye, *Electrifying America: Social Meanings of a New Technology* (Cambridge: MIT Press, 1990), 85–89.

[25]National Public Radio, Morning Edition, "After Solyndra Loss, U.S. Energy Loan Program Turning a Profit," Nov 13, 2014.

[26]Benjamin Storrow, "Coal's Decline Continues with 13 Plant Closures Announced in 2020: The fuel is increasingly uncompetitive with cheaper natural gas and renewable energy," *Scientific American* May 27, 2020.

will move forward even when resisted. In its most extreme forms, this argument equated mastery of energy technologies with evolutionary advance. In 1949 Leslie White argued in his influential book *The Science of Culture* that "the degree of civilization of any epoch, people, or group of peoples, is measured by ability to utilize energy for human advancement or needs."[27] The idea that control of more energy made possible advanced civilization is no longer accepted in an age confronting global warming.

More recent theorists have developed nuanced models of energy transitions. One that is widely discussed is the theory of deep transitions presented in in the work of Johan Schot, John Grin, Jan Rotmans, Laur Kanger, and others.[28] Their meta-theory of socio-technical change argues that the first "deep transition" began in Britain and spread to other nations. A series of four "surges" followed, one of which was electrification. Taken together the "First Deep Transition" required more than two centuries. They further argue that the world is now entering a "Second Deep Transition" that is in part a response to the environmental crisis, and in part due to the emergence of alternative energies and other new technologies. Because this "Second Deep Transition" lies mostly in the future, it cannot be precisely described, but it too will consist of a series of surges. This research group has developed a "Multilevel Perspective," (MLP) which describes the interaction of "niches, socio-technical regimes and sociotechnical landscapes." The regimes are "sets of rules or routines directing the behavior of actors" within "a specific socio-technical system." In contrast, in the "niches" alternative ideas and emergent technologies gradually develop until they can challenge a dominant regime. The interactions between regimes and niches occur within a surrounding context, which MLP scholars refer to as the landscape, defined as "the exogenous environment shaping both niches and regimes. Landscape pressures involve trends such as globalization, urbanization, and climate change, but also events such as wars, natural disasters, and economic crises." This metaphorical use of "landscape" in MLP terminology means less a physical place than a congeries of historical events and social trends which constitute the context within which surges, regime changes, and transitions occur. However, some MLP publications also hold on to the more physical sense of landscape as a space shaped by a socio-technical infrastructure such as a railroad system, with all its attendant services and concentrations of energy, manufacturing, and consumption.

As a general theorization of historical change since c. 1770, MLP seems to provide a way to understand the emergence and development of industrial society. However, there are some serious problems. First, MLP scarcely is interested in geographical or cultural differences, but argues instead for uniformity in the rhythms

[27] Leslie White, *The Science of Culture* (New York: Grove Press, 1949), 368.

[28] Johan Schot, Laur Kanger, "Deep Transitions: Emergence, Acceleration, Stabilization and Directionality," *Research Policy* 47:6 (July 2018): 1045–1059. See also John Grin, Jan Rotmans, Johan Schot, *Transitions to sustainable development: new directions in the study of long-term transformative change* (London: Routledge, 2010); and Johan Schot, "Confronting the Second Deep Transition through the Historical Imagination," *Technology and Culture* 57:2 (2016): 445–456.

of historical change. Despite occasional rhetorical salutes to contingency and choice that suggest the model can accommodate a more pluralistic history, the argument (and the diagrams and charts that illustrate it) presents a deterministic chain of events. In this model, the adoption of energy systems, notably steam power and electrification, occurred in every society in much the same way and meant much the same thing. Ideology and politics do not have an important place in the argument. The MLP model does not conceptualize much resistance to surges of change, as its rules, regimes, and techno-social landscapes roll over ideologies and cultural differences, creating a universal experience of modernity.

In contrast, many historians of energy foreground politics and ideology, as represented in this volume by Cyrus C. M. Mody on the oil industry in the turbulent 1970s and by Stathis Arapostathis and Yannis Fotopoulos on Greek electrification from 1940 to 2010.[29] They remind us that energy systems often serve state purposes. Governments subsidize and regulate energy markets to achieve political and economic ends. States also promote particular kinds of energy (and models of energy development) to achieve their goals, as illustrated by Abby Spinak's chapter on how rural electrification in Ecuador was promoted by the American Peace Corps during the Cold War.[30]

Such work does not fit well within MLP theory, where actors are de-emphasized and often not even named. In it, inventors, entrepreneurs, workers, politicians, and consumers are caught up in vast structural changes over which they have little control. In MLP, history consists of socio-technical surges, each of which takes the better part of a lifetime, and each of which creates a regime that defines dominant rules for economic behavior and social development. Historical change is explained by abstract forces linked like a series of intersecting gears, driven by a "deep transition" gear that rumbles inexorably beneath the surface of events, carrying history forward for centuries. Then it gives way to a second, deep transition that is required because of the "accumulated social and ecological challenges" created by the first deep transition, which necessitate not only a transformation of existing socio-technical systems but also the development and imposition of new rules.

The MLP model seems more cogent to social scientists than to historians, who see not a universal pattern but rather variegated regional differences that overwhelm attempts to generalize. Roger Fouquet from the London School of Economics concluded in 2016 that "at present, no formal economic theory of how energy transitions unfold exists."[31] Only if one looks at an unrepresentative selection of nations does a theory seem within reach. To construct a theory in which energy

[29] Stathis Arapostathis and Yannis Fotopoulos, "Between material dependencies and the politics of electricity transitions: Networks of power in Greece, 1940–2010," this volume; Cyrus C. M. Mody, "Energy Modernism: The Oil Industry and the Energy Infrastructure Scramble of the Long 1970s," this volume.

[30] Abby Spinak, "Co-ops Against Castroism: American-style Rural Electrification in the Global Countryside," this volume.

[31] Roger Fouquet, "Historical energy transitions: Speed, prices, and system transformation," *Energy Research and Social Science* 22 (2016): 11.

transitions converge in uniform systems, it would be helpful if nations started with similar patterns of energy production and consumption and then went through similar processes of change, ending up with similar energy regimes. But nations do not begin or end transitions with similar energy systems. For a little more than a century the reliance on fossil fuels made energy systems converge more than they had before. This was because oil, coal, and gas could be shipped long distances and stored for long periods, but even so differences in geography and social systems prevented full homogeneity. Alternative energies are far more dependent on local weather and geography and as they are adopted energy systems will diverge.

There are great variations in regional and national electrical systems, which make it difficult to substantiate a deterministic theory of energy transitions. One might expect neighboring countries with similar political systems to have similar energy profiles. They do not. Scandinavia provides a good example. Norway produces almost 100% of its electricity from hydropower. Sweden produces about half its electricity with hydro but the other half comes from nuclear reactors. Denmark has almost no hydropower, and it has refused to develop nuclear power. Instead, it burns natural gas from its North Sea oil fields, and supplies more than half its electricity with wind turbines. Iceland has taken yet another path, developing thermal power. Its largest electrical station is located on the edge of a volcano and extracts boiling water from fifty wells that are between 1000 and 2000 meters deep. The four Scandinavian nations have similar political systems, religious beliefs, and social welfare institutions, but they have entirely different systems of electrical generation. In part, this is due to national geography, as Denmark does not have mountains like Norway or volcanos like Iceland. But cultural choices are also salient, notably in the varying levels of resistance to nuclear power and in the level of interest in wind turbines.

Such differences in energy systems are not the exception but the norm. Germany has decided to abandon nuclear power and is rapidly adopting wind and solar energy. Next door, France is heavily committed to nuclear power, which provides the lion's share of its electricity and has become a symbol of the nation's technical prowess. New Zealand has extensive waterpower, while Australia has developed coal fields. Such national differences are compounded by the taxes, subsidies, incentives, and regulations that governments use to shape energy markets, for example by making gasoline cheaper or more expensive, or by discouraging or stimulating adoption of electric cars. One nation builds co-generation plants that provide electricity and steam heating for entire towns, while a more individualistic nation relies on stand-alone heating systems for each household. The universalism of MLP theory does not accord well with so much national variation in energy systems. The divergence is pronounced, and it is not diminishing.

The third kind of argument, developed as a model in the second half of this essay, does not present energy transitions as a deterministic process leading to uniformity. Rather, each society selects between various technical possibilities to shape an energy system based on its geography, history, and politics. The systems that result are therefore neither homogeneous nor inevitable but heterogeneous and contingent. Uniformity lies not in the energy systems nations select but rather in the six stages that every technology in a transition undergoes: invention, introduction, resolution of technical problems, expansion, technological momentum, and market domination. The sequence of stages is a constant, but they lead not to uniformity but to diversity. This model of energy transitions synthesizes two earlier traditions, the work of economic historians and historians of technology. Before going into the details of this argument, however, some general characteristics of energy transitions need to be made specific.

2.2.2 A Typology of Transitions

The rejection of deterministic theories such as the MLP conception of a linear development driven by deep structural forces demands a reconception of energy transitions that includes politics and culture. An adequate theory of energy transitions that accounts for differences must begin by noting the various kinds of possible transitions. There appear to be at least six, each of which can occur in different historical circumstances and cultural contexts.

(1) An *energy production and consumption transition* involves a wide range of actors. As an example, consider the shift from horses, oxen, and mules (or muscle power) to electricity and gasoline motors. As late as 1890 horses were the primary source of power on farms and in cities, including the most industrialized areas of Europe and the United States. Huge investments were tied up in feeding, housing, and providing medical care for horses, and a large workforce was required for that alone. On the consumption side, almost every adult had to know how to deal with horses. Consumption of saddles, riding clothes, harnesses, wagons, buggies, and other horse-drawn equipment was an important part of every local economy. The energy transition to motors destroyed many businesses and made both ownership of horses unnecessary and knowledge of them obsolete. That was a total energy transition, where one system replaced another, making nearly all of its stables, feed stores, equipment, and practical knowledge obsolete. Few skills were transferable. Consumers had to acquire new skills to operate and maintain an automobile. Such complete transitions have widespread effects on employment, land use, and the organization of everyday life.

(2) In contrast, *production transitions* replace the source of energy but require few changes in consumption. Wind and solar electrical generation requires a replacement of production technologies and modification of transmission, and storage, but they have little effect on consumption. The consumer retains the same wiring system and appliances as before and needs to make few adjustments, unless he or she wants to install generating equipment and become self-sufficient. An electric automobile is somewhat different from a gasoline car, but the steering wheel, tires, brakes, signaling, locking systems, and driving experience remain much the same. Likewise, the technology of electric motors is well-known. The shift to

renewable electrical energy is primarily a production transition that is less complicated and less expensive for the consumer than the earlier transition from horse drawn vehicles to cars and trucks.

(3) There are also energy *consumption transitions* in which the production of energy does not change, but a new or larger public begins to use a form of energy. The result is by no means always rapid or automatic. As Ruth Sandwell has demonstrated, Canadian consumers adopted electricity differently than Americans. They long used it primarily for lighting and had fewer electrical appliances,[32] exemplifying how neighboring nations with shared cultural characteristics nevertheless may have different energy transitions.

(4) There are also *redeployment transitions* in which an energy source does not disappear when it is replaced in one market because it expands into another market. This redeployment may also entail technical improvements and the development of new resources. A good example is the case of gas. When eclipsed in the street-lighting market during the early twentieth century, it expanded into home heating and cooking. By 1930, although it had disappeared from most city streets, more gas was being sold than before. This expansion occurred along with a shift away from manufactured coal-gas to natural gas, transported by pipelines. Such *redeployment transitions* have some similarities to *production transitions*, but they are based less on new technologies than on comparative costs and the forces of supply and demand. *Redeployment transitions* typically occur as a knock-on effect during the introduction of a new energy source. Another example is the shift in the primary use of coal, from driving steam engines and heating buildings to generating electricity.

(5) Some energy technologies exemplify *post-transition niche persistence*. An older technology may persist because the workforce has not made the necessary skills transition to use a new energy system, or because consumers are more comfortable with familiar methods. There are also niche markets where an old technology persists in places that newer technologies cannot easily serve or that are too expensive. Examples include wood-burning stoves in remote cabins and vacation homes; or the use of electric vehicles in warehouses and golf courses after 1930 when the electric car had disappeared from highways. *Niche persistence* may also be based less on economic than on cultural factors. Examples include the use of gas-lighting in historic districts to recreate the ambiance of the past, or the Amish rejection of steam and electric power in favor of horses. Fossil fuels seem likely to persist in the niche market of aviation after being phased out for automobiles.

(6) One can also distinguish between *pioneering transitions* which encounter new technical problems, and *imitative transitions* in which a region or nation catches up by installing technologies developed elsewhere. In such cases, the problems

[32] Ruth Sandwell, "Heating and cooking in rural Canada: home energy in transition, 1850–1940," *History of Retailing and Consumption* 4:1, pp. 64–80.

of technology transfer amd adoption are less technical than they are political and cultural. Who is making the transition? In a *colonial imitative transition,* a governing elite may be seeking to change an indigenous culture. The colonizing power may empty a landscape and define it as a sacrifice zone that will be flooded or mined in the name of progress. Such a process can provoke resistance. There will be winners, losers, and environmental effects. As Rob Nixon argues, in such cases some people become developmental refugees or "uninhabitants."[33] In this volume, Nathan Kapoor provides one example of such an imitative transition in New Zealand, which developed hydroelectric dams to benefit the colonizers.[34] The creation of the Panama Canal offers another example. The United States took control of a strip of land between the Atlantic and Pacific oceans, defined it as a "canal zone" and then developed it for commercial and military purposes. The canal required a hydroelectric dam to regulate the water levels in the canal, to provide the electricity to open and close the locks, to pump water into locks, to drive the locomotives towing ships through the canal, and to light the facility throughout the night in order to maximize traffic flow. Electricity was also essential to provide comfortable buildings for the American engineers and administrators. For a generation, the Canal Zone was a popular tourist destination, and it was widely praised as a progressive utopia where modern technology had mastered nature.[35] However, the military controlled the Canal Zone. It commanded a small army of mostly Caribbean laborers to construct and maintain the facilities, and it subordinated the local population who were defined as outsiders, or unihabitants of the Zone. In such imitative transitions, outsiders impose new technologies, and the problems are often less technological than social and political.

This typology of possible energy transitions is only preliminary, but it suggests the heterogeneity of possible systems, some only partially developed, others persistent in niches long after they have become technically obsolete, and still others just emerging. This variety further underscores the inability of the MLP model to capture the complexity of historical change. The same could be said for any model that posits an inexorable succession of dominant regimes to describe the history of energy. The third model recognizes political, geographical, and cultural influences on energy systems, making each somewhat unique, but at the same time locates a core set of unavoidable stages in development. It also makes clear why transitions usually require roughly half a century.

---

[33] Rob Nixon, *Slow Violence and the Environmentalism of the Poor* (Cambridge: Harvard University Press, 2011), 150–174.

[34] Nathan Kapoor, "Colonial Self-Sufficiency: Electrifying the 'Britain of the South,' 1880–1914," this volume.

[35] Alexander Missal, *Seaway to the Future* (Madison: University of Wisconsin Press, 2008).

2.3 A Model for Heterogeneous Electrical Energy Transitions

The six-stage model for electrical energy transitions proposed in this section draws on two research traditions, that of economic historians and that of historians of technology.[36] Both of these research traditions are useful because they do not share MLP theory's assumption that development is inexorable. Instead, both economic and technological historians have focused a good deal on blockages and rigidities that interfere with energy transitions. Economic historians speak of "lock in" and "path dependence," while historians of technology are concerned with "reverse salients" and "technological momentum." In different ways, these research traditions seek to explain incomplete transitions and failures of rationality. The economic historians seek to explain why the most efficient form of a technology often is not adopted. For example, the typewriter keyboard is not the most efficient possible. Technological historians study obstacles that had to be overcome in order to implement a new system, such as the need for reliable and efficient electric motors in the 1880s, or the search for ways to transmit electricity efficiently over long distances. Only when such problems were overcome could electricity become fully competitive with existing technologies based on steam and gas. Neither group has paid much attention to the other's scholarship, but their work can be combined and modified to suggest an alternative model to MLP, an alternative that is more deeply grounded in empirical examples and that does not assume different nations converge on a uniform system of production or similar patterns of consumption. Technologies set limits to what is possible, but within those limits there are considerable national and regional variations.

2.3.1 Thomas Hughes and Technological Momentum

In his study of the development of the electrical industry, *Networks of Power,* Thomas Hughes laid out the stages in an energy production and consumption transition. In his theory, a transition culminates when it achieves "technological momentum," a concept that Hughes also applied to other large-scale systems, such as the railway. Hughes did not use the terms "lock in" or "path dependence" in his five stages of system development. In the case of US electrification, the first stage began in the 1870s with invention and early development of practical electrical lighting in a few locations (1875–1882). Second came technology transfer to other regions (1882–1890). Third came growth (1890–1900), which was only possible after solving major technological problems, which Hughes calls "reverse salients." Hughes adopted this term from military tactics, in which a salient is a bulge in the

[36] Sovacool, "How long will it take?" 202–207.

trench lines between two opposing forces. The protrusion of one army's lines forward into the trenches of the enemy threatens to halt or reverse its progress. For example, at Verdun the bulge in the French lines was "a reverse salient" that the Germans had to eliminate if they wanted to advance. Hughes adopted this term and used it to define sticking points or bottlenecks in a technological system: "Networked systems consisting of a number of heterogeneous components or firms evolve like a shifting front. The components and firms that fall behind, I name 'reverse salients'; those that move ahead are 'salients'. Reverse salients in networked systems need to be corrected in order for the systems to continue to evolve. An example of a reverse salient in early electric power systems was the absence of a satisfactory motor for alternating current systems."[37]

There were many such technical problems that constituted reverse salients in the early electrical industry. As these difficulties were overcome, the industry could develop the infrastructures of production that lowered unit costs. As the system became more efficient after 1900 it entered the crucial fourth stage that Hughes calls "technological momentum," when electricity became a preferred standard source of light, heat, and power. Finally, came the mature stage (after. c. 1910) where the problems faced by management required financiers and consulting engineers as problem solvers.[38] Beginning not with the first demonstration of an arc light in 1808, but with the first practical arc lighting systems, these five stages took 40 years (1875–1915). Nevertheless, in 1915 only 15 percent of US homes had electricity, and only 20 percent of factory power was electric.[39] Hughes's model requires a sixth stage when electricity's market share rose to over 50 percent in factories, homes, and businesses, in a process that required an additional 25 years after 1915. Altogether, the full adoption of the electrical system in the US required about 65 years (1875–1940). Most other historians agree. Sovacool notes that a survey of fourteen historical transitions concluded that they take a minimum of 40 years and in some cases require centuries.[40]

The Hughes model is not deterministic, and his fourth stage of "technological momentum" is crucial. Before then, a technology may falter and fail to take hold. The success of a new energy technology after it enters the market remains uncertain for 20 to 25 years. Technological momentum only arises after considerable development. The design is not "locked in" early but rather late, in contrast to one of the governing ideas of economic historians, who usually see "path dependence" beginning at an earlier stage.[41] Only some systems achieve technological momentum,[42]

[37]Thomas Parke Hughes, "Afterword," *Annales Historiques de l'électricité,* (juin 2004): 174.

[38]Hughes, *Networks of Power*, 14–17.

[39]Nye, *Electrifying America*, 261, 187.

[40]Sovacool, "How long will it take?" 207.

[41]David E. Nye, "Electricity and Culture: Conceptualizing the American Case," in *Annales Historiques de l'électricité* (juin 2004): 125–138.

[42]Thomas Parke Hughes, "Technological Momentum: Hydrogenation in Germany, 1900–1933," *Past and Present* (Aug. 1969): 106–132.

and they vary from one culture to another. The bicycle achieved technological momentum in Denmark and the Netherlands, and for three generations before c. 1960 these countries relied more on bicycles than automobiles. They developed an infrastructure of paved trails, urban cycle lanes, and traffic lights for bicycles, as well as extensive repair services and specialized equipment. In contrast, in the United States, the automobile achieved technological momentum by 1920 and became the center of a socio-technical system that almost entirely displaced the bicycle for adult transportation.

Hughes defined "network system momentum as the inertia of a mass in motion." He argued that,

> . . .the mass of an electric utility holding company, for example, consists of the following: vested capital, especially in the technical core of its power systems; the system-specific skills or knowledge of the workers, engineers, scientists, and managers; dedicated physical structures such as dams and the housing of generating plants; and task-oriented organizational bureaucracy. The direction of the company I associate metaphorically with its strategic policies and the characteristics of its products. The velocity arises from the growth and evolutionary change over time. As systems become larger and more complex, they gather momentum; they become less shaped by, and more the shaper of, their context or environment.[43]

Once technological momentum has been achieved, a new kind of management is needed. In the first three stages inventors are prominent. In the fourth, fifth, and sixth stages manager-entrepreneurs become central figures. In the case of the electrical industry, Hughes presented Samuel Insull as an exemplary figure. Leaders of the mature stage seldom were technicians because once technological momentum had been achieved, the central problems of utilities changed. Financial entrepreneurs with political skills were needed. With the emergence of such figures, the integration of Hughes's theory with Path Dependency Theory can begin.

Before turning to that discussion, note how useful the concept of technological momentum is for understanding large systems. These have some flexibility in their initial phases, when each region or nation can move in different directions. However, once technical specifications are widely adopted, and the system is run by a bureaucracy with thousands of workers, it is less flexible and resists outside pressures.[44] Despite the wishful thinking of policy analysts and political scientists, once a socio-technical system achieves momentum, it cannot be rapidly transformed. When a system such as electrification is woven into society, it becomes naturalized as part of everyday life. People become deeply dependent upon it, and after a generation they can scarcely imagine their lives without it. The same applies to many individual elements of an energy system. For example, in 1950 scarcely any American houses had air conditioning. By 1970 it had become common, and by 2000 it was installed in almost 80% of American homes. A similar process of

[43] Hughes, "Afterword," 175–176.

[44] See also Thomas Parke Hughes, "Technological Momentum," in Merritt Roe Smith and Leo Marx, *Does Technology Drive History?* (Cambridge: The MIT Press, 1994), 111; and Thomas Parke Hughes, *American Genesis* (New York: Penguin, 1989), 460.

naturalization occurred for televisions, clothes dryers, dishwashers, microwave ovens, and personal computers. In 1950 virtually no one had these devices. Two generations later, doing without them was considered a hardship. The naturalization of each additional appliance further increased the technological momentum of the electrical system. By 2001 lighting was only 8.8% of total electricity use. More electricity was used on space heating (10.1%), water heating (9.1%), air conditioning (16%), or family freezers and stoves (17.2%).[45] The ever-lengthening list of domestic appliances that have become "natural" and "necessary" exemplifies the momentum of the system and suggests how difficult it can be to shift to a new energy regime, especially if it is a full transition that demands new technologies for both producers and consumers.

2.3.2 Path Dependence Theory

Studies of path dependence examine why corporations and nations persist in patterns of behavior, or keep to the same path, when superior alternatives are available. Because inefficient patterns persist, the economists W. Paul David, Brian Arthur, and others have argued that history and culture must be taken into account to explain some economic behavior.[46] David used the QWERTY typewriter keyboard as an example of path dependence. This internationally familiar keyboard is not a random arrangement of letters. It was designed to minimize the jamming of keys in the early typewriters. The arrangement was based in part on a knowledge of how frequently individual letters are used in English. (Notably, France did not adopt QWERTY.) This keyboard was based not on mathematical analysis but experience and some guesswork. The original typewriter keyboard did not become the standard, and the standardization of typewriter design as a whole did not occur until 1899.[47]

In 1936 August Dvorak introduced a more ergonomically correct keyboard layout, but it failed to win acceptance. Some historians have argued that there is insufficient proof that Dvorak's arrangement definitely was superior, but it seems indisputable that another keyboard that is easier to use could be created. However, as David argues, path dependence (Hughes would call it inertia) has prevented the shift to a new layout. That change would include not only replacing all existing keyboards but also retraining millions of people who are accustomed to the QWERTY layout.[48] The implications of such an argument are important: technological designs persist,

[45] Statistics from US Department of Energy, "Table US-1. Electricity Consumption by End Use in U.S. Households, 2001," http://www.eia.doe.gov/emeu/reps/enduse/er01_us_tab1.html

[46] W. Brian Arthur, "Competing technologies, increasing returns, and lock-in by historical events," *Economic Journal* vol. 99, (March 1989): 116–131.

[47] James Utterback, *Mastering the Dynamics of Innovation* (Boston: Harvard Business School Press, 1994), 1–11.

[48] Paul A. David, "Clio and the Economics of QWERTY," *American Economic Review* 75 (May 1985): 332–337. For a discussion, see Utterback, Ibid., 3–13.

even when inefficient, long after the necessity for them has disappeared. When typewriters became electronic, jamming keys ceased to be a problem and any keyboard layout was possible. Yet, QWERTY remained. Techno-social arrangements become entrenched when they are widely adopted and become parts of many systems.

David's analysis of QWERTY has become something of a founding myth for the path dependence school.[49] However, it is not an ideal example, since the typewriter keyboard is not peculiar to one particular firm or nation. The importance of path dependence lies in whether it can explain why some consumers, some firms and some nations fail to adapt to new systems, rather than why, at times, almost everyone sticks to an antiquated but familiar standard. QWERTY may exemplify path dependence in a general sense, but it is somewhat divorced from the problems of corporate (or national) competition, growth, and decline. An even more serious problem with the QWERTY example, from the point of view of historians of technology, is that the keyboard arrangement is only minimally a technological problem. Compare it, for example, to Hughes's "reverse salient," of the car battery that thwarted the electric car industry for a century. Edison and many others after him sought a car battery that is light-weight and can be quickly recharged. Because there was no remedy for this reverse salient between 1890 and 1915, the electric car fell decisively behind the gasoline automobile and retreated into a few niche markets, such as warehouses and golf courses. Edison wrestled more successfully with another reverse salient in the late 1870s, that of finding a light bulb filament that glowed long enough and was bright enough to make it commercially viable. In these examples, the choices inventors and entrepreneurs made are less "accidental" or arbitrary than the QWERTY example suggests. However, economic historians seem to prefer arbitrary choices, such as the width of railway gauge[50] or the arrangement of letters on a keyboard.

Path development economists are keen to demonstrate that "lock in" is fortuitous. They look for accidents. The earlier in the cycle of product development an accidental choice occurs, and the longer lasting its effects, the better for their argument. In contrast, historians of technology study experimentation and invention. They do not see the design of the light bulb or the battery as historical accidents, but as painstaking social and technical constructions. To write about the history of the electrical system in the spirit of path dependency, one might focus on "accidental" features of the system that became permanent, such as the American adoption of the 110-standard voltage vs. 220 for continental Europe.[51] But historians of electrification focus far less on such issues than on the problems faced by inventors and

[49] Paul A. David, "Path Dependence; A foundational concept for social science," *Cliometrica* 1:2 (May 2007): 91–114.

[50] On railway gauge, see D. J. Puffert, "Path dependence in spatial networks: the standardization of railway track gauge," *Explorations in Economic History* 39:3 (July 2002): 282–314.

[51] On Europe vs. the United States, see David E. Nye, "Path Insistence: Comparing European and American Attitudes Toward Energy," *Journal of International Affairs*, 53:1 (Fall 1999): 129–148.

entrepreneurs, how these were solved, and the process of refining an invention into a marketable product. Success is only possible (though by no means inevitable) if reverse salients, such as a longer-lasting and quick charging car battery can be overcome. Path dependence theory uses the metaphor of being "locked in" to characterize the moment when a design becomes fixed. This is a less happy word choice than Hughes's metaphor of technological momentum. Being "locked in" suggests immobility. In contrast, the momentum of a large, moving system is a less static and more accurate metaphor.

Economic historians focus less on invention than on what happens after lock-in, when, as Paul Krugman explains, "increasing returns and cumulative processes are pervasive and give an often-decisive role to historical accident."[52] Gavin Wright explained path dependency's larger implications this way: "As nations continue to differ from each other in corporate structures, in labour relations, in educational systems, in economic policies, and in many other ways, so we should expect them to continue to differ in the technologies they generate and adopt." In contrast to MLP theory, this statement recognizes the persistence of cultural differences which are sustained by external structures, policies, and cultural systems. For many economic historians, however, technologies remain black boxes, which magically appear in response to needs. In short, path dependency tends to assume that necessity is the mother of invention. This approach is seldom acceptable to technological historians.[53] Wright concluded, "the broadest lesson is that if economics really wants to take technology seriously, economics will have to become a more historical discipline."[54] A historian of technology would agree but add that economists will have to take more seriously the processes of invention and early product development.

Another prominent proponent of path dependence theory, Douglass C. North, has also argued that social and legal institutions should be part of economic analysis. He declared: "Path dependence is more than the incremental process of institutional evolution in which yesterday's institutional framework provides the opportunity set for today's organizations and individual entrepreneurs (political or economic). The institutional matrix consists of an interdependent web of institutions and consequent political and economic organizations that are characterized by massive increasing returns."[55] In such statements, the word technology does not even appear.

---

[52] Paul Krugman, "History and Industry Location: The Case of the Manufacturing Belt," *The American Economic Review* 81:2 (May 1991): 82.

[53] See John Staudenmaier, "Rationality versus Contingency in the History of Technology," in Smith and Marx, *Does Technology Drive History?* (Cambridge: The MIT Press, 1994), 260–273; Nye, *Technology Matters*, 15–31.

[54] Gavin Wright, "Towards a More Historical Approach to Technological Change," *The Economic Journal* 107: 444 (Sept. 1997): 1565.

[55] These returns include both profits and lowered transaction costs. Douglass C. North, "Institutions," *Journal of Economic Perspectives* 5:1 (Winter, 1991): 109.

These economists published after Hughes did, but they made no reference to his *Networks of Power*.[56] Yet if neither North, Wright, Krugman, David, nor any other path dependence theorist seems to have engaged the concept of technological momentum, the two approaches do have commonalities. In each case, technology is understood to be inflected by its cultural context. In each case, this context plays a particularly important role in the early stages of system development until it matures, both technically and organizationally. The path dependence economists speak of an "institutional matrix," Hughes sees a "socio-technical system." In each case, the authors conceive of an interdependent web of institutions that is inextricably woven together and that becomes less flexible over time. In each case, the mature form of the system resists radical change. Finally, both Hughes and the path dependence economists describe feedback mechanisms that reinforce the chosen technological form as it is institutionalized. The economists find that there are "increasing returns and cumulative processes" as an institution increases its profits, by incrementally improving set routines in production and distribution. Hughes concluded in *Networks of Power* that "the system builders' efforts were usually directed to increasing the size of the systems incrementally."[57]

The concepts of "technological momentum" and "path dependence" both argue that culture shapes technology in ways that persist over long periods. Krugman observed that, "If there is one single area of economics in which path dependence is unmistakable, it is economic geography—the location of production in space."[58] He argues that path dependence explains why the industrial manufacturing belt in the United States long remained in the region that first industrialized, i.e. the area bounded by Boston and Baltimore in the east, and St. Louis and Chicago to the west. Long after US population and resource development had shifted toward the south and west, the manufacturing commitment to this region persisted. A similar kind of argument can be made concerning the electrical infrastructure, which was first thoroughly developed in this same region. Hughes devotes much of his argument to the New York, Pittsburg, and Chicago areas, where electrical utility systems were invented and developed, and where Westinghouse, General Electric, and other electrical manufacturers concentrated.

Yet despite these affinities, the two concepts are by no means identical. Hughes conceived technological momentum as a fruitful way to avoid technological determinism without embracing a relativistic social constructionism. In contrast, the path dependence school developed its ideas to escape from present-minded neoclassical economic theory and to find a way to discuss contingencies that cannot be explained through the usual market forces of supply, demand, cost/benefit ratios, and the like. Because these concepts arose in quite different fields, the implicit focus of each

[56] Few seem to read outside their field. An exception is Robert McAdams, *Paths of Fire: An Anthropologist's Inquiry into Western Technology* (Princeton: Princeton University Press, 1996), 23–25.

[57] Hughes, *Networks of Power*, 4, 65.

[58] Krugman, "History and Industry Location," 80.

concept is not the same. Hughes was primarily interested in the history of technology, and therefore economics is ultimately secondary. Path dependence theorists are interested in explaining how historical contingencies deflect market forces. For them, technologies are examples whose illustrative value increases if they are arbitrary or accidental. The ideal example for path dependency theory seems to be an inefficient technological arrangement, arrived at haphazardly, that is widely adopted and persists in the marketplace. One major theme of path dependency is that inefficient economic behavior becomes deeply embedded in society, contrary to the tenets of neoclassical theory. This means that history and culture matter because they explain otherwise irrational behavior. The path dependence approach to materials, machines, and invention therefore may seem naïve to historians of technology.

These differences do not prevent synthesis between the two concepts. Hughes is most concerned with the technical and social processes from which technological momentum emerges. In contrast, the path dependent school are mostly concerned with economic behavior after lock-in, or, as Hughes would have it, after technological momentum. Hughes's book ends with system maturity, that is, at the moment when economists have a well-defined case of path dependence to study. A dialogue between Hughes's theory and path dependence would strengthen both. The economists would have to open the black box of each technology that they study. In doing so, they would need to acknowledge that the early choices often are less arbitrary or accidental than QWERTY suggests. Indeed, they would need to rethink their myth of origins, and recognize, with Hughes, that the early stages of invention and diffusion are not as powerfully shaping as they think. Rather, they might learn from historians of technology that the invention, refinement, and introduction of a socio-technical system takes 20–25 years before reaching technological momentum and the moment of "lock in." For their part, historians of technology have begun to pay more attention to the mechanisms that reinforce systems and become a part of a dynamic feedback that sustains technological momentum. Finally, the combination of these two research traditions can help us to think more clearly about the crises that systems undergo, at the moments when they meet external challenges. These systemic crises typically have multiple causes that are economic, cultural, and technological.

There is still much to be done before we can merge these approaches into a theory of energy transitions, but one can sketch how this might be accomplished. Hughes's model of system development provides the basis for a division of labor between these concepts. His model is useful for the years of invention and diffusion of a new technology, when it is socially constructed, and also for discussions of the middle stages, when the system faces technical problems (or reverse salients) that must be overcome in order to reach the stage of technological momentum. Only at this point does path dependence 'lock in' and not, as economic historians think, during the earlier stages.[59]

---

[59] QWERTY was not the layout of the first typewriter. See Utterback, *Mastering*, 3–13.

Even QWERTY was not 'locked in' immediately. Various patents for writing machines were granted during the first half of the nineteenth century, but none were manufactured. What would become the modern typewriter took several years to develop and was patented in 1868. It then took six years to perfect a machine for the marketplace, and during this time different keyboards were tried. The first typewriters were on sale in 1874, and manufacturers with different keyboards entered the competition. Only a few thousand machines were sold in the first three years. In 1878 Underwood offered an improved machine that had both upper- and lowercase letters, and it then entered the third stage of Hughes's model, when a product begins to penetrate a larger market. However, in 1880 few offices used typewriters, and technological momentum did not emerge until c. 1890. Even then, substantial market penetration still lay in the future. Path dependency with regard to the keyboard emerged among the consumers, who did not want to learn a new typing system each time they upgraded to a newer machine.

Path dependence theory explains the history of the typewriter or the electrical energy system only in their later stages. Hughes's theory is needed to understand the sequence of events from invention to technological momentum, including the identification of reverse salients and their resolution. Path dependency seldom emerges early, nor is it accidental; rather it emerges around the middle of a socio-technical system's development. Indeed, Hughes theory needs a sixth stage, of growth to market dominance, and this, along with more attention to marketing, can be adopted from economic history.

2.4 The Model and the Decarbonizing Energy Transition

The six stages of this combined theory can be applied to the decarbonizing energy transition now underway. The first stage of wind and solar energy product development resembles the first stage of the creation of the electrical system. In the 1870s, many inventors in several countries simultaneously worked on electric arc lighting and power generation, and by c. 1878 the first electrical arc-lighting systems were displayed in New York, London, Moscow, and Paris. In the early 1880s these systems entered a second stage, and were being installed in additional high-profile locations. However, the cost remained far higher than gas lighting, which remained predominant in the United States until c. 1900 and for almost a generation longer in Britain.[60]

Almost exactly a century later, wind and solar power underwent similar development. In the 1980s, wind turbines went online in California and a few other high-profile locations, but they were not cost competitive during the 1980s or 1990s. Research and development gradually brought the cost down, but consumers still had to pay a premium for alternative energies during the second stage. In stage two,

[60] Nye, *American Illuminations*, 44–59.

during the 1880s, electric arc lights could only be seen on a few streets in major cities, and electric motors supplied little factory power and ran almost no streetcars.

During the 1890s came the third stage of electrification, when many reverse salients were overcome, notably the development of alternating current and improvements in the motors used in street railways and factory machinery. Yet in 1900 less than 5 percent of American homes had electricity, and factory electrification was only starting to become important. A century later solar and wind power overcame their reverse salients, becoming more reliable and less expensive technologies. They also began to realize manufacturing efficiencies as they scaled up production of wind turbines and solar panels. But in 2000, these alternative energies were still more expensive than fossil fuels and supplied only a tiny fraction of the energy of the United States.

Fourth, the electrical system attained technological momentum in a few nations, notably Germany and the United States, between 1900 and 1910. Something similar occurred with alternative energies between 2000 and 2015, when they gradually became price competitive, leading the World Economic Forum to conclude by 2017 that solar and wind power had become good investment opportunities.[61] Global sales of photovoltaic cells grew by 40% in 2007 alone, with 2.3 gigawatts of newly installed capacity. Europeans owned eighty percent of that generation, focused largely in Germany and Spain. In contrast, the United States, with huge areas of sunny desert better suited to generating solar energy than Germany, had installed only one third as much capacity.[62] A study published in *Scientific American* in 2008 estimated that two-thirds of all US electricity needs could be supplied from solar installations in Arizona and New Mexico.[63] In 2007, Germany had built fifteen of the world's twenty largest solar arrays, and its solar equipment manufacturing industry employed 40,000 people.[64] Between 2000 and 2010 wind power was also developed in Germany and Denmark. By 2015 the Danes produced 56 percent of their electricity from alternative energies, primarily from wind turbines.

Yet the technological momentum for alternative energies was not a global phenomenon. In 2015, less than 2% of the world's energy was supplied by solar and wind power. However, as these energy sources became cost competitive with coal, large-scale investment and technology transfer rapidly emerged. In the European Union wind power provided more energy than nuclear plants in 2013,

[61] World Economic Forum, *Renewable Infrastructure Investment Handbook: A Guide for Institutional Investors*. 2017, 5–8. www.weforum.org

[62] John J. Berger, *Charging Ahead: The Business of Renewable Energy and What it Means to America* (New York: Henry Holt, 1997). In the 1990s German engineers created a system of renewable energy sources that supplied 100% of one community's power. Their system linked eleven wind energy farms, four biogass generators, twenty photovoltaic plants, and one pumped storage facility.

[63] Ken Zweibel, James Mason and Vasilis Fthenakis, "A Solar Grand Plan," *Scientific American* 298:1 (Dec. 2008): 64–73.

[64] Craig Whitlock, "Cloudy Germany a Powerhouse in Solar Energy," *Washington Post*, May 5, 2007, A1.

more than hydroelectric power in 2015, and more than coal in 2016. The American Energy Information Administration predicted that in the U.S. alternative energies would surpass coal in 2030, but this estimate may have been too conservative. As Bloomberg News noted, the cost of solar power fell by almost 40 percent between 2012 and 2016, and it continued to decline after that. The rapid shift to solar and wind power resembled what occurred with electricity in the US between 1910 and 1930. In short, wind and solar power have now entered stage four of Hughes's model. The technologies are reliable and cost competitive. In 2020, they are in the position that the US electrical industry had reached in 1910, when only one home in 10 had electrical wiring, and only 18.7% of factory power was electric. In the next quarter century, home electrification boomed, reaching almost every urban home by 1930, and factory electrification reached 78%.[65]

Electrification was a comprehensive transition that included both production and consumption, reaching into all aspects of daily life. The pace of change varied between different countries, but the succession of stages was similar for each of the technologies involved, including the same reverse salients for lighting, street railways, long-distance transmission, motors, and various home appliances. Lighting was the entering wedge of change in most nations, starting c. 1880. But electric lighting in the public sphere proceeded more slowly in Europe than in the U.S. and was less intense when it was installed. Nor, between 1900 and 1920, did electric advertising penetrate all European countries in the same ways or to the same degree as it did the US. In contrast, electric streetcars were only widespread in American cities for a generation and most disappeared in the 1930s. Trams remained more common in Europe until well after World War II. The pace of adoption and the subsequent retention of electrical technologies varied. But underlying these variations were the technological stages of invention, demonstration, adoption in high profile venues, expansion until reaching technological momentum, and growth to dominate markets. The many transitions had the same structures but heterogeneous results.

2.5 What Is Different in the Decarbonization Transition?

While there are many similarities between the current energy transition to alternative energies and the transition to electricity between 1875 and 1940, there are also important differences. Like the process of electrification, some energy transitions require construction of entirely new infrastructures for both production and consumption. Coal had to be mined, coal canals and railroads had to be constructed, and delivery to consumers required vast coal yards, horses, wagons, and so on.[66] After 1860, oil required an entirely different system of extraction, refinement, and

[65] Nye, *Electrifying America*, 187.

[66] Christopher F. Jones, *Routes of Power* (Cambridge: Harvard University Press, 2016).

delivery. Gas demanded yet another. When electricity emerged, some of the coal infrastructure could be repurposed to deliver fuel to power plants, but enormous sums were invested in hydroelectric dams and transmission lines, connected through another complex invention, the electrical grid.[67] Every factory, business, and home also had to be wired for electricity. In contrast, electricity from wind turbines and solar panels can use an expanded grid to reach the same wiring. Even so, as Julie Cohn explains in this volume, new high-tension lines must be built to link new wind and solar generation with consumers, and locating these towers and wires in the landscape often becomes a political issue.[68]

Most nations have heterogeneous power systems based on a mix of hydroelectric dams, steam power plants burning gas or coal, solar panels, wind turbines, and energy exchanges with other regions. As wind and solar electricity becomes a larger part of the supply, they also require larger storage capacity to deal with windless or sunless days. Incorporating electric automobiles into that storage system is an attractive idea, but it presents reverse salients. Elsewhere in this volume, Matthew Eisler examines the obstacles to linking vehicles into the grid, "including competing charging standards, uncertain economics, and a mismatch of design to application."[69] Utilities are accustomed to linking a range of different generation sources into larger grids, however, and eventually the reverse salients Eisler identifies may well be overcome, though not necessarily in the same ways that are equally useful to every society.

Some consumers will install alternative generation systems, notably solar panels, but they are not required to buy new equipment in order to purchase alternative energies. Most people scarcely notice whether their electricity is from wind turbines, solar panels, nuclear plants, or steam engines. Denmark produces more than half of its electricity from alternative sources, but this change has not demanded any adjustment from consumers. Similarly, in order to drive an electric car the consumer faces almost no problems compared to the transition from horses to automobiles. Driving a team of horses did not prepare one to get behind the wheel of a Model T. A driver has most of the skills needed to own and drive an electric car, and the roads and traffic laws are unchanged. Metaphorically speaking, one does not need to learn another keyboard in order to drive a hybrid or electric car. The alternative energy transition demands much more from producers than consumers.

Previous transitions were always part of an expansion in energy production and consumption. The energy systems being displaced often were redeployed. When electric lighting replaced gas lighting between c. 1880 and 1910, the amount of gas being produced and sold actually increased, but it was used for cooking and heating instead of illumination. Likewise, when steam increased its market share of factory

---

[67] Julie A. Cohn, *The Grid: Biography of an American Invention* (Cambridge: MIT Press, 2017).

[68] Cohn, "Large-scale Renewables and Local Gatekeepers: Moving Wind and Solar Power Across the Landscape," this volume.

[69] Matthew N. Eisler, "Fragile Sustainability: V2G and Grid Management in the Renewable Energy Era," this volume.

power, the amount of power generated by water wheels in the United States also increased.[70] Eventually, much of this energy system was repurposed to generate electricity. In short, during previous transitions older energy forms were not abandoned but repurposed, and often they did not diminish. Society added new layers of energy systems to those it already had. As David remarked concerning the shift from steam to electricity, the "overlaying of one technical system upon a preexisting stratum is not unusual during historical transitions."[71] In the past, systems often coexisted for generations, during a gradual but never complete substitution.[72]

This is not the situation in 2021. To halt global warming, fossil fuels must be largely abandoned, not repurposed. Yet utilities resist shutting down their coal and gas power plants until they have amortized their investments. They may retrofit a few power plants, substituting gas for coal, but they will often be reluctant to throw away their investment. They are accustomed to purchase new systems only after equipment has depreciated. As Smil summarized the problem: "The inertia of existing massive and expensive energy infrastructures and prime movers and the time and capital investment needed for putting in place new convertors and new networks make it inevitable that the primary energy supply of most modern nations will contain a significant component of fossil fuels for decades to come."[73] Slow growth or falling demand makes an energy transition more difficult. When demand grew rapidly during previous transitions, old systems were still needed. Chinese and Indian energy demand is rising, and they therefore can add wind and solar power without decommissioning fossil fuel plants. In contrast, the EU and the US are making this transition amid stagnant or falling demand, and their utilities will have to abandon investments in fossil fuel plants.

Consumers also play a role in reducing energy use. Since 1950 US corporations have become sensitive to energy costs, and they have reduced their share of the national electricity bill, but consumers have increased their share. Between 1940 and 2001 the average US household increased its electrical consumption by 1300%. A typical family in 2020 used more electricity every month than grandparents did during an entire year. This increased energy use has occurred despite greater efficiencies. Since the 1980s, much has been accomplished by increasing energy efficiency. More GNP is being produced with less energy. Houses are kept warm using less fuel. Washing machines use less electricity. Cars go further with the same amount of electricity or gasoline. The old incandescent lighting is being abandoned in favor of bulbs that are at least four times as efficient. As a result, the United States, even though it never signed the Kyoto Accords, has made progress. George W. Bush

[70] Nye, *Consuming Power*, 82.

[71] Paul David, "The Dynamo and the Computer: An Historical Perspective on the Modern Productivity Paradox," *The American Economic Review* 80:2 (May 1990): 357.

[72] Nye, *American Illuminations*, 35–60.

[73] Smil, *Energy Transitions*, chapter four.

refused to acknowledge the reality of global warming, but by 2008 the US per capita use of energy was falling, and this process accelerated under President Obama.[74]

Until c. 2015 the US increased the energy efficiency of its fossil fuel consumption, but otherwise the shift toward decarbonizing was slow. Since then, solar and wind power have achieved technological momentum. But fossil fuels are so intertwined with the US economic and political system that they retain their technological momentum as well. Other nations where fossil fuels have been economic drivers of the economy also have moved slowly toward the green production transition. In contrast, nations without domestic fossil fuels extraction industries, such as Chile and Portugal, quickly embraced alternative energies.[75] Yet as has been the case during previous transitions, change will accelerate in the later stages, driven by cost advantages.

Finally, climate change is an external driver. Previous transitions took little account of the environmental consequences of rising energy use. They were driven by increasing consumption, cost advantages, convenience, and flexibility. But the decarbonizing transition requires energy systems that also eliminate of CO_2 emissions, reduce consumption, and recycle. Nations will not make this transition in the same way. Some will have more success in reducing CO_2, others in recycling. Some will take less interest in making the transition than in making short-term profits from fracking for oil and natural gas.

Given geographical and cultural differences, there is little reason to expect that energy systems will converge after a transition to alternative energies. France will persist with nuclear power. Norway will still produce electricity with hydroelectric installations, supplemented by wind turbines. Germany will maximize solar and wind power, as well as recycling and energy efficiency, for example in zero emission housing. Portugal will install more alternative energy, slashing its fossil fuel imports. Chile with its long windy coastline, hot deserts where it almost never rains, and volcanic regions, will become self-sufficient and a potential powerhouse for Latin America. Given this heterogeneity, there seems to be little evidence that deep transitions are driving the adoption of a new homogeneous energy regime.

Diversity in national production will be accompanied by equally diverse choices in consumption. Each nation has its own energy culture, just as each has a distinctive food culture, musical tradition, and political system. Some will become less dependent on outside suppliers, others more dependent. Some systems will be highly centralized, others not. Energy efficiency will also vary. CO_2 emissions will fall, but not uniformly. Each energy culture will have its own style, and the components of its energy systems will achieve momentum at different historical moments. But most nations will shift from fossil fuels toward alternative forms of generation that are in accord with their geography, resources, culture, technology, and political systems. As Thomas Hughes put it, "power supply problems are governed completely by

[74] See David E. Nye, "The United States and Alternative Energies Since 1980: Technological Fix or Regime Change?" *Theory, Culture & Society* 31:5 (June 2014): 103–125.

[75] "Portugal runs for 4 days straight on renewable energy alone." *The Guardian*, May 18, 2016.

local conditions. Out of these come contrasting utility styles."[76] Energy technologies set the boundaries for possible transitions, and their introduction follows a sequence of stages. Yet they dictate neither the form or nor the cultural meanings of an electrical system.

Bibliography

Aklin, M., and J. Urpelainen. 2018. *Renewables: The Politics of a Global Energy Transition*. Cambridge: MIT Press.

Arapostathis, S., and Y. Fotopoulos. this volume. Between Material Dependencies and the Politics of Electricity Transitions: Networks of Power in Greece, 1940–2010. In *Electrical Conquest: New Approaches to the History of Electrification*, ed. W. Bernard Carlson and Erik M. Conway. Cham: Springer.

Arthur, W. Brian. 1989. Competing Technologies, Increasing Returns, and Lock-in by Historical Events. *Economic Journal* 99: 116–131.

Berger, John J. 1997. *Charging Ahead: The Business of Renewable Energy and What it Means to America*. New York: Henry Holt.

Bloomberg, Michael, and Carl Pope. 2017. *Climate of Hope*. New York: St. Martin's Press.

Bromley, Peter Sircom. 2016. Extraordinary Interventions: Toward a Framework for Rapid Transition and Emission Reductions in the Energy Space. *Energy Research & Social Science* 22: 165–171.

Calvin, William. 2008. *Global Fever*. Chicago: University of Chicago Press.

Cohn, Julie A. this volume. Large-scale Renewables and Local Gatekeepers: Moving Wind and Solar Power Across the Landscape. In *Electrical Conquest: New Approaches to the History of Electrification*, ed. W. Bernard Carlson and Erik M. Conway. Cham: Springer.

———. 2017. *The Grid: Biography of an American Invention*. Cambridge: MIT Press.

David, Paul A. 1985. Clio and the Economics of QWERTY. *American Economic Review* 75: 332–337.

———. 2007. Path Dependence, A Foundational Concept for Social Science. *Cliometrica* 1 (2): 91–114.

———. 1990. The Dynamo and the Computer: An Historical Perspective on the Modern Productivity Paradox. *The American Economic Review* 80 (2): 355–361.

Eisler, Matthew N. this volume. Vehicle-to-Grid, Regulated Deregulation, and the Energy Conversion Imaginary. In *Electrical Conquest: New Approaches to the History of Electrification*, ed. W. Bernard Carlson and Erik M. Conway. Cham: Springer.

Emmett, Robert S., and David E. Nye. 2017. *The Environmental Humanities*. MIT Press.

Fouquet, Roger. 2016. Historical Energy Transitions: Speed, Prices, and System Transformation. *Energy Research and Social Science* 22: 11–12.

Grin, John, Jan Rotmans, and Johan Schot. 2010. *Transitions to Sustainable Development: New Directions in the Study of Long-Term Transformative Change*. London: Routledge.

Hecht, Gabrielle. 1998. *The Radiance of France*. Cambridge: The MIT Press.

Hughes, Thomas P. 1994. Technological Momentum. In *Does Technology Drive History?* ed. Merritt Roe Smith and Leo Marx. Cambridge: The MIT Press.

———. 2004. Afterword. *Annales Historiques de l'électrcité* 2: 173.

[76]Hughes, "Afterword," 176. See also Thomas P. Hughes, "Regional Technological Style," in *Technology and its Impact on Society*, ed. Sigvard Strandh (Stockholm: Tekniska Museet, 1979): 211–234.

———. 1979. Regional Technological Style. In *Technology and its Impact on Society*, ed. Sigvard Strandh, 211–234. Stockholm: Tekniska Museet.

———. 1969. Technological Momentum: Hydrogenation in Germany, 1900–1933. *Past and Present* 44: 106–132.

———. 1989. *American Genesis*. New York: Penguin.

———. 1983. *Networks of Power*. Baltimore: Johns Hopkins University Press.

Jones, Christopher F. 2016. *Routes of Power*. Cambridge: Harvard University Press.

Kapoor, Nathan. this volume. Colonial Self-Sufficiency: Electrifying the 'Britain of the South,' 1880–1914. In *Electrical Conquest: New Approaches to the History of Electrification*, ed. W. Bernard Carlson and Erik M. Conway. Cham: Springer.

Krugman, Paul. 1991. History and Industry Location: The Case of the Manufacturing Belt. *The American Economic Review* 81 (2): 80–83.

McAdams, Robert. 1996. *Paths of Fire: An Anthropologist's Inquiry into Western Technology*. Princeton: Princeton University Press.

McShane, Clay, and Joel Tarr. 2011. *The Horse in the City: Living Machines in the Nineteenth Century*. Baltimore: Johns Hopkins University Press.

Missal, Alexander. 2006. *Seaway to the Future*. Madison: University of Wisconsin Press.

Mody, Cyrus C.M. this volume. Energy Modernism: The Oil Industry and the Energy Infrastructure Scramble of the Long 1970s. In *Electrical Conquest: New Approaches to the History of Electrification*, ed. W. Bernard Carlson and Erik M. Conway. Cham: Springer.

National Public Radio. 2014. *Morning Edition, After Solyndra Loss, U.S. Energy Loan Program Turning a Profit*, Nov 13.

Nixon, Rob. 2011. *Slow Violence and the Environmentalism of the Poor*. Cambridge: Harvard University Press.

North, Douglass C. 1991. Institutions. *Journal of Economic Perspectives* 5 (1): 97–112.

Nye, David E. 2004. Electricity and Culture: Conceptualizing the American Case. *Annales Historiques de l'électrcité* 2: 125–138.

———. 1999. Path Insistence: Comparing European and American Attitudes Toward Energy. *Journal of International Affairs*. 53 (1): 129–148.

———. 2014. The United States and Alternative Energies Since 1980: Technological Fix or Regime Change? *Theory, Culture & Society* 31 (5): 103–125.

———. 2018. *American Illuminations*. Cambridge: The MIT Press.

———. 1998. *Consuming Power: A Social History of American Energies*. Cambridge: The MIT Press.

———. 1990. *Electrifying America: Social Meanings of a New Technology*. Cambridge: The MIT Press.

———. 2006. *Technology Matters: Questions to Live With*. Cambridge: The MIT Press.

Parsons, R.H. 1940. *The Early Days of the Power Station Industry*. Cambridge: Cambridge University Press.

Puffert, D.J. 2002. Path dependence in spatial networks: The standardization of railway track gauge. *Explorations in Economic History* 39 (3): 282–314.

Sandwell, Ruth W. 2018. Heating and cooking in rural Canada: Home energy in transition, 1850–1940. *History of Retailing and Consumption* 4 (1): 64–80.

Schot, Johan, and Laur Kanger. 2018. Deep Transitions: Emergence, Acceleration, Stabilization and Directionality. *Research Policy* 47 (6): 1045–1059.

Schot, Johan. 2016. Confronting the Second Deep Transition through the Historical Imagination. *Technology and Culture* 57 (2): 445–456.

Smil, Vaclav. 2017. *Energy Transitions: Global and National Perspectives*. 2nd ed. New York: Praeger.

———. 2016. Examining Energy Transitions: A Dozen Insights Based on Performance. *Energy Research and Social Science* 22: 194–197.

Sovacool, Benjamin K. 2016. How Long Will it Take? Conceptualizing the Temporal Dynamics of Energy Transitions. *Energy Research and Social Science* 13: 202–215.

Spinak, Abby. this volume. *Co-ops Against Castroism: American-style Rural Electrification in the Global Countryside*. In *Electrical Conquest: New Approaches to the History of Electrification*, ed. W. Bernard Carlson and Erik M. Conway. Cham: Springer.

Staudenmaier, John. 1994. Rationality Versus Contingency in the History of Technology. In *Does Technology Drive History?* ed. M.R. Smith and L. Marx, 260–273. Cambridge: The MIT Press.

Storrow, Benjamin. 2020. *Coal's Decline Continues with 13 Plant Closures Announced in 2020*. https://www.scientificamerican.com/article/coals-decline-continues-with-13-plant-closures-announced-in-2020/

US Department of Energy. 2001. *Table US-1. Electricity Consumption by End Use in U.S. Households*. http://www.eia.doe.gov/emeu/reps/enduse/er01_us_tab1.html

Utterback, James. 1994. *Mastering the Dynamics of Innovation*. Boston: Harvard Business School Press.

White, Leslie. 1949. *The Science of Culture*. New York: Grove Press.

World Economic Forum. 2017. *Renewable Infrastructure Investment Handbook: A Guide for Institutional Investors*. www.weforum.org

Wright, Gavin. 1997. Towards a More Historical Approach to Technological Change. *The Economic Journal* 107: 444.

Zweibel, Ken, James Mason, and Vasilis Fthenakis. 2008. A Solar Grand Plan. *Scientific American* 298 (1): 64–73.

David E. Nye is a Senior Research Fellow at the University of Minnesota and a by-fellow of Churchill College, Cambridge University. His publications include *American Illuminations, Electrifying America, Consuming Power,* and *When the Lights Went Out*. He received the Leonardo da Vinci Medal from the Society for the History of Technology in 2005 and was knighted by the Danish Queen in 2014.

Chapter 3
Surveying the Landscape: The Oil Industry and Alternative Energy in the 1970s

Cyrus C. M. Mody

Abstract The multi-level perspective (MLP) is one of the most widely-used frameworks for understanding the conditions and choices underpinning energy transitions. Although this volume contains several cogent critiques of the concept of an energy transition, it is still important for historians to be able to address the transitions studies community on its terms—which means being able to convey our historical cases in the language of MLP. At the same, we can use the rich and ambiguous historical record to point out—and possibly correct—shortcomings in the MLP framework. In this chapter I show that MLP's vocabulary offers useful heuristics for understanding the energy debates of the 1970s—the "landscape" changes that destabilized the fossil fuel "regime" and stimulated alternatives such as solar and nuclear to "break out" from their "niches." However, I show that the oil industry's role in the 1970s debates complicates the MLP framework, in that the era's most important landscape shifts were not exogenous to the oil industry, while oil firms exerted substantial control over niche alternatives. By the mid-1980s, the fossil fuel regime re-stabilized and alternative energy technologies were forced into even smaller niches than before—a reversal that is in tension with the more optimistic and linear stories usually found in transitions studies.

Keywords Sustainability · Transitions studies · Multi-level perspective · Resource scarcity · Environmentalism · The "long" 1970s · Nuclear energy · Solar energy

---

C. C. M. Mody (✉)
Maastricht University, Maastricht, Netherlands
e-mail: c.mody@maastrichtuniversity.nl

W. B. Carlson, E. M. Conway (eds.), *Electrical Conquest*, Archimedes 67,
https://doi.org/10.1007/978-3-031-44591-0_3

3.1 Introduction: Energy Transitions and Electrical History

The 2015 Paris climate accords committed the world's nations to limiting "global warming to well below 2, preferably to 1.5 degrees Celsius, compared to pre-industrial levels."[1] Various means of meeting that goal have been proposed, but the main route imagined at Paris was widespread replacement of fossil fuels by alternative energy technologies—commonly known as an "energy transition" or, more broadly, a "sustainability transition."[2]

Other chapters in this volume critique the energy transition concept. The relative prevalence of different energy sources continually changes over time and across contexts of use; conceptualizing this flux as "a transition" ignores the chaotic social life of energy and will likely lead to ineffective policies. A particular weakness of transition talk is that it pictures a single energy source replacing others in *relative* terms, but neglects *absolute* quantities. As Odinn Melsted has shown, for instance, Iceland underwent an energy transition from coal to oil in the 1940s and from oil to hydropower and geothermal in the 1970s—and yet Iceland uses as much coal now as ever, and several times as much oil as it did before the "transition" to renewables.[3] Furthermore, energy transition discourse narrows attention to high technologies and away from low-tech social and cultural changes that could have more immediate effect.[4]

Yet despite these critiques, scholarly communities have arisen in the past two decades to better understand and encourage energy and sustainability transitions. The questions this chapter asks, therefore, are: which of these communities' ideas and findings can be retained, and how can those ideas and findings incorporate critiques of the transition concept? I approach these questions by complicating—and suggesting ways to refine—one of the most popular frameworks in the energy/sustainability transitions research community, the multi-level perspective (MLP). I'll do so by examining some factors affecting energy consumption, production, and discourse during the "long 1970s" (roughly 1968–1986), primarily but not only in the United States. The long 1970s were one of the most turbulent decades in energy history, with two oil "shocks" and a counter-shock, plus a global debate about

---

[1]United Nations Framework Convention on Climate Change, "The Paris Agreement," https://unfccc.int/process-and-meetings/the-paris-agreement/the-paris-agreement

[2]See the essays in Geert Verbong and Derk Loorbach (eds.), *Governing the Energy Transition: Reality, Illusion, or Necessity?* (New York: Routledge, 2012).

[3]Odinn Melsted, Icelandic Energy Regimes: Imported Fossil Fuels, Renewable Resource Development and the Making of a Low-Carbon Energy Balance, 1940–1980, PhD dissertation, University of Innsbruck, 2020.

[4]Elizabeth Shove and Gordon Walker, "Governing Transitions in the Sustainability of Everyday Life," *Research Policy* 39, no. 4 (2010): 471–476.

resource scarcity that grabbed headlines and shaped individual practices.[5] I'll show that MLP provides useful heuristics for understanding those events—but also that MLP's disaggregation of distinct "levels," and its neglect of interactions among those levels, prevents it from explaining why an energy transition did *not* occur in the 1970s. In particular, I look at how the oil industry actively shaped conditions for both adoption and rejection of alternative energy in ways that stymie simple models of energy transitions.

3.2 The Multi-Level Perspective

The multi-level perspective is a theory of technological choice/change that combines ideas from science and technology studies (STS), evolutionary economics, and neo-institutional theory.[6] The version of STS it builds on is the social construction of technology (SCOT) and especially Thomas Hughes' large technological systems (LTS) framework.[7] Hughes' work is a touchstone for this volume and might therefore provide a link between MLP and "new electrical history." However, MLP has not fully absorbed developments since the late 1990s in SCOT or LTS, much less STS more broadly. Although MLP's most prominent adherents—particularly Frank Geels and Johan Schot—still travel in STS and history of technology circles, they are more associated today with the sustainability transitions studies community.

MLP is "multi-level" because it imagines three distinct fields that influence technological choice in different ways. Hughes' LTS corresponds to MLP's "regime," an entrenched system characterized by Hughesian technological momentum, reverse salients, and heterogeneous elements such as technological artifacts, intellectual property laws, research organizations, supplies of materials, media, user communities, etc. But regimes have to come from somewhere; no Large Technological System starts out "large." MLP identifies "niches"—explicitly analogized to ecological niches—as the breeding ground for potential future regimes. In the niche, alternative technologies are protected from competition with the established regime

[5]Elisabetta Bini, Guiliano Garavini, and Federico Romero, *Oil Shock: The 1973 Crisis and Its Economic Legacy* (London: I. B. Tauris, 2016); Duccio Basosi, Giuliano Garavini, and Massimiliano Trentin (eds), *Counter-Shock: The Oil Counter-Revolution of the 1980s* (London: I. B. Tauris 2018); Paul Sabin, *The Bet: Paul Ehrlich, Julian Simon, and Our Gamble over Earth's Future* (New Haven: Yale University Press, 2013).

[6]Frank W. Geels, "Micro-Foundations of the Multi-Level Perspective on Socio-Technical Transitions: Developing a Multi-Dimensional Model of Agency through Crossovers between Social Constructivism, Evolutionary Economics, and Neo-Institutional Theory," *Technological Forecasting and Social Change* 152 (2020).

[7]For MLP the canonical expressions of these programs are: Wiebe E. Bijker, *Of Bicycles, Bakelites, and Bulbs: Towards a Theory of Sociotechnical Change* (Cambridge, MA: MIT Press, 1995) for SCOT; and Thomas P. Hughes, *Networks of Power: Electrification in Western Society, 1880–1930* (Baltimore: Johns Hopkins University Press, 1983) for LTS.

and can mature until they could join or challenge existing regimes. A classic example is the transition from wooden sailing ships to metal steamships. As Geels relates, the latter were initially too dangerous and expensive for wide adoption, but possessed advantages for "niche" applications such as military gunships.[8] With military sponsorship metal ships could develop in the protected niche until they were competitive with wooden vessels, at which point metal designs could "break out" and trigger a general transition from wood to metal construction.

For breakout to occur there must be changes at the third, "landscape," level that destabilize the regime. Notably, MLP pictures the landscape as *exogenous* to regimes and niches. The landscape *sets* conditions for the viability of regimes and niches. Changes to the landscape may make a regime less viable, allowing a niche technology to gain adherents who can either insinuate the technology into an existing regime (possibly with help from incumbent actors) or promote it as the basis for a new, alternative regime.[9] MLP analogizes the resources and selection pressure offered by the landscape to that of the environment in evolutionary theory—although a possible critique of MLP is that in evolutionary theory the environment is not truly exogenous since organisms can shape it as it shapes them.

There are a number of other criticisms of MLP in the literature: that it attends too much to structure and not agency[10]; that social actors do not experience discontinuous "levels" but rather operate in non-hierarchical "arenas;"[11] that MLP is too focused on technological innovation and not social changes such as bathing and driving less[12]; that regimes are treated as homogeneous and regime/niche and transition/no-transition as binary.[13] In response, Geels and co-authors have offered several refinements.[14]

[8]Frank W. Geels, "Technological Transitions as Evolutionary Reconfiguration Processes: A Multi-Level Perspective and a Case-Study," *Research Policy* 31, no. 8/9 (2002): 1257–1274.

[9]Frank W. Geels et al., "The Enactment of Socio-Technical Transition Pathways: A Reformulated Typology and a Comparative Multi-Level Analysis of the German and UK Low-Carbon Electricity Transitions," *Research Policy* 45, no. 4 (2016): 896–913.

[10]Adrian Smith, Andy Stirling, and Frans Berkhout, "The Governance of Sustainable Socio-Technical Transitions," *Research Policy* 34, no. 10 (2005): 1491–1510; Audley Genus and Anne-Marie Coles, "Rethinking the Multi-Level Perspective of Technological Transitions," *Research Policy* 37 (2008): 1436–1445.

[11]Ulrik Jørgenson, "Mapping and Navigating Transitions: The Multi-Level Perspective Compared with Arenas of Development," *Research Policy* 41, no. 6 (2012): 996–1010.

[12]Shove and Walker, "Governing Transitions."

[13]Lea Fuenfschilling and Bernhard Truffer, "The Structuration of Socio-Technical Regimes: Conceptual Foundations from Institutional Theory," *Research Policy* 43, no. 4 (2014): 772–791.

[14]Frank W. Geels, "The Multi-Level Perspective on Sustainability Transitions: Responses to Seven Criticisms," *Environmental Innovation and Societal Transitions* 1, no. 1 (2011): 24–40; Frank W. Geels, "Disruption and Low-Carbon System Transformation: Progress and New Challenges in Socio-Technical Transformations Research and the Multi-Level Perspective," *Energy Research and Social Science* 37 (2018): 224–231; Benjamin K. Sovacool and Frank W. Geels, "Further Reflections on the Temporality of Energy Transitions: A Response to Critics," *Energy Research and Social Science* 22 (2016): 232–27.

It's unclear whether those refinements would make MLP acceptable to historians such as those in this volume (myself included). True, MLP is a theory of change over time, so its adherents often offer historical examples. Yet those historical episodes are deployed to reinforce or refine the framework without being *historicized*. Steamships are like municipal water systems are like wheat fields, and the nineteenth century is like the 20th is like the 21st—lessons from one are transported rather directly to the next.[15] Messiness, contingency, ambivalence, non-linearity, multicausality—the things historians love—are for MLP obstacles that can be overcome either by ignoring them or by adding complex gadgets to the original framework.

Yet MLP is widely used in the sustainability transitions community, so any contribution historians make to discussions of transitions (even if only to critique the concept) should acknowledge what MLP's adherents have to say. Because MLP draws on LTS and historical case studies, historians could engage with MLP more easily than other frameworks that are popular in the sustainability transitions community, such as the Technological Innovation Systems approach.[16] In my view, MLP offers useful heuristics for teaching and research, but the same simplicity that makes it helpful also hinders its ability to grasp complex, multi-causal situations where the boundaries among its elements are blurred.

Take, for example, solar and nuclear energy. MLP's vocabulary is useful for understanding how a "landscape" feature—the Cold War—opened up protected "niches" for nuclear power (e.g., nuclear submarines) and solar energy (e.g., satellites). That vocabulary draws our attention to why these niches existed and how they were maintained. MLP's welcome emphasis on path-dependence also spurs us to ask what imprint the national-security landscape left on the solar and nuclear niches. For instance, designs for early civilian nuclear power plants that were influenced by national-security considerations translated poorly to attempts to create a nuclear energy regime.[17] Similarly, photovoltaic solar power was incubated in the national-security niche and thus had powerful allies who attempted to promote it to regime status at the expense of alternatives such as photothermal systems.

With respect to the energy debates of the 1970s, MLP's heuristics capture some dynamics well: in that decade landscape changes (social movements, embargoes, etc.) destabilized the coal and oil regimes, leading many to think that niche

---

[15] Frank Geels, "Co-Evolution of Technology and Society: The Transition in Water Supply and Personal Hygiene in the Netherlands (1850–1930): A Case Study in Multi-Level Perspective," *Research Policy* 27, no. 3 (2005): 363–397; Cameron Roberts and Frank W. Geels, "Conditions for Politically Accelerated Transitions: Historical Institutionalism, the Multi-Level Perspective, and Two Historical Case Studies in Transport and Agriculture," *Technological Forecasting and Social Change* 140 (2019): 221–240.

[16] Anna Bergek et al., "Analyzing the Functional Dynamics of Technological Innovation Systems: A Scheme of Analysis," *Research Policy* 37, no. 3 (2008): 407–429.

[17] Robin Cowan, "Nuclear Power Reactors: A Study in Technological Lock-In," *The Journal of Economic History* 50, no. 3 (1990): 541–567; Gabrielle Hecht, *The Radiance of France: Nuclear Power and Nuclear Identity after World War II* (Cambridge, MA: MIT Press, 1998).

alternative energy technologies could break out to become energy sources for a new regime. Yet MLP's adherents have so far said little about the 1970s, perhaps for two main reasons. First, the 1970s undermine MLP's optimism that more sustainable technologies *will* break out and become predominant if the landscape destabilizes an incumbent regime. The fossil fuel regime underwent an existential crisis in the 1970s, but it exited that crisis more entrenched than ever; whereas the promising alternatives of the 1970s not only didn't "break out" but even retreated to smaller niches than before. And second, the 1970s show that MLP's disaggregated levels are in fact deeply entangled; the rest of this chapter will expand on that point. Yet I would argue that MLP's adherents should confront the 1970s head-on; even if that decade highlights some shortcomings in the MLP framework, addressing those shortcomings offers opportunities to make MLP a more truly historical perspective that can accommodate the ambiguities and contradictions present in the historical record.

In what follows, I survey the major shifts of the 1970s—primarily in the United States—that MLP would categorize as landscape changes, and then identify the main energy technologies that vied for regime status. I will particularly elucidate ways that the *incumbent* fossil fuel regime—especially the oil industry—complicates MLP's picture of an exogenous landscape and protected niches. MLP's canonical landscape shifts include "oil prices, economic growth, wars, emigration, broad political coalitions" as well as "environmental and demographic change, new social movements, shifts in general political ideology, broad economic restructuring, emerging scientific paradigms, and cultural developments."[18] *None* of these were exogenous to the oil industry in the 1970s. Oil firms manipulated oil prices, fostered wars, meddled in political coalitions and the formation of political ideologies, contributed both to environmental change and the environmental movement, undertook Nobel Prize-winning scientific research, sponsored cultural institutions, and promoted economic globalization. At the same time, the oil industry was involved with various alternative energy technologies. MLP does acknowledge the possibility that incumbent regime actors might foster niche alternatives, but oil-alternative energy interactions in the 1970s go far beyond what MLP has so far imagined.

Grappling with that complexity should sap MLP adherents' optimism that a sustainability transition can be accelerated or managed.[19] The transition that many actors thought was taking shape in the 1970s did not occur—at least not in a way that anyone today understands as a sustainability transition—despite vigorous efforts to manage that transition. Any attempt to apply MLP in substituting alternative energy for fossil fuels today is likely to fall foul of the same non-linear entanglements

[18] Geels, "Technological Transitions," 1260; Adrian Smith, Jan-Peter Voß, and John Grin, "Innovation Studies and Sustainability Transitions: The Allure of the Multi-Level Perspective and Its Challenges," *Research Policy* 39 (2010): 435–448.

[19] Roberts and Geels, "Conditions for Politically Accelerated Transitions." For further skepticism that transitions can be managed, see Elizabeth Shove and Gordon Walker, "Caution! Transitions Ahead: Politics, Practice, and Sustainable Transition Management," *Environment and Planning A* 39, no. 4 (2007): 763–770.

among incumbent fossil fuel actors, societal ("landscape") changes, and the energy sources most commonly promoted as underpinning a new energy regime.

What follows is not exclusively a "new approach to the history of electrification," though it is that. Some of the alternative energy technologies I discuss (particularly solar photovoltaics and nuclear fission) are mainly—though not exclusively—designed to generate electricity. Others, however, are mainly designed to deliver heat or motive power, though they are sometimes adapted to provide electricity. Oil, of course, is mostly used as a source of motive power and, in some regions, heat (putting aside petrochemicals and other non-energy oil products) and *not* electricity. However, coal *is* central to electrical history, and in the 1970s the oil industry captured much of the North American coal industry. Oil firms were also at the forefront in developing nuclear, solar, and other alternative sources of electricity. So any electrical history set in the 1970s should include the oil industry, even if oil itself is not very "electrical."

Yet if oil firms were important to the electric power regime in the 1970s, they also cultivated ties to other "regimes" such as the food and transportation systems. This is, indeed, one of the weaknesses of the MLP—it isn't always clear which regime an actor is an incumbent in, and therefore which niches that actor might be threatened by or seek to coopt. For my purposes that heterogeneity is an important finding: by placing oil at the center of my story I can trace connections between electrical history and other topics that might not be visible if we place electricity itself at the center of the story.

3.3 Landscape Level: Coming Unstuck

So the MLP model fits the 1970s well at first glance: landscape changes destabilized the incumbent fossil fuel regime, which stimulated hopes that alternative energy technologies could break out of their niches and either join or replace the incumbent regime. But classical MLP misses that the incumbent fossil fuel regime exerted enormous influence over both the shifting landscape *and* the niche alternatives. In this section, I'll step through the most important landscape shifts that seemingly threated the incumbent fossil fuel regime; in the next section, I'll show how the oil industry shaped *all* of those landscape shifts; and in the section after that, I'll show that the oil industry also shaped all of the niche alternatives that *could* have broken out thanks to the destabilization of the fossil fuel regime.

3.3.1 Environmentalism

Obviously, the growing environmental movement was one of the most important sources of (almost literal) landscape pressures on the incumbent fossil fuel regime. Notably, spectacular failures of that regime—such as the Santa Barbara oil spill of

1969—galvanized support for environmental initiatives: Earth Day in 1970, the creation of the EPA and Council on Environmental Quality that same year, and passage of the Clean Water Act of 1972 and supplementary legislation later in the decade.[20] Although the environmental movement was a heterogeneous mix of sometimes competing factions, in the 1970s those strands largely came together when confronting *visible* environmental threats such as oil spills and smog generated by burning fossil fuels.[21] There was more disagreement on abstract or invisible issues such as overpopulation or rising carbon dioxide levels; yet the areas of agreement were, temporarily, large enough that environmentalism possessed a mass appeal that spurred governments, firms, and social movements to combat certain forms of pollution and encourage development of supposedly cleaner sources of energy.

3.3.2 Counterculture

The environmental movement and related social movements such as Appropriate Technology also overlapped with another politically heterogeneous movement: the youth counterculture. Only a small portion of the counterculture deliberately aimed to destabilize the fossil fuel regime—a prominent example being Denis Hayes, the Earth Day organizer who went on to run the Solar Energy Research Institute under the Carter administration. Much more common were what Fred Turner terms the "New Communalists"—young people who sought independence from all large technological systems.[22] Thus, the communes temporarily offered a protected niche for consumption and even development of some alternative energy technologies, from wood-burning stoves to solar ponds.[23] Yet when the commune movement collapsed after a few years, the ensuing wave of disillusionment pushed many conservative and libertarian communards, including some leaders of the movement (most notably Stewart Brand), toward technologies such as nuclear power and space "colonies" beaming solar power to earth—technologies that could only be viable if integrated into gigantic, new energy regimes.

[20] Adam Rome, "'Give Earth a Chance': The Environmental Movement and the Sixties," *Journal of American History* 90, no. 2 (2003): 525–554; Teresa Sabol Spezio, *Slick Policy: Environmental and Science Policy in the Aftermath of the Santa Barbara Oil Spill* (Pittsburgh: University of Pittsburgh Press, 2018).

[21] J. Merritt McKinney, Air Pollution, Politics, and Environmental Reform in Birmingham, Alabama, 1940–1971, PhD dissertation, Rice University, 2011.

[22] Fred Turner, *From Counterculture to Cyberculture: Stewart Brand, the Whole Earth Network, and the Rise of Digital Utopianism* (Chicago: University of Chicago Press, 2006).

[23] Henry Trim, "'We Are as Gods': The Green Technical Fix," *RCC Perspectives* 4 (2016): 55–60.

3.3.3 *Decolonization*

The mention of space "colonies" should remind readers that European empires were still giving way in the 1970s to newly sovereign states. Decolonization of oil-producing nations and nationalization of their oil reserves was viewed by the major multinational oil firms (and politicians in their home countries) as an existential threat to the oil regime.[24] Their response was, in part, more colonization. That's one way to think about the scramble to get oil out of the North Slope of Alaska by any means, regardless of people and animals living there.[25]

We can also interpret the extension of oil drilling further offshore in this way. The technology of offshore drilling improved dramatically over the long 1970s, *in tandem* with the legal frameworks that facilitated moving further offshore—most notably the 1982 UN Convention on the Law of the Sea (UNCLOS).[26] Oil companies helped craft the US position in UNCLOS negotiations, as a result of which the United States gained sovereignty over Exclusive Economic Zones amounting to 120% of the nation's continental footprint.[27] That's a truly impressive imperial expansion that the oil industry and its allies believed was necessitated by the retreat of American imperial influence in other parts of the world.

3.3.4 *National Security*

If decolonization represented a new phase in North-South relations, it also took place in the context of the evolving East-West confrontation. In MLP terms, Cold War-inspired protection of niches for alternative energy technologies bore regime-level fruit in that decade. Moreover, the Cold War was such a pervasive context that when other landscape changes destabilized the fossil fuel regime, solutions to the ensuing crisis largely built on a national-security state infrastructure. That's evident,

---

[24] Giuliano Garavini, *The Rise and Fall of OPEC in the Twentieth Century* (Oxford: Oxford University Press, 2019).

[25] The polar bear quote is from Edward Teller to the Alaska Chamber of Commerce, as recounted in Scott Kirsch, *Proving Grounds: Project Plowshare and the Unrealized Dream of Nuclear Earthmoving* (New Brunswick: Rutgers University Press, 2005). The trajectory from nuclear earthmoving to pipeline is recounted in Peter A. Coates, *The Trans-Alaska Pipeline Controversy: Technology, Conservation, and the Frontier* (Bethlehem, PA: Lehigh University Press, 1991).

[26] Tyler Priest, *The Offshore Imperative: Shell Oil's Search for Petroleum in Postwar America* (College Station: Texas A&M University Press, 2007).

[27] An interesting player in this was Hollis Hedberg, head of R&D at Gulf. See, for instance, Hollis Hedberg, "Energy in the Nineties – Key Technologies." Box 27, Folder D "Energy in the Nineties," Hollis Hedberg Papers, Spencer Research Library, University of Kansas. On EEZs, see Ruth Oldenziel, "Islands: The United States as a Networked Empire," in *Entangled Geographies: Empire and Technopolitics in the Global Cold War*, ed. Gabrielle Hecht (Cambridge, MA: MIT Press, 2011), 13–42.

for instance, in the transformation of the US Atomic Energy Commission (i.e., the complex centered on nuclear weapons production) into the Energy Research and Development Administration in 1974 and then the Department of Energy (1977). The AEC legacy meant that ERDA and DoE were dominated by former Pentagon and weapons-complex staff, and hence solar energy and energy conservation were sidelined, while nuclear fission and even nuclear fusion projects prospered and the oil industries' interests were protected.[28]

3.3.5 Resource Scarcity

Entangled with these other factors were worries about the long-term availability of various resources. Scarcity wasn't new in the 1970s, of course, but that decade saw a vigorous global conversation about resource scarcity, stimulated by new predictive techniques and concerned elite networks.[29] Books offering alarming predictions of societal collapse resulting from over-consumption, such as *The Population Bomb* and *Limits to Growth*, became global bestsellers. Those books also elicited a backlash from both the left and right, but not before worries about impending scarcity pushed young people into the environmental movement and the communes while simultaneously driving oil companies into the oceans, tundra, and deserts.

3.4 The *Mutual* Shaping of Regime and Landscape

Now let's review the landscape shifts listed above, but this time with an eye to how the incumbent fossil fuel regime facilitated those shifts rather than simply reacting to them: i.e., in this section I show that the landscape that shaped energy technology choices in the 1970s was not—as MLP would have it—exogenous to the incumbents.

3.4.1 Environmentalism

The oil industry has earned its poor environmental reputation through more than a century of spills, effluent streams, emissions, opposition to environmental

[28] Ray Reece, *The Sun Betrayed: A Report on the Corporate Seizure of U.S. Solar Energy* (Montreal: Black Rose Books, 1979).

[29] Thomas Turnbull, From Paradox to Policy: The Problem of Energy Resource Conservation in Britain and America, 1865–1981, PhD dissertation, University of Oxford, 2017.

legislation, and organized stoking of climate denialism.[30] Industry attempts to appear proactive on environmental issues have been rightly critiqued as greenwashing and as a strategy of cooptation, misinformation, and delay. Yet there were also fossil fuel actors who were—apparently sincerely—involved in environmentalist networks in the 1970s, particularly the circle around Robert O. Anderson (founder of Atlantic Richfield) and the Canadian oilman-diplomat Maurice Strong. That circle largely centered on the Aspen Institute, which Anderson took control of in the early 1960s and used as a forum for elite discussion of resource scarcity and coordination of global environmental governance. Anderson and his hand-picked director of Aspen, Joseph Slater (another former oil executive), worked closely with Strong and other allies to prepare the way for the 1972 UN Conference on the Human Environment in Stockholm, which Strong presided over.[31] One of those allies, the economist Barbara Ward, co-wrote the main preparatory text for the Stockholm meeting, *Only One Earth*; at Anderson's invitation she then became first director of the International Institute for Environmental Development, a think tank founded with Anderson's seed money.

Anderson and Strong weren't the only oil actors to forge similar ties with environmentalists. For example, George Mitchell, the "father of fracking," was also an early and active proponent of sustainable development.[32] In the late 1970s and through the 1980s he held a series of Limits to Growth conferences that brought together figures from all sides of the environmental and resource scarcity debates. Similarly, both of the co-founders of the oil (and later uranium) company Kerr-McGee presented themselves as environmental spokesmen for their industry; one, Robert Kerr, even has an EPA facility named after him.[33] For other environmentally-minded oil actors, the Worldwide Fund for Wildlife served as a conclave where they could meet-and-greet prince consorts, the leaders of newly independent African

---

[30] Hugh S. Gorman, *Redefining Efficient: Pollution Concerns, Regulatory Mechanisms, and Technological Change in the U.S. Petroleum Industry* (Akron: University of Akron Press, 2001). For a more micro study see Raechel Lutz, Crude Conservation: Nature, Pollution, and Technology at Standard Oil's New Jersey Refineries, 1870–2000, PhD dissertation, Rutgers University, 2018. On denialism see Naomi Oreskes and Erik M. Conway, *Merchants of Doubt: How a Handful of Scientists Obscured the Truth on Issues from Tobacco Smoke to Global Warming* (London: Bloomsbury, 2010).

[31] Or at least that's the narrative relayed in Kenneth Harris, *The Wildcatter: A Portrait of Robert O. Anderson* (New York: Weidenfeld & Nicolson, 1987), 122–125.

[32] Loren C. Steffy, *George P. Mitchell: Fracking, Sustainability, and an Unorthodox Quest to Save the Planet* (College Station: Texas A&M Press, 2019); Jurgen Schmandt, *George P. Mitchell and the Idea of Sustainability* (College Station: Texas A&M Press, 2010); Jon Gertner, "George Mitchell, Father of Fracking," *New York Times Magazine* (December 21, 2013).

[33] See Dean McGee's remarks opening a panel at the Growth with Environmental Quality conference which he was invited to moderate because of his reputation as one of the oil industry's experts on the topic: Dean McGee, "Summary of Opening Remarks," Energy Panel of the National Forum on Growth with Environmental Quality, September 23–26, 1973. In Box 137, Folder "Growth with Environmental Quality," Dean McGee papers, Spencer Research Library, University of Kansas.

nations, and other influential people while promoting environmental causes.[34] Notably, the WWF started giving out a J. Paul Getty Wildlife Conservation Prize in 1974.

3.4.2 Counterculture

Getty, Anderson, and Kerr were decidedly strait-laced, but some oil barons came to environmentalism via the counterculture. Perhaps the most bizarre example is Ed Bass, an heir of Richardson and Bass Oil, who has sunk billions into various environmental and ecological organizations as well as the ill-fated Biosphere 2![35] Along with Biosphere 2, Bass also funded the Synergia Ranch and Institute of Ecotechnics—all apparently at the urging of an itinerant group of actor/environmentalists—the Theater of All Possibilities—led by the eccentric metallurgist and ecologist John Allen (aka Johnny Dolphin). Perhaps surprisingly, Royal Dutch Shell also took an interest in Biosphere 2.[36] Or perhaps not so surprising, since Shell was where Pierre Wack's scenario planning group applied countercultural ideas to predicting the future.[37] Maurice Strong, meanwhile, often associated with countercultural figures such as the New Alchemists and Francisco Varela, and lent his Colorado ranch to New Age icons such as Shirley Maclaine and the Lindisfarne Association.[38] Likewise, George Mitchell's circle included the art-sociology-futurism power couple John and Magda McHale (whose writings also

---

[34] Alexis Schwarzenbach, *Saving the World's Wildlife: WWF – the First 50 Years* (London: Profile Books, 2011); Wilfried Huismann, *PandaLeaks: The Dark Side of the WWF* (Bremen: Nordbook, 2014).

[35] Joe Nick Patoski and Bill Crawford, "The Long, Strange Trip of Ed Bass," *Texas Monthly* (June 1989).

[36] Andrew Kirk, *Counterculture Green: The Whole Earth Catalog and American Environmentalism* (Lawrence: University Press of Kansas, 2007), 212 talks about how "Biosphere II hosted the first Shell Oil Learning Conference, sponsored by Shell and Volvo, and 'brought together what would become the Global Business Network (GBN)'." The link appears to have been Peter Schwartz, who was part of the scenario planning group at Shell before joining Stewart Brand to found the GBN.

[37] Bretton Fosbrook, How Scenarios Became Corporate Strategies: Alternative Futures and Uncertainty in Strategic Management, PhD dissertation, York (Canada) University, 2017.

[38] Julia Rubin, "Colorado Site Called 'a Place of Power': Spiritualists, Environmentalists Find Haven in the Baca," *Los Angeles Times* (August 20, 1989). According to Rubin, Robert Anderson also invested in the Strongs' ranch. Some sense of the debt owed to Maurice and Hanne Strong (and to Laurance Rockefeller) as well as the wider networks of the Lindisfarne Association can be gleaned from the blog of its founder, William Irwin Thompson, https://williamirwinthompsonblog.wordpress.com/. The Schumacher Center for a New Economics (one of the surviving institutions of the Appropriate Technology movement) owns the Lindisfarne Tapes, which give a similar sense of the intersecting countercultural, environmental, and industrial networks that I'm pointing to. For a list of the figures who appear in the tapes see https://centerforneweconomics.org/envision/legacy/lindisfarne-tapes/

influenced Strong).[39] In other words, though many corners of the oil industry remained aloof from the counterculture, some oil actors gave it their full support.

3.4.3 Resource Scarcity

Similarly, while the stereotypical oilman did not give much credence to predictions of resource depletion, some oilmen, such as Mitchell, Anderson, and Strong, put money and effort into better understanding and managing scarcity. Those efforts took many forms, but perhaps the most controversial was the *Limits to Growth* report that Mitchell's conferences were named after. People associated with *Limits* were everywhere in this period: one of its main authors, Dennis Meadows, headlined Mitchell's event; his co-author and wife Dana was one of the main consultants behind Biosphere 2 before being purged when Ed Bass brought in (future Donald Trump strategist) Steve Bannon to run the facility. *Limits* was sponsored by the Club of Rome; Anderson, Mitchell, Strong, Slater, and others in their network all worked closely with the Club and were members of its US branch, as were other executives.[40] And while the literature has depicted *Limits* as an outgrowth of computer modeling of urban systems, early drafts were saturated with references to King Hubbert's non-digital forecast of peak oil (first articulated in a presentation to the American Petroleum Institute in 1956, when Hubbert was an employee of Shell Oil).[41]

3.4.4 Decolonization and National Security

The 1970s, and indeed the whole postwar era, were also strongly shaped by a web of interactions among oil firms, the US foreign policy establishment, and philanthropic foundations. It's clear that there has long been a tight relationship between oil

---

[39] I'm indebted to Simone Schleper for much of my thinking about Strong, including the inspiration he took from the McHales. Simone Schleper, *Planning for the Planet: Environmental Expertise and the International Union for Conservation of Nature and Natural Resources, 1960–1980* (New York: Berghahn, 2019).

[40] Peter Moll, *From Scarcity to Sustainability: Futures Studies and the Environment: The Role of the Club of Rome* (Frankfurt: Peter Lang, 1991); Matthias Schmelzer, "'Born in the Corridors of the OECD': The Forgotten Origins of the Club of Rome, Transnational Networks, and the 1970s in Global History," *Journal of Global History* 12 (2017): 26–48.

[41] Kevin T. Baker, World Processors: Computer Simulation, the Limits to Growth, and the Birth of Sustainable Development, PhD dissertation, Northwestern University, 2019, 318–319; Tyler Priest, "Hubbert's Peak: The Great Debate over the End of Oil," *Historical Studies in the Natural Sciences* 44.1 (2014): 37–79.

companies and the US national-security state's imperial adventures.[42] It's also long been known that philanthropies—especially the Rockefeller Foundation, which was founded with oil money—had a similarly tight relationship with US foreign policy and military strategy.[43] Yet as far as I know those facts haven't been brought into the same analytical frame. I don't mean that the Rockefeller Foundation advocated modernization theory or counter-insurgency tactics as part of a larger oil conspiracy. But I do mean that Rockefeller Foundation officers generally shared an oil-centric worldview that also underlay modernization theory's obsession with, say, tractors and mechanized agriculture.

I'm also pointing to the typical career paths of people who shaped American foreign policy during decolonization: figures like Joseph Slater who first worked in the oil industry (often in the Global South—Slater worked for Standard Oil's subsidiary in Venezuela, Creole Petroleum), then passed through the revolving door to philanthropic foundations and/or government (both, in Slater's case). Something similar happened at a global level with figures like Maurice Strong moving back-and-forth between oil/energy jobs and international diplomacy, in the process hammering out post-colonial policy frameworks such as "sustainable development" (with which Strong is closely associated).[44] Nor were Slater or Strong unusual. From George H. W. Bush (CIA director, then president) to James Webb (administrator of NASA) to George McGhee (US ambassador to Turkey and Germany) to, more recently and internationally, Liz Truss (briefly prime minister of the United Kingdom) and Sigrid Kaag (formerly Foreign, later Finance, Minister of the Netherlands), ex-employees of oil companies are ubiquitous in the upper reaches of politics, especially in areas relating to national security and diplomacy.

3.5 The Fossil Fuel Regime and Niche Alternatives

So the oil industry profoundly shaped the landscape that MLP paints as exogenous to technological regimes. Now let's turn to the oil industry's influence—both broad and deep—over the various niche technologies that seemed ready to "break out" thanks to destabilization of the fossil fuel regime in the 1970s. The oil industry was implicated in virtually all of the new, futuristic energy sources of that era: geothermal, nuclear fusion, synfuels (in various guises), hydrogen fuel cells, and especially

[42] This is almost common knowledge that needs no citation. It is, at any rate, so well known that it suffuses even highly conventional oil history narratives, e.g. Daniel Yergin, *The Prize: The Epic Quest for Oil, Money, and Power* (New York: Free Press, 1991).

[43] Mark Solovey, *Shaky Foundations: The Politics-Patronage-Social Science Nexus in Cold War America* (New Brunswick: Rutgers University Press, 2013).

[44] For a study of Strong, the Club of Rome, and the path to "sustainable development" (but with almost no acknowledgement of Strong's or the Club's oil ties), see Stephen J. Macekura, *Of Limits and Growth: The Rise of Global Sustainable Development in the Twentieth Century* (Cambridge, UK: Cambridge University Press, 2015).

solar energy and nuclear fission. At the time, many politicians and members of the public thought that oil companies were trying to gain monopoly control of these nascent industries in order to stifle them.[45] No doubt there were oilmen wily enough to think they could pull that off. But the oil industry was simply too heterogeneous for that to be a universal explanation. A more plausible explanation is that oil actors understood that the landscape changes going on around them could destabilize fossil fuel energy regimes; they therefore tried to steer the landscape in a direction that would re-stabilize their industry while also cultivating various niche technologies that could break out over the medium- and/or long-term, either to assist in restabilizing the incumbent regime or to be prepared for the emergence of a new regime. Oil actors of the 1970s were as certain as MLP adherents today that regime change would happen eventually; but they also interfered with the factors encouraging regime change in ways that MLP struggles to grasp. That twin strategy—bet on regime change while tilting the playing field against it—becomes clearer if we step through the oil industry's entanglements with each of the era's most plausible alternatives:

3.5.1 Conventional Nuclear

The oil industry's involvement with things nuclear goes back almost to the beginning. Chemical companies, most of them future petrochemical companies, dominated contracting for the Manhattan Project, and several of them—most prominently Dupont—continued as contractors for Atomic Energy Commission sites after the war.[46] Because of the Manhattan Project's need for rapid construction of large, technically complex facilities, companies with experience building chemical refineries and oilfield platforms—Stone & Webster, M.W. Kellogg (the K of KBR), Bechtel—were also central both to the wartime bomb program and to the postwar AEC. Oil production companies such Humble Oil and Standard Oil of New Jersey also contributed personnel to the Manhattan Project (such as the head of centrifuge development, Esso's Eger Murphree).[47] After the war some of these companies got

[45]Probably the most vocal promoter of this view at the time was Senator James Abourezk. See James G. Abourezk, *Advise and Dissent: Memoirs of South Dakota and the U.S. Senate* (New York: Lawrence Hill Books, 1989). Meg Jacobs shows how the wider public shared Abourezk's suspicion that oil companies were gaining ownership of alternatives in order to strangle them. Meg Jacobs, *Panic at the Pump: The Energy Crisis and the Transformation of American Politics in the 1970s* (NY: Macmillan, 2016).

[46]Dupont's involvement with the Manhattan Project is well known, for example from Pap Ndiaye, *Nylon and Bombs: Dupont and the March of Modern America* (Baltimore: Johns Hopkins University Press, 2007). Despite the vast literature on the bomb program, however, we do not yet have a *business* history of the Manhattan Project as a whole. If we did, the contributions of oil, oilfield services, and petrochemical companies besides Dupont would be more evident.

[47]In the meantime, we can infer what those contributions were from passing mentions in, for example, Richard Rhodes, *The Making of the Atomic Bomb* (New York: Simon and Schuster, 1986).

out of nuclear but others replaced them. Phillips Petroleum, for instance, ran the Materials Testing Reactor for the AEC from 1952–1970; while Kerr-McGee leveraged its experience in oil exploration to insert itself into the uranium procurement and processing business.[48] In the Netherlands, Shell steered government policy on nuclear energy and got into centrifuge development early on.[49]

From the mid-1960s onward oil companies ramped up involvement with nuclear power. Atlantic Richfield, for instance, moved into nuclear waste processing in 1967.[50] Gulf purchased General Atomics in the same year; Shell bought a 50/50 share in 1973 but sold it back to Gulf in 1982.[51] Ashland Oil made a failed attempt to take over United Nuclear in 1969.[52] And so on; as Wesley Cohen argues, once Exxon formed Jersey Nuclear in 1969, the rest of the oil industry—which was accustomed to taking its cues from Exxon—followed suit for fear of missing out.[53]

Not many oil companies were directly producing electricity from nuclear power in the 1970s, but many did insert themselves into the nuclear fuel chain, usually at multiple points—exploration, mining, and refining of uranium, plus processing of nuclear waste. A 1976 Federal Trade Commission report found that twelve of the top 25 uranium mining and milling companies in the US were also (or were owned by) oil firms, including five of the top ten.[54] By 1981 "about 45% of all US uranium [wa] s produced by oil companies."[55]

---

[48] See, e.g., Phillips Petroleum Company Atomic Energy Division, *Fundamentals in the Operation of Nuclear Test Reactors,* U.S. Atomic Energy Commission, contract AT(10–1)-205, 1963. Kerr-McGee moved into uranium mining and processing in 1952. Dean A. McGee, "The Kerr-McGee Story," speech for the Oklahoma State Chamber of Commerce, November 13, 1970. In Box 176, Folder H – "The Kerr-McGee Story," Dean McGee papers, Spencer Research Library, University of Kansas. Note that the plant where Karen Silkwood worked was operated by Kerr-McGee.

[49] Abel Streefland, Jaap Kistemaker en uraniumverrijking in Nederland, 1945–1962, PhD dissertation, Universiteit van Amsterdam, 2017.

[50] ARCO was mostly a contractor at the Hanford site, and apparently not always a careful one. Wallace Turner, "Atom-Waste Blast Contaminates Ten," *New York Times* (August 31, 1976).

[51] A brief overview of these transactions is recounted in Keith E. Asmussen (General Atomics) to Merrit Baker (Nuclear Regulatory Commission), re: Request to Terminate General Atomics' U.S. NRC Special Nuclear Materials License, October 31, 2018, available at https://www.nrc.gov/docs/ML1830/ML18309A038.pdf

[52] Joel Dean, "Some Causes and Consequences of Conglomerate Merger – A Statement," in *Economic Concentration, part 8 The Conglomerate Merger Problem,* hearings before the Subcommittee on Antitrust and Monopoly of the Committee on the Judiciary, United States Senate, 91st Congress, November 1969 to February 1970.

[53] Wesley Marc Cohen, Firm Heterogeneity, Investment, and Industry Expansion: A Theoretical Framework and the Case of the Uranium Industry, PhD dissertation, Yale University, 1981.

[54] Bureaus of Competition and Economics, *Federal Energy Land Policy: Efficiency, Revenue and Competition* (Washington: US Government Printing Office, 1976), 684A.

[55] P. Bauder and A.-E. Wagner, "Uranium: Market, Reserves and Industry," *Energy Exploration and Exploitation 1981* (New York: Sage, 1981), 5–28.

3.5.2 Solar

Oil firms' long involvement with nuclear power was aided by affinities and complementarities between the two energy sources: both require mining scarce raw materials, sometimes from adjacent regions; both produce problematic waste through accidents as well as normal use; and both had the support of the national-security state. Yet while there are fewer affinities between oil and solar power, solar was as popular with oil companies as nuclear: most of the majors and a few mid-size and independent firms formed solar units, invested in solar start-ups, and/or purchased solar subsidiaries.

Some oil companies, such as Atlantic Richfield, were big players in solar. ARCO invested heavily in one of the most prominent solar firms, Stanford Ovshinsky' ECD.[56] ECD also had upfront money from Standard Oil of Ohio and some backdoor financing from Exxon and Mobil.[57] ARCO had its own solar subsidiary, too, which at one point made it the world's largest manufacturer of photovoltaic panels before it sold to Siemens in 1990.[58] Likewise, Shell had substantial solar operations in the Netherlands and Indonesia, and Shell US was the main funder of Solar Energy Systems, a start-up associated with the University of Delaware. Chevron also seems to have put money into SES.[59] And so on: Mobil had Mobil-Tyco, BP had BP Solar, Phillips had Acurex Solar, etc. One of the bigger and certainly more surprising efforts was at Exxon which had *two* solar subsidiaries, Solar Power Corporation and Daystar.

Texas Instruments—originally (and at that point still in part) an oilfield services company—also had a solar system for residential heating and electricity that it was preparing to move into production until the declining price of oil scuttled the idea.[60] Notably, when TI sought partners or buyers for the project, the only promising targets were oil companies or a Saudi prince.[61] Likewise, when RCA was looking for

---

[56] Lillian Hoddeson and Peter Garrett, *The Man Who Saw Tomorrow: The Life and Inventions of Stanford R. Ovshinsky* (Cambridge, MA: MIT Press, 2018).

[57] The backdoor financing appears to have been channeled through a $3.4 million investment in ECD from a Japanese syndicate. Robert Metz, "New Products, New Hopes," *New York Times* (June 1, 1978).

[58] The rest of the paragraph is based on Geoffrey G. Jones and Loubna Bouamane, "'Power from Sunshine': A Business History of Solar Energy," Harvard Business School Working Paper, No. 12–105, May 2012.

[59] Ralph Flood, "Big Oil Reaches for the Sun," *New Scientist* 92.1279 (12 November 1981): 426–428.

[60] Cyrus C. M. Mody, "After the IC: Jack Kilby's Solar Misadventures," *IEEE Spectrum* 53, no. 10 (2016): 50–55.

[61] Prince Mohammed Faisal. See "Texas Instruments Takes a Partner," undated (probably 1977), Box 2, Record Group 4, Texas Instruments Records, A2005.0025, DeGolyer Library Special Collections, Southern Methodist University. Another Saudi royal, Crown Prince (later King) Fahd, also associated himself with solar energy, e.g., by investing in a trial solar system for the Terraset Elementary School in Reston, Virginia in 1977. The Saudi embassy in the US was an enthusiastic supporter of "Sun Day," the not very auspicious sequel to Earth Day held on May 3, 1978 in support of solar initiatives around the US. See "SunDay A Special Day... A Special Relationship" [ad from the Royal Embassy of Saudi Arabia], *New York Times* (May 3, 1978): A26.

a partner or buyer for its solar project in the early '80s, the prime candidates were Amoco, Sohio, Texaco, and Total.[62] In 1983 RCA sold to Solarex—which was owned by Amoco (and which had earlier bought Exxon's Solar Power Corporation). Any history of solar energy, thus, needs to acknowledge that the oil industry was *the* obligatory point of passage, not just in the US but globally, in the late '70s and early '80s.

3.5.3 The Rest

Indeed, the oil industry played that role for almost all energy technologies in that period, both fossil fuels and "alternatives." Oil firms became major coal producers in the 1970s, and were largely responsible for shifting the center of US coal production to the West and Midwest and away from Appalachia. Oil firms (especially Exxon and Tosco) were also prominent partners in the Carter administration's synfuels initiative, which poured hundreds of millions of dollars into "western oil shale, tar sand, coal liquefaction and gasification, peat, unconventional gas (western tight gas sands, eastern Devonian gas shales, methane from coal seams, and methane from geopressured aquifers), and fuel ethanol."[63]

Meanwhile, Exxon as well as Shell and Gulf established nuclear fusion projects, and one of the most promising fusion companies of the era, KMS Industries, was 20% owned by Burmah Oil (Texas Gas Commission also had a stake in KMS).[64] Geothermal energy, too, saw considerable oil investment. By 1975, "Union Oil Company operate[d] the only geothermal field now producing electric power on a commercial basis and ... other oil companies, such as Chevron, Getty, Phillips Petroleum and Shell have leased thousands of acres of public land for geothermal efforts."[65] Chevron was at one point the world's largest private producer of geothermal energy, with fields operating in the Philippines, Oregon, and elsewhere.[66]

[62] This list is from the documents in Box 1, Brown Williams papers, Hagley Library.

[63] Robert M. Reed (ed.) *Preparation of Environmental Analyses for Synfuel and Unconventional Gas Technologies*, Environmental Sciences Division Publication No. 1843 (Oak Ridge National Laboratory, 1982). There have been a couple dissertations on the Synthetic Fuels Corporation but it merits further study. Hervey Amsler Priddy, United States Synthetic Fuels Corporation: Its Rise and Demise, PhD dissertation, University of Texas at Austin, 2013; Joshua Kevin Kundert, Fluid Technologies: Synthetic Fuels, Petroleum, and the American Political Context, PhD dissertation, University of Wisconsin – Madison, 2009.

[64] On Shell, see Bron, The Sun on Earth. On Burmah and the Texas Gas Commission see Jeff Hecht and Dick Teresi, *Laser: Light of a Million Uses* (Mineola, NY: Dover, 1998). There's a mention of the Exxon program in Victor K. McElheny, "Change of Goals in Exxon Research," *New York Times* (April 12, 1978).

[65] "Abourezk Seeks Oil-Concern Curb," *New York Times* (January 30, 1975).

[66] Alexander Richter, "Chevron Geothermal Philippines Celebrates 40 Years of Operation," *Think Geoenergy*, https://www.thinkgeoenergy.com/chevron-geothermal-philippines-celebrates-40-years-of-operation/

A few oil firms also positioned themselves in biofuels: Shell Forestry, for instance, bought up huge holdings in New Zealand and South America, and several oil firms invested in—or even wholly owned—leading plant biotechnology start-ups.

Oil companies were also at the forefront of R&D in energy storage, nominally with an eye to combining new storage technologies with cleaner energy generation. The Atlantic Richfield Foundation, for instance, sponsored the World Hydrogen Energy Conferences in cooperation with the University of Miami's Clean Energy Research Institute.[67] Shell had a methanol fuel cell project for powering vehicles that started in the early '70s on contract with the British military but then continued until 1980 as an independent effort.[68] In 2019, Stanley Whittingham won a share of the Nobel Prize in Chemistry for his invention of the rechargeable lithium ion battery while at Exxon Research in 1977.[69] Also in the late 1970s, Exxon Enterprises—Exxon's corporate venture capital arm—invested in energy storage and other companies related to long-range plans to enter the electric car business.[70] Later, Whittingham moved to the Schlumberger-Doll research center—Schlumberger being a major oilfield services company with fingers in nuclear power, microelectronics, and biotech as well.

3.6 Rushing in and Rushing Out Again

So why did oil companies dabble in multiple niche and regime technologies in the 1970s, in ways that seem counter-intuitive today? One conspiratorial explanation is that oil firms were trying to undermine competitors; another conspiracy theory says that Anderson, Strong, et al. aimed to create a socialist/eugenicist world government.[71] Even if such conspiracy theories hold a few grains of truth, though, they're inadequate. Oil firms needed to secure the cooperation of non-oil actors like Barbara

[67] T. Nejat Veziroğlu (ed.), *Hydrogen Energy Processes IV* (Oxford: Pergamon Press, 1982).

[68] Rob van Veen, personal communication.

[69] Interview with M. Stanley Whittingham conducted by Arne Hessenbrugh and Bernadette Bensaude-Vincent, October 30, 2000, https://authors.library.caltech.edu/5456/1/hrst.mit.edu/hrs/materials/public/Whittingham_interview.htm; Matthew Eisler, "Cold War Computers, California Supercars, and the Pursuit of Lithium-Ion Power," *Physics Today* 69, no. 9 (2016): 30–36.

[70] "The New Diversification Oil Game," *Business Week*.

[71] William P. Hoar, "New World Order," *American Opinion* 20 (1977); Rod Adams, "Robert O. Anderson – Banking Heir, Oil Wildcatter, Big Oil Exec, Financier of Antinuclear Movement," https://www.ans.org/news/article-1392/robert-anderson-antinuclear-financier/; William Engdahl, "Independent Oilmen Honor Anderson," *Executive Intelligence Review* 8.49 (December 22, 1981): 14-; Sanjeev Sabhlok, "How Arch-Socialist Maurice Strong Steered the World towards Precautionary Principle and Climate Panic – Sanjeev Sabhlok's Blog," August 16, 2019, https://www.sabhlokcity.com/2019/08/how-arch-socialist-maurice-strong-steered-the-world-towards-precautionary-principle-and-climate-panic/; informarmy, "The Green Genocide Agenda: Saving the Earth by Killing Humans | L'informazione Di INFORMARMY," informarmy.net, July 29, 2009, http://www.informarmy.net/2009/07/green-genocide-agenda-saving-earth-by/

Ward and Stanford Ovshinsky who required non-conspiratorial explanations for their cooperation to be plausible.

And there were plenty of alternative explanations. Oil was becoming more expensive and harder to access in this period—a good enough reason for oil companies to look for cheaper feedstocks and energy sources. Some oil actors, such as Maurice Strong, seem to have sincerely believed that helping build a more sustainable future would make up for their industry's "sins."[72] Other oil actors saw alternative energy more instrumentally. Solar energy, for instance, is a great way to power oil drilling operations in the desert or ocean where there isn't a grid.[73] Oil firms may also have put money into alternatives to win goodwill: from Congress, the public, and even from their own employees (especially younger, greener researchers).

The more interesting question to me is why oil companies retreated from these entanglements in the 1980s. The destabilization of the fossil fuel regime that many actors believed was underway in the 1970s either was reversed or didn't lead to the break-out of any niche alternatives. Thus, oil companies went back to being oil companies and gave up most—though not all—of their forays into nuclear, solar, and other emerging energy technologies. So how did collective memory largely forget that the oil industry was once a key player in those niches, and that oil industry executives were once publicly confident that those niches would break out to form a new regime?

A common answer is that improvements in the technology of oil pipelines and platforms in the 1970s alleviated the scarcity of oil and gas, in particular by opening up Alaska and the North Sea.[74] For MLP these improvements would count as clearing up reverse salients in the established energy regime. At the same time, the technological limitations of niche alternatives became more apparent: it became clearer that both nuclear power plants and the coalitions that support them break down, sometimes catastrophically; and consensus was reached that solar power wasn't going to be cheap any time soon.[75] Synfuels were a bust (at least until the current shale boom); the profitable applications of biotech turned out to be in GMOs and pharma rather than energy; fusion is still the perpetual energy source of the future; and fuel cells (to borrow from Matt Eisler) have yet to meet their "overpotential."

But that technological story doesn't satisfactorily explain why oil companies retreated from their 1970s investments. It was known at the beginning of the

[72]The first line of Strong's obituary in the *New York Times* was "Maurice Strong, a former industrialist and confessed ecological sinner." Sam Roberts, "Maurice Strong, Environmental Champion, Dead at 86," *New York Times* (December 1, 2015).

[73]John Perlin, *From Space to Earth: The Story of Solar Electricity* (Cambridge, MA: Harvard University Press, 1999), 57–66.

[74]Hans Veldman and George Lagers, *50 Years Offshore* (Delft: Foundation for Offshore Studies, 1997).

[75]For a contemporary lament see "In Exotic New Fuels: More Hope than Heat," *New York Times* (January 4, 1976).

1970s that Alaskan and North Sea oil and gas would come online eventually. Likewise, the risks associated with nuclear power were apparent before Three Mile Island. Few people in the 1970s denied that solar, fusion, the hydrogen economy, synfuels, and other emerging energy technologies would take until the 1990s or even (much) longer to become viable. Yet in the 1970s, the oil industry was willing to invest in these other energy sources for the long-term—whereas by the mid-1980s it was not. A fuller explanation, therefore, has to center on the decline of long-term thinking, rather than the "facts" that oil really wasn't scarce while other forms of energy really weren't ready. And for a change like the decline of long-termism, you need cultural and political explanations.[76]

Here I don't want to join the chorus that blames neoliberalism for everything—but neoliberalism was an important factor. In the late 1970s, the Carter administration in the US and the Wilson/Callahan governments in the UK liberalized price and supply controls and forced oil and gas companies to compete on price. Then in the 1980s, the elections of Margaret Thatcher and Ronald Reagan and their allies (with the enthusiastic support of many oil executives) squelched the global conversation about scarcity and state planning.[77] Their administrations became unreliable partners for development of alternatives such as solar power and synfuels, leaving oil companies little choice but to exit the partnerships they formed in the 1970s. Shell and Gulf, for instance, were willing to put money into General Atomics' fusion efforts so long as their investments were balanced by US federal support for fusion R&D; but once that balance was removed, Shell and Gulf weren't willing to bear the burden alone.

Neoliberalism also changed business as much as it changed government. Indeed, it turned much of government *into* business. Especially in Europe, the 1980s saw a wave of privatization of state (or state-supported) enterprises, especially in the energy sector.[78] Now, companies like BP answered only to investors, rather than to government ministers and, by extension, voters. Neoliberalism also changed firms' habits with respect to pricing and investment. The pre-neoliberal generation of oilmen had reacted to the destabilization of fossil fuel regimes in the 1970s by diversifying into new technologies to help their industry weather the storm; they could do so because the price of oil was propped up and they could therefore invest substantial cash reserves in long-term projects. From the late 1970s onward, though, price controls were removed, and in the 1980s the oil glut wiped out oil firms' ready cash. Also in the 1980s, aggressive investors (such as T. Boone Pickens, Carl Icahn, and Ivan Boesky) gained legal rights to override firms' long-term strategies.[79] These

---

[76] Jo Guldi and David Armitage, *The History Manifesto* (Cambridge, UK: Cambridge University Press, 2014).

[77] Jeroen Touwen, "De impact van het *Rapport van de Club van Rome*," in *Wereldgeschiedenis van Nederland*, ed. Lex Heerma van Voss et al. (Amsterdam: Ambo|Anthos, 2018), 659–664.

[78] Jacob William Ward, Information and Control: Inventing the Communications Revolution in Post-War Britain, PhD dissertation, University College London, 2018.

[79] Brian R. Cheffins (ed.), *The History of Modern U.S. Corporate Governance* (Cheltenham: Edward Elgar, 2011).

investors' sole priority was getting money out of their investments sooner rather than later, with little regard for long-term consequences.

Thus, from a variety of directions, oil companies faced enormous pressure to get rid of their long-range investments in solar and nuclear projects.[80] Investor and market pressure also meant oil companies were forced to move oil and gas to market as quickly as possible and had little capacity to spare for side projects. Neoliberal-inspired changes to anti-trust regulation also removed constraints on oil firms' ability to grow through mergers rather than non-oil acquisitions.[81] The 1980s and 1990s therefore saw a wave of ever-larger mergers between oil firms, while those same firms also shed solar, nuclear, and other non-oil units that were no longer central to their growth strategies. In MLP terms, the landscape now re-stabilized the incumbent regime, while removing many of the protections that formerly shielded niche alternatives—hardly the optimistic, linear story about niches overthrowing regimes that is offered by much of the sustainability transitions literature.

The declining price of oil and investors' pressure to focus on core competences foreclosed the possibility that fossil fuel incumbents would aid the breakout of niche alternatives; instead, short-termism stoked climate denialism. There was, true, denialism even earlier.[82] Yet climate change is mainly an existential problem for the *fossil fuel* industry, whereas in the long 1970s climate change was not necessarily an existential problem for companies like Shell or even Exxon that were busy acquiring the capacity to enter non-fossil fuel markets. Such firms could simply say that climate change wouldn't be a problem for a long time, and that by then the energy supply would be diversified enough to cope.

But once oil firms chose or were forced to focus on their core competence in fossil fuels and retreat from long-range plans to add solar, nuclear, etc., then the existential threat posed by climate change became unavoidable, and other energy sources became competitors. To keep stock prices high and remove any obstacle to getting oil out of the ground, climate change had to be defanged as a public topic of concern and alternative energy had to be discredited. Oilmen like Anderson and Strong who acknowledged climate change lost any influence they once had in the industry, while their opponents in the 1970s scarcity debate were enrolled to promote climate denialism. This wasn't an overnight switch: Oreskes and Conway have shown that institutions of climate denialism formed gradually over the course of the 1980s, while the journalist Jelmer Mommers and others have shown that Shell and other oil companies acknowledged—and even promoted public knowledge of—climate change until the early 1990s.[83]

---

[80] Mario Shao, "Oil Firms' Capital, Exploration Outlays Are Expected to Be Flat or Decline in 1983," *Wall Street Journal* (December 27, 1982): 20.

[81] This particular point, and the general thrust of this section, has benefited greatly from conversation with Erik Conway and the other authors in this volume.

[82] Benjamin Franta, "Early Oil Industry Disinformation on Global Warming," *Environmental Politics* (2021).

[83] Jelmer Mommers, "Shell erkent al dertig jaar het gevaar van klimaatverandering (en deze film bewijst dat)," *De Correspondent* (February 28, 2017).

The tipping point in that process seems to have been the Rio summit of 1992—Strong's last hurrah as an environmental diplomat.[84] In the run-up to Rio, for instance, Shell paid the chemist Frits Böttcher—a founding member of the Club of Rome but a *Limits to Growth* skeptic—to build a European climate denialist network similar to that constructed by the "merchants of doubt" in the US.[85] Like Strong and his Club of Rome colleagues, Böttcher believed that innovation and ingenuity can overcome limits. The difference was that that view led Strong toward the concept of "sustainable development;" whereas for Böttcher the same view occasioned denial of limits talk in the 1970s and then denial of climate change in the 1990s.

One way to see the disparate paths of Böttcher and Strong is as an intra-industry battle over how the fossil fuel regime should shape its landscape—with an eye to what the reshaped landscape would imply for the regime's relationship to niche alternatives. Strong (along with Anderson, Mitchell, Slater, etc.) led an arm of the oil industry that wanted to shift the global conversation in a way that would legitimate oil firms' elevation of niche alternatives such as nuclear and solar. Böttcher and his corporate allies, meanwhile, tried to move the landscape-level conversation in a direction that would force the oil industry to abandon those niches.

The lesson for adopters of the multi-level perspective—and for all of us—is to pay closer attention to how regime (and niche) actors shape the landscape that, in turn, shapes them. As hinted in my summary of MLP at the beginning of this chapter, this is one way in which MLP could be more faithful to evolutionary theory than its current debt to evolutionary economics allows. Evolutionary biology is full of examples where organisms radically modified landscapes that in turn exerted selection pressure on those same organisms—from early microorganisms' creation of an oxygen-rich atmosphere to our more recent ancestors' invention of agriculture. Oil actors seek to modify the "landscape" of society as much as anyone; and the size and wealth of their industry allows them to do so more effectively than other actors. Of course, the size and heterogeneity of the oil industry also means there will be intra-industry disagreements about *how* to shape that landscape: divides between big and small firms, public and private ones, European and North American, producers and oilfield services. Some firms—e.g., Exxon—have tried to forestall destabilization of the fossil fuel regime for as long as possible; while others—e.g., Atlantic Richfield and to some extent its current owner, BP—have positioned themselves to take a role in whatever new energy regime emerges after destabilization of the current one.

For electrical historians this point will seem familiar or even banal. The publications in which Thomas Hughes articulated the characteristics of large technological systems are replete with examples of electrical actors fighting over how to shape various societal landscapes in ways that would allow the electrical regime to grow.

---

[84]Nathaniel Rich, "Losing Earth: The Decade We Almost Stopped Climate Change," *New York Times* (August 1, 2018); Marten Boon, "A Climate of Change? The Oil Industry and Decarbonization in Historical Perspective," *Business History Review* 93, no. 1 (2019): 101–125.

[85]Alexander Beunder, "Hoe Frits Böttcher met steun van tientallen bedrijven de basis legde voor de klimaatscepsis in Nederland," *Volkskrant* (February 22, 2020).

The most successful electrical actors—an Edison or Insull or Lilienthal—grew the regime to the point where it was almost indistinguishable from the societal landscape. Indeed, Lilienthal headed a project, the Tennessee Valley Authority, that radically changed the literal landscape of parts of the US South. As the other chapters in this volume show, electrical history *is* landscape history—the construction of electrified societies was shaped (though not determined) by physical landscapes that were themselves remade by electrification. It is probably non-controversial to say that the same is true of the relationship between electrification and the figurative societal "landscapes" that appear in MLP.

Given the co-production of landscapes and electrification, the case for electrical history to speak more directly to environmental history is strong—a theme that runs through much of this volume. Much of this volume, and particularly this chapter, also makes a case for electrical history to speak more directly to the broader field of *energy* history. Of course, for historians (if not for thermodynamics), different forms of energy are not convertible and electrical history cannot therefore be submerged in an undifferentiated history of "energy" in the abstract: food energy plays a different role in history than nuclear fission or geothermal heating. But this chapter has tried to show that the histories of different forms of energy are woven together: the history of food includes the history of nuclear power and *vice versa*, and so on. The relationships among different energy histories are a tangled thicket. However, I've tried to show that for the 1970s, especially in the US, most branches of that thicket ran through the oil industry in one way or another. Putting the oil industry at the center of electrical history is counter-intuitive, given that that industry is not very involved in generating or transmitting electricity; but I've tried to show that any electrical history of the 1970s should take account of oil firms' multiple entanglements with electrical technologies. Indeed, putting oil at the center of "new approaches to the history of electrification" in the 1970s uncovers many connections to domains that are not usually understood as part of electrical history and thus broadens the relevance of the electrical history literature.

Electrical history also has much to contribute to discussions of energy and sustainability transitions, even if many historians are uncomfortable with the concept of "transitions." I've tried to show that the one of the more popular (and more history-friendly) transitions frameworks, the multi-level perspective, offers some heuristics for understanding what was driving energy actors of the 1970s: societal changes with the potential to destabilize fossil fuel regimes and the emergence of niche technologies with the potential to displace all or part of those regimes. Still, MLP's clear categories don't do justice to the complexity of that era: on inspection, niches, regimes, and landscapes were too mixed up for us to speak of distinct "levels." That's an analytic point, but one that also has implications for action in the present. Attempts to curb climate change, such as the 2015 Paris accords, will fall apart if different energy technologies are imagined to reside in distinct domains, or if incumbent energy regimes are pictured as only passively reactive to (rather than actively shaping) their landscape. Energy is a social practice, and social life is too multivalent—and too much an emergent property of heterogeneous networks—to be managed in reductionist fashion; and ignoring the landscape-shaping political power of incumbents would be naive to the point of ineffectiveness.

Bibliography

Abourezk, James G. 1989. *Advise and Dissent: Memoirs of South Dakota and the U.S. Senate*. New York: Lawrence Hill Books.

Adams, Rod. 2013. *Robert O. Anderson – Banking Heir, Oil Wildcatter, Big Oil Exec, Financier of Antinuclear Movement*. https://www.ans.org/news/article-1392/robert-anderson-antinuclear-financier/.

Anonymous. 1975. Abourezk Seeks Oil-Concern Curb. *New York Times*. January 30.

———. 1976. In Exotic New Fuels: More Hope than Heat. *New York Times*. January 4.

———. The New Diversification Oil Game. *Business Week*.

Baker, Kevin T. 2019. *World Processors: Computer Simulation, the Limits to Growth, and the Birth of Sustainable Development*. PhD dissertation, Northwestern University, pp 318–319.

Basosi, Duccio, Giuliano Garavini, and Massimiliano Trentin, eds. 2018. *Counter-Shock: The Oil Counter-Revolution of the 1980s*. London: I. B. Tauris.

Bauder, P., and A.-E. Wagner. 1981. Uranium: Market, Reserves and Industry. In *Energy Exploration and Exploitation 1981*, 5–28. New York: Sage.

Bergek, Anna, et al. 2008. Analyzing the Functional Dynamics of Technological Innovation Systems: A Scheme of Analysis. *Research Policy* 37 (3): 407–429.

Beunder, Alexander. 2020. Hoe Frits Böttcher met steun van tientallen bedrijven de basis legde voor de klimaatscepsis in Nederland. *Volkskrant*. February 22.

Bijker, Wiebe E. 1995. *Of Bicycles, Bakelites, and Bulbs: Towards a Theory of Sociotechnical Change*. Cambridge, MA: MIT Press.

Bini, Elisabetta, Guiliano Garavini, and Federico Romero. 2016. *Oil Shock: The 1973 Crisis and Its Economic Legacy*. London: I. B. Tauris.

Boon, Marten. 2019. A Climate of Change? The Oil Industry and Decarbonization in Historical Perspective. *Business History Review* 93 (1): 101–125.

Bureaus of Competition and Economics. 1976. *Federal Energy Land Policy: Efficiency, Revenue and Competition*. Washington: US Government Printing Office.

Cheffins, Brian R., ed. 2011. *The History of Modern U.S. Corporate Governance*. Cheltenham: Edward Elgar.

Coates, Peter A. 1991. *The Trans-Alaska Pipeline Controversy: Technology, Conservation, and the Frontier*. Bethlehem, PA: Lehigh University Press.

Cohen, Wesley Marc. 1981. *Firm Heterogeneity, Investment, and Industry Expansion: A Theoretical Framework and the Case of the Uranium Industry*, PhD dissertation, Yale University.

Cowan, Robin. 1990. Nuclear Power Reactors: A Study in Technological Lock-In. *The Journal of Economic History* 50 (3): 541–567.

Eisler, Matthew. 2016. Cold War Computers, California Supercars, and the Pursuit of Lithium-Ion Power. *Physics Today* 69 (9): 30–36.

Engdahl, William. 1981. Independent Oilmen Honor Anderson. *Executive Intelligence Review* 8 (49): 14.

Flood, Ralph. 1981. Big Oil Reaches for the Sun. *New Scientist* 92 (1279): 426–428.

Fosbrook, Bretton. 2017. How Scenarios Became Corporate Strategies: Alternative Futures and Uncertainty in Strategic Management, PhD dissertation, York (Canada) University.

Franta, Benjamin. 2021. Early Oil Industry Disinformation on Global Warming. *Environmental Politics*. 30 (4): 663–668.

Fuenfschilling, Lea, and Bernhard Truffer. 2014. The Structuration of Socio-Technical Regimes: Conceptual Foundations from Institutional Theory. *Research Policy* 43 (4): 772–791.

Garavini, Giuliano. 2019. *The Rise and Fall of OPEC in the Twentieth Century*. Oxford: Oxford University Press.

Geels, Frank W. 2018. Disruption and Low-Carbon System Transformation: Progress and New Challenges in Socio-Technical Transformations Research and the Multi-Level Perspective. *Energy Research and Social Science* 37: 224–231.

———. 2020. Micro-Foundations of the Multi-Level Perspective on Socio-Technical Transitions: Developing a Multi-Dimensional Model of Agency through Crossovers between Social Constructivism, Evolutionary Economics, and Neo-Institutional Theory. *Technological Forecasting and Social Change* 152.

———. 2002. Technological Transitions as Evolutionary Reconfiguration Processes: A Multi-Level Perspective and a Case-Study. *Research Policy* 31 (8/9): 1257–1274.

———. 2011. The Multi-Level Perspective on Sustainability Transitions: Responses to Seven Criticisms. *Environmental Innovation and Societal Transitions* 1 (1): 24–40.

Geels, Frank W., et al. 2016. The Enactment of Socio-Technical Transition Pathways: A Reformulated Typology and a Comparative Multi-Level Analysis of the German and UK Low-Carbon Electricity Transitions. *Research Policy* 45 (4): 896–913.

Geels, Frank. 2005. Co-Evolution of Technology and Society: The Transition in Water Supply and Personal Hygiene in the Netherlands (1850–1930): A Case Study in Multi-Level Perspective. *Research Policy* 27 (3): 363–397.

Genus, Audley, and Anne-Marie Coles. 2008. Rethinking the Multi-Level Perspective of Technological Transitions. *Research Policy* 37: 1436–1445.

Gertner, Jon. 2013. George Mitchell, Father of Fracking. *New York Times Magazine*. December 21.

Gorman, Hugh S. 2001. *Redefining Efficient: Pollution Concerns, Regulatory Mechanisms, and Technological Change in the U.S. Petroleum Industry*. Akron: University of Akron Press.

Guld, Jo, and David Armitage. 2014. *The History Manifesto*. Cambridge, UK: Cambridge University Press.

Harris, Kenneth. 1987. *The Wildcatter: A Portrait of Robert O. Anderson*, 122–125. New York: Weidenfeld & Nicolson.

Hecht, Gabrielle. 1998. *The Radiance of France: Nuclear Power and Nuclear Identity after World War II*. Cambridge, MA: MIT Press.

Hecht, Jeff, and Dick Teresi. 1998. *Laser: Light of a Million Uses*. Mineola, NY: Dover.

Hoar, William P. 1977. *New World Order*. American Opinion 20.

Hoddeson, Lillian, and Peter Garrett. 2018. *The Man Who Saw Tomorrow: The Life and Inventions of Stanford R. Ovshinsky*. Cambridge, MA: MIT Press.

Hughes, Thomas P. 1983. *Networks of Power: Electrification in Western Society, 1880–1930*. Baltimore: Johns Hopkins University Press.

Huismann, Wilfried. 2014. *PandaLeaks: The Dark Side of the WWF*. Bremen: Nordbook.

Informarmy. 2009. The Green Genocide Agenda: Saving the Earth by Killing Humans | L'informazione Di INFORMARMY. infromarmy.net, July 29, 2009, http://www.informarmy.net/2009/07/green-genocide-agenda-saving-earth-by/.

Jones, Geoffrey G. and Loubna Bouamane. 2012. 'Power from Sunshine': A Business History of Solar Energy. *Harvard Business School Working Paper*, No. 12–105, May

Jørgenson, Ulrik. 2012. Mapping and Navigating Transitions: The Multi-Level Perspective Compared with Arenas of Development. *Research Policy* 41 (6): 996–1010.

Kirk, Andrew. 2007. *Counterculture Green: The Whole Earth Catalog and American Environmentalism*. Lawrence: University Press of Kansas.

Kirsch, Scott. 2005. *Proving Grounds: Project Plowshare and the Unrealized Dream of Nuclear Earthmoving*. New Brunswick: Rutgers University Press.

Kundert, Joshua Kevin. 2009. *Fluid Technologies: Synthetic Fuels, Petroleum, and the American Political Context*, PhD dissertation, University of Wisconsin – Madison.

Lutz, Raechel. 2018. *Crude Conservation: Nature, Pollution, and Technology at Standard Oil's New Jersey Refineries, 1870–2000*, PhD dissertation, Rutgers University.

Macekura, Stephen J. 2015. *Of Limits and Growth: The Rise of Global Sustainable Development in the Twentieth Century*. Cambridge, UK: Cambridge University Press.

McElheny, Victor K. 1978. Change of Goals in Exxon Research. *New York Times*. April 12.

McKinney, J. Merritt. 2011. *Air Pollution, Politics, and Environmental Reform in Birmingham, Alabama, 1940–1971*, PhD dissertation, Rice University.

Melsted, Odinn. 2020. *Icelandic Energy Regimes: Imported Fossil Fuels, Renewable Resource Development and the Making of a Low-Carbon Energy Balance, 1940–1980*, PhD dissertation, University of Innsbruck.
Metz, Robert. 1978. New Products, New Hopes. *New York Times*. June 1.
Mody, Cyrus C.M. 2016. After the IC: Jack Kilby's Solar Misadventures. *IEEE Spectrum* 53 (10): 50–55.
Moll, Peter. 1991. *From Scarcity to Sustainability: Futures Studies and the Environment: The Role of the Club of Rome*. Frankfurt: Peter Lang.
Mommers, Jelmer. 2017. Shell erkent al dertig jaar het gevaar van klimaatverandering (en deze film bewijst dat). *De Correspondent*. February 28.
Ndiaye, Pap. 2007. *Nylon and Bombs: Dupont and the March of Modern America*. Baltimore: Johns Hopkins University Press.
Oldenziel, Ruth. 2011. Islands: The United States as a Networked Empire. In *Entangled Geographies: Empire and Technopolitics in the Global Cold War*, ed. Gabrielle Hecht, 13–42. Cambridge, MA: MIT Press.
Oreskes, Naomi, and Erik M. Conway. 2010. *Merchants of Doubt: How a Handful of Scientists Obscured the Truth on Issues from Tobacco Smoke to Global Warming*. London: Bloomsbury.
Patoski, Joe Nick, and Bill Crawford. 1989. The Long, Strange Trip of Ed Bass. *Texas Monthly*. 1989.
Perlin, John. 1999. *From Space to Earth: The Story of Solar Electricity*, 57–66. Cambridge, MA: Harvard University Press.
Priddy, Hervey Amsler. 2013. *United States Synthetic Fuels Corporation: Its Rise and Demise*, PhD dissertation, University of Texas at Austin.
Priest, Tyler. 2014. Hubbert's Peak: The Great Debate over the End of Oil. *Historical Studies in the Natural Sciences* 44 (1): 37–79.
———. 2007. *The Offshore Imperative: Shell Oil's Search for Petroleum in Postwar America*. College Station: Texas A&M University Press.
Reece, Ray. 1979. *The Sun Betrayed: A Report on the Corporate Seizure of U.S. Solar Energy*. Montreal: Black Rose Books.
Rhodes, Richard. 1986. *The Making of the Atomic Bomb*. New York: Simon and Schuster.
Rich, Nathaniel. 2018. Losing Earth: The Decade We Almost Stopped Climate Change. *New York Times*. August 1.
Roberts, Cameron, and Frank W. Geels. 2019. Conditions for Politically Accelerated Transitions: Historical Institutionalism, the Multi-Level Perspective, and Two Historical Case Studies in Transport and Agriculture. *Technological Forecasting and Social Change* 140: 221–240.
Roberts, Sam. 2015. Maurice Strong, Environmental Champion, Dead at 86. *New York Times*. December 1.
Rome, Adam. 2003. 'Give Earth a Chance': The Environmental Movement and the Sixties. *Journal of American History* 90 (2): 525–554.
Sabhlok, Sanjeev. 2019. *How Arch-Socialist Maurice Strong Steered the World towards Precautionary Principle and Climate Panic – Sanjeev Sabhlok's Blog*. August 16 https://www.sabhlokcity.com/2019/08/how-arch-socialist-maurice-strong-steered-the-world-towards-precautionary-principle-and-climate-panic/.
Sabin, Paul. 2013. *The Bet: Paul Ehrlich, Julian Simon, and Our Gamble over Earth's Future*. New Haven: Yale University Press.
Spezio, Sabol, and Teresa. 2018. *Slick Policy: Environmental and Science Policy in the Aftermath of the Santa Barbara Oil Spill*. Pittsburgh: University of Pittsburgh Press.
Schleper, Simone. 2019. *Planning for the Planet: Environmental Expertise and the International Union for Conservation of Nature and Natural Resources, 1960–1980*. New York: Berghahn.
Schmandt, Jurgen. 2010. *George P. Mitchell and the Idea of Sustainability*. College Station: Texas A&M Press.

Schmelzer, Matthias. 2017. 'Born in the Corridors of the OECD': The Forgotten Origins of the Club of Rome, Transnational Networks, and the 1970s in Global History. *Journal of Global History* 12: 26–48.
Schwarzenbach, Alexis. 2011. *Saving the World's Wildlife: WWF – the First 50 Years*. London: Profile Books.
Shao, Mario. 1982. Oil Firms' Capital, Exploration Outlays Are Expected to Be Flat or Decline in 1983. *Wall Street Journal* 27: 20.
Shove, Elizabeth, and Gordon Walker. 2007. Caution! Transitions Ahead: Politics, Practice, and Sustainable Transition Management. *Environment and Planning A* 39 (4): 763–770.
———. 2010. Governing Transitions in the Sustainability of Everyday Life. *Research Policy* 39 (4): 471–476.
Smith, Adrian, Andy Stirling, and Frans Berkhout. 2005. The Governance of Sustainable Socio-Technical Transitions. *Research Policy* 34 (10): 1491–1510.
Smith, Adrian, Jan-Peter Voß, and John Grin. 2010. Innovation Studies and Sustainability Transitions: The Allure of the Multi-Level Perspective and Its Challenges. *Research Policy* 39: 435–448.
Solovey, Mark. 2013. *Shaky Foundations: The Politics-Patronage-Social Science Nexus in Cold War America*. New Brunswick: Rutgers University Press.
Sovacool, Benjamin K., and Frank W. Geels. 2016. Further Reflections on the Temporality of Energy Transitions: A Response to Critics. *Energy Research and Social Science* 22: 232–227.
Steffy, Loren C. 2019. *George P. Mitchell: Fracking, Sustainability, and an Unorthodox Quest to Save the Planet*. College Station: Texas A&M Press.
Streefland, Abel. 2017. *Jaap Kistemaker en uraniumverrijking in Nederland, 1945–1962*, PhD dissertation, Universiteit van Amsterdam.
Touwen, Jeroen. 2018. De impact van het *Rapport van de Club van Rome*. In *Wereldgeschiedenis van Nederland*, ed. Lex Heerma van Voss et al., 659–664. Amsterdam: Ambo|Anthos.
Trim, Henry. 2016. 'We Are as Gods': The Green Technical Fix. *RCC Perspectives* 4: 55–60.
Turnbull, Thomas. 2017. *From Paradox to Policy: The Problem of Energy Resource Conservation in Britain and America, 1865–1981*, PhD dissertation, University of Oxford.
Turner, Fred. 2006. *From Counterculture to Cyberculture: Stewart Brand, the Whole Earth Network, and the Rise of Digital Utopianism*. Chicago: University of Chicago Press.
Turner, Wallace. 1976. Atom-Waste Blast Contaminates Ten. *New York Times*. August 31.
United Nations Framework Convention on Climate Change. *The Paris Agreement*. https://unfccc.int/process-and-meetings/the-paris-agreement/the-paris-agreement.
Veldman, Hans, and George Lagers. 1997. *50 Years Offshore*. Delft: Foundation for Offshore Studies.
Verbong, Geert, and Derk Loorbach, eds. 2012. *Governing the Energy Transition: Reality, Illusion, or Necessity?* New York: Routledge.
Veziroğlu, T. Nejat, ed. 1982. *Hydrogen Energy Processes IV*. Pergamon Press: Oxford. 1982.
Ward, Jacob William. 2018. *Information and Control: Inventing the Communications Revolution in Post-War Britain*, PhD dissertation, University College London.
Yergin, Daniel. 1991. *The Prize: The Epic Quest for Oil, Money, and Power*. New York: Free Press.

Cyrus C. M. Mody is Professor of the History of Science, Technology, and Innovation at Maastricht University. He is an historian of late Cold War physical and engineering sciences, particularly industrial and industry-supported research. His contribution to this volume is part of a project funded by the Dutch Research Council, "Managing Scarcity and Sustainability: The Oil Industry, Environmentalism, and Alternative Energy in the Age of Scarcity."

Chapter 4
"We Have no Niagara": Electrifying the "Britain of the South"

Nathan Kapoor

Abstract To explore the relationship between New Zealand's electrification and its settler-colonial past, this chapter recasts one of New Zealand's most notable electric power projects, the Lake Coleridge Power Station. Typically, the power station is associated with New Zealand's becoming a Dominion and it is also often seen as the origin of the country's expansive, state-run hydroelectric infrastructure. While the station is fundamentally linked to the transition from colony to Dominion and the country does continue to maintain a sizeable hydroelectric base, it is far more important to understand how the dam did not constitute the beginning of something new. The power station was built on decades of settler colonial rhetoric and legislation that sought to make New Zealand a successful and self-sufficient British country. The legislation and language of the Public Works Department and Christchurch City Council make clear how the power station figured into that settler vision and embedded those practices into the country's electric power infrastructure.

Keywords New Zealand · Public works administration · Hydroimperialism · Settler colonialism · Infrastructure · British empire · Technopolitics

4.1 Introduction

During the first decades of the twentieth century, journalists and engineers lamented New Zealand's lack of an industrial scale electric power supply by comparing its small power stations to Niagara.[1] But they also noted the hydroelectric power potential of the country's waterways. Beyond the aspirations connected to international competitiveness, parties interested in electrical power found that creating such a power source could be most easily achieved by continuing colonial practices of

[1] "A Valuable Asset," *Wairarapa Age*, July 6, 1910, 4, https://paperspast.natlib.govt.nz/newspapers/TS19141126.2.86

N. Kapoor (✉)
Illinois State University, Normal, IL, USA
e-mail: nnkapo1@ilstu.edu

W. B. Carlson, E. M. Conway (eds.), *Electrical Conquest*, Archimedes 67,
https://doi.org/10.1007/978-3-031-44591-0_4

centralizing control of water and electric power bureaucracies. For example, on the Southern Island, the establishment of New Zealand's first national electric power supply, which began with the Lake Coleridge Power Station, exercised New Zealand's rebranded colonial authority under the auspices of the young Dominion government. As intimated by other chapters in this volume and authors elsewhere, the process of electrification did not follow a predetermined path, nor did it serve as the natural next energy transition following the Industrial Revolution.[2] New Zealand did not have to create a nationalized power system, nor were they destined to build a primarily hydroelectric infrastructure over the twentieth century. The various parties interested in electrification, the electrical stakeholders, namely Christchurch and Public Works Department officials, imagined and designed electric power systems that exemplified their industrial and modern aspirations for New Zealand. Borrowing from Sheila Jasanoff and Sang-Hyun Kim, the creation of these systems can be better understood as a sociotechnical imaginary or a:

> .... collectively held, institutionally stabilized, and publicly performed visions of desirable futures, animated by shared understandings of forms of social life and social order attainable through, and supportive of, advances in science and technology.[3]

The legislation and technologies involved in constructing the Lake Coleridge Power Station show how a few individuals' shared ideas led to the communal adoption of a vision for electric power and what ought to be the dominant energy source. For them, electric power performed their visions and expectations for a modern New Zealand – a New Zealand modeled on the original British expectation for the colony as a space for white settlers, a valuable Pacific resource hub, and a British-styled government.

Indeed, as many of the chapters in this volume demonstrate, stakeholders in existing or proposed electric power systems envision those systems to achieve political and economic ends. During the late-nineteenth century, the electric power transition required a complex web of socio-cultural and political alignments. Even within a single state, the pattern of and justification for electrification varied significantly. New Zealand's electrification demonstrates how the motivation for electrification differed from other colonial spaces and the metropole itself. New Zealand serves as an excellent case study for connecting electrification and colonialism. Still, it is worth noting that electrification projects in settler and subject colonies often adopted different rhetoric based on overt racism.[4] As much as the Colonial Office and many New Zealanders liked to frame the settlement as the "Britain of the South" or "England of the Pacific," New Zealand's electrification did not follow the British

---

[2]Nathan Kapoor, "Who Has Seen the Wind: Imagining Wind Power for the Generation of Electricity in Victorian Britain," *Technology and Culture* 60, no. 2 (2019): 467–493; Ute Hasenöhrl and Jan-Henrik Meyer, "The Energy Challenge in Historical Perspective," *Technology and Culture* 61, no. 1 (2020): 295–306.

[3]Sheila Jasanoff and Sang-Hyun Kim, eds., *Dreamscapes of Modernity: Sociotechnical Imaginaries and the Fabrication of Power* (Chicago: University of Chicago Press, 2015), 4.

[4]Moses Chikowero, "Subalternating Currents: Electrification and Power Politics in Bulawayo, Colonial Zimbabwe, 1894–1939," *Journal of Southern African Studies* 33, no. 2 (2007): 287–306; Animesh Chatterjee, "'New Wine in New Bottles': Class Politics and the 'Uneven Electrification' of Colonial India," *History of Retailing and Consumption* 4, no. 1 (2018): 94–108.

Fig. 4.1 Lake Coleridge power station. Head, Samuel Heath, 1868–1948: (Negatives. Ref: 1/1–007272-G. Alexander Turnbull Library, Wellington, New Zealand)

style. Like most urban elites, they saw electrification as a means of claiming their status as a modern and progressing city. Furthermore, they insisted that manufacturing and industrial development needed cheap electricity. Unlike Britain's locally regulated electric power network at the turn of the century, which Thomas Hughes described as "disordered and small scale," electric power stakeholders in New Zealand saw electrification as both a national and imperial project (Fig. 4.1).[5]

Contrary to some "tools of empire" and "technology transfer" narratives that either misrepresent or ignore how British colonial spaces electrified, I maintain that electrification took on new meanings in colonial spaces like New Zealand because electric power infrastructures served as mechanisms for expanding and adapting colonial authority. The historiography of electric power in New Zealand, while useful, portrays electrification as national or corporate history that does not connect the electrical network to either the colony or Dominion.[6] In both colonial and

[5]Thomas P. Hughes, *Networks of Power: Electrification in Western Society, 1880–1930* (Baltimore: Johns Hopkins University Press, 1983), 227.

[6]Neil Rennie, *Power to the People: 100 Years of Public Electricity Supply in New Zealand* (Wellington: Electricity Supply Association of New Zealand, 1989); John Martin, *People, Politics, and Power Stations: Electric Power Generation in New Zealand* (Wellington: Bridgett Williams Books, 1991); Hellen Reilly, *Connecting the Country: New Zealand's National Grid* (Wellington: Steele Roberts, 2008).

post-colonial spaces, electrification does not neatly align with the Euro-American models.[7] As illustrated by other chapters in this volume, stakeholders often imagine electric systems will accomplish long-running goals or sometimes upend existing structures. One thing that makes New Zealand's electrification distinctive is that stakeholders developed the first large power stations, such as Lake Coleridge, with a large-ubiquitous national system in mind. This was not because they envisioned an electrical future. Rather, they understood the utility and authority conferred by the control of the land and resources required to generate power.

Therefore, to achieve their imagined electric power future, the Colonial and subsequent Dominion governments appropriated indigenous land, established national electric-power regulations, and supported creating a centralized Public Works Department with the authority and resources to develop hydroelectric infrastructures.

> May the new Dominion flourish With the progress time will bring, And a knot of heroes nourish For the Empire and the King.[8] – Anon. "The New Dominion" (1907)

This is the final stanza of a poem, written by an anonymous citizen from Auckland, read in the New Zealand Parliament following deliberations over the establishment of the Dominion. In 1907 the Colony of New Zealand became the Dominion of New Zealand. Celebrations abounded, but the most publicized and photographed exhibition happened at the parliament buildings in Wellington. Electric lights adorned the building and formed the words, "Advance New Zealand," "Colony 1840," and "Dominion 1907." The use of electric lights at political and industrial exhibitions typified many turn-of-the-century celebrations around the world.[9] Electric lights symbolized technological capability and modernity; however, for New Zealand, the relationship between progress and electricity had, for decades, been associated with New Zealand's settler-colonial mission to create a model colony.[10] During the 1880s, electric power bolstered New Zealand's plans for self-sufficiency and economic productivity in service to the Empire. In the early years of the country's transition to electric power, hydroelectricity proved an effective means of achieving financial gain and establishing colonial infrastructures, such as lighting

---

[7] Elizabeth Chatterjee, "The Asian Anthropocene: Electricity and Fossil Developmentalism," *Journal of Asian Studies* 79, no. 1 (2020): 3–24.

[8] Anon., "The New Dominion," *Parliamentary Debates* 139, July 12, 1907, 387–388.

[9] David E. Nye, "Electrifying Exhibitions: 1880–1939," in *Fair Representations: World Fairs and the Modern World*, eds. Robert Rydell and Nancy Gwinn (Amsterdam: Free University Press, 1994), 140–156.

[10] Graeme Gooday, *Domesticating Electricity: Technology, Uncertainty, and Gender 1880–1914* (London: Pickering and Chatto, 2008): 121–122. This chapter builds on and uses research from my forthcoming book *Empowering Colonialism: Electrification in New Zealand* under contract with Pittsburgh University Press, expected 2024.

frontier outposts, mining gold in the frontier building long-distance telegraphic communication.[11] Across the Empire, the British enacted regulations to control settler populations and marginalize indigenous peoples, like the Māori of New Zealand.[12] The colony's government introduced new infrastructures, like state-run hydroelectric dams, to manage water, land, and people more efficiently.

Even though Dominion Day's signage and rhetoric touted independence, New Zealand remained part of the British Empire. British colonialism does not end with the creation of a Commonwealth. The continued military relationship, the high tax rate, and exploitation of the indigenous population highlight this relationship.[13] Nineteenth-century British colonialism in New Zealand is how the Colonial Office extracted resources, established a settler community, and attempted to subjugate the Māori. Modern colonialism is the extraction of tribute, goods, and wealth coupled with indigenous people's removal, assimilation, and economic reorganization, which draws the colonized and colonizers into a complicated relationship.[14] The New Zealand Parliament, Public Works Department, and Christchurch City Council adopted colonial rhetoric and practices, such as centralized control over natural resources or the utilization of stolen land, to advocate for and produce state-run electric power. The transition to electric power systems had fulfilled visions of a self-reliant colony; the expansion and demand for electric power gave officials a means to facilitate a more centralized government that would continue to pursue a colonial agenda under a new banner. The machinations of government, utilization of lands, stratification of people, and imagination for the future developed during the British settlement of New Zealand remained firmly in place.

[11] Eric Pawson and Neil C. Quigley, "The Circulation of Information and Frontier Development: Canterbury 1850–1890," *New Zealand Geographer* 38, no. 2 (1982): 65–76; Peter Petchey, "New Zealand's Technological Participation in the International Goldfields: The Example of the Stamper Battery," *Technology and Culture* 60, no. 2 (2019): 494–522.

[12] H. Hoag, "Turning Water into Power: Debates Over the Development of Tanzania's Rufiji River Basin," *Technology and* Culture 49, no. 3 (2008): 624–651; T. Tvedt, "Hydrology and Empire: The Nile, Water Imperialism and the Partition of Africa," *Journal of Imperial and Commonwealth History* 39, no. 2 (2011): 173–94; H. Hungerford and S. Smiley, "Comparing Colonial Water Provision in British and French Africa," *Journal of Historical Geography* 52 (2016): 74–83; L. Bruce Railsback, "Rain, Riches, and Empire: The Relationship between Nations Ruling Distant Lands, Nations of Great Wealth, and Regions of Regular Moderate Atmospheric Precipitation," *Weather, Climate, and Society* 9, no. 3 (2017): 455–469.

[13] Avner Offer, "The British Empire, 1870–1914: A Waste of Money?," *Economic History Review* 46, no. 2 (1993): 215–238.

[14] Ania Loomba, *Colonialism/Postcolonialism 3rd ed.* (New York: Routledge, 2015), 21.

4.2 Lake Coleridge and an Industrial New Zealand

> More favourably [sic] than any other water supply... the Christchurch and tributary districts could be very effectively supplied, that the installation would be simple and that the initial installation would not be enormously costly, as the plant could be enlarged as needed.[15]

This excerpt from the *Evening Post* is one of many exciting newspaper articles touting the potential of a Christchurch-based hydroelectric power station on the Rakaia River watershed, which included Lake Coleridge, in the early twentieth century. Construction on the dam and power station commenced in 1911 and began supplying power in 1914. Over the next few decades, the dam added more generators and expanded its supply network. *Engineering New Zealand*'s interpretation of the Lake Coleridge Power Station, like most histories of electrification in New Zealand, treats the dam as a simple electrification success story of the young state:

> Through the sale and promotion of Lake Coleridge power, most of the boroughs and counties near Christchurch were induced to become electricity supply authorities. The making of numerous contracts with relatively small-scale consumers or groups became unwieldy and led to the Government passing the Electric Power Boards Act 1918, which was later consolidated as the 1925 Act, setting the groundwork for the orderly distribution of electric power in New Zealand.[16]

Stories of the dam and the country's electrification always conclude with later developments in the twentieth century. During the mid-twentieth century, the Public Works Department added new river sources into the lake to increase output. The government replaced generators and linked it to other power boards. Today the dam remains an important power station in the TrustPower network on the South Island.[17] While creating a national electric power supply in New Zealand parallels most industrial states' electrification, it is worth re-examining how this project succeeded and work to move electrification narratives away from arguments that suggest such a grid was the plan all along. Like many other projects worldwide, New Zealand's electric power stakeholders founded these systems on colonial policies that hinged on centralized control over natural resources aimed at profit.

The story of how the city of Christchurch surveyed, justified, and ultimately constructed the dam clarifies how the power station carried colonial purpose. The technology did not merely flow from other industrial states; New Zealand maintained its own rationale. Even though many earlier electric projects relied on hydropower, the Lake Coleridge scheme harnessed hydropower on a larger scale and at a higher cost. The transition to Dominion represents the completion of the initial

---

[15] "Electrical Transmission of Power," *Evening Post*, October 31, 1904, 2. https://paperspast.natlib.govt.nz/newspapers/EP19041031.2.3

[16] Engineering New Zealand, "Lake Coleridge," accessed August 7, 2020, https://www.engineeringnz.org/our-work/heritage/heritage-records/lake-coleridge-power-station/

[17] "Our History, TrustPower, accessed December 29, 2020, https://www.trustpower.co.nz/Getting-To-Know-Us/Our-History

push for settlement because the colony achieved local self-government, one of the supposedly cherished aims of British colonialism. Nationalized electrification demonstrates the duplication of colonial infrastructures within New Zealand's new national government.

Understanding Lake Coleridge's role in local enactments of colonial power, how New Zealand's government adhered to the Colonial Office's emphasis on centralized governance is aided by placing the project in the context of the broader global transition to hydroelectric power, especially in British colonies and territories. Hydroelectric infrastructures and the accompanying political institutions (like the Public Works Department in New Zealand), expanded rapidly in the British Empire, especially in Canada, at the turn of the twentieth century.[18] Historians and sociologists classify this type of technopoltics as *hydro-imperialism*, or the usage and control of water resources in a colonial setting.[19] In studies of hydroelectric systems, dams, such as the Lake Coleridge Power Station, exist within an imperialist dynamic, in which New Zealanders reimagined the river as an industrial resource in service to the state and Great Britain. Across the Empire, the British enacted water regulations to control settler populations, marginalize indigenous peoples by stripping away their ancestral claims, and justify the introduction of new infrastructures.[20] Even though the dam was designed and built after Dominion, Lake Coleridge's construction still depended on imperial power dynamics and exploitation. Late-nineteenth and early-twentieth-century politicians and intellectuals knew the power potential of water.[21] As John Miller Dow Meiklejohn (1836–1902), a Scottish academic textbook author, clearly underscores in one of his geography texts:

> Divine Providence, brought it about that the people of Great Britain are now the traders and news carriers for the whole world. These functions have given us another office, have forced upon us another mission. This is to keep the Water-ways of the world– the waterways in ocean, sea, lake, river, and canal. Great Britain is, therefore, the Guardian of Water-ways.[22]

By the turn of the twentieth century, New Zealand politicians and settlers, particularly those on the West Coast of the southern island, had substantial experience altering the environment around waterways for electric power stations to create

[18] Dorotea Gucciardo, "The Powered Generation: Canadians, Electricity, and Everyday Life," (Ph.D. thesis, University of Western Ontario, 2011): 49–50.

[19] Sara Pritchard, "From Hydroimperialism to Hydrocapitalism: 'French' Hydraulics in France, North Africa, and Beyond," *Social Studies of Science* 42, no. 4 (2012): 591–615.

[20] Ute Hasenöhrl,"Rural Electrification in the British Empire," *History of Retailing and Consumption* 4, no. 1 (2018): 10–27; M. Lewis, "The Personal Equation: Political Economy and Social Technology on India's Canals, 1850–1930," *Modern Asian Studies*, 41 (2007): 967–994; Daniel Macfarlane and Peter Kitay, "Hydraulic Imperialism: Hydroelectric Development and Treaty 9 in the Abitibi Region," *American Review of Canadian Studies* 46, no. 3 (2016): 380–397.

[21] John Broich, "Engineering the Empire: British Water Supply Systems and Colonial Societies, 1850–1900," *Journal of British Studies* 46 (2007): 346–365.

[22] J. M. D. Meiklejohn, *The British Empire: Its Geography, Resources, Commerce, Land-ways, and Water-ways*, 6th ed. (London: Alfred M. Holden, 1899): 16.

settlements and enhance economic productivity in the service of colonial interests, especially mining. And while the power boards operated locally, water rights had to be obtained from the colonial government. Placing the construction of Lake Coleridge in the context of hydro-imperialism situates New Zealand's transition to electric power within the broader frame of imperial utilization of water as means of centralizing national government control over indigenous and settler populations to assert control over colonized places.

4.3 Dominion and the Public Works Department

New Zealand's electrification cannot be separated from the transition to Dominion. For New Zealanders, especially politicians in the Canterbury region, both electrification and Dominion signaled progress and the country's ability to function as a self-sufficient state. For historians of electricity, the merging of this political ideology with the rationalization of a large-scale centralized state electric power system serves as a useful reminder that we cannot assume that economies of scale were obvious. Instead, the design of these larger-scale systems must be seen as immersed in the technical and political context in which it appeared. British politicians and writers celebrated New Zealand's status as a model colony during the nineteenth century, and they subsequently portrayed the transition to Dominion as the definitive conclusion of the settler-colonial mission. The Dominion shift represents the political reproduction of British settler colonialism. A modern state in New Zealand could only be achieved through the guidance of settler (British) government, industry, and technology. During that transition, the Public Works Department, especially electric power and communication networks, provided the physical framework that facilitated more centralized authority within New Zealand. The idea for a national electric power supply began in the country's earliest public works legislation.

Throughout the late nineteenth-century, New Zealand continued to favor self-rule policies; the Public Works Department and its antecedents played a critical role in upholding New Zealand's ideas about self-sufficiency and demonstrated its ability to operate with little support independently. In June 1870, Julius Vogel, Colonial Treasurer, pushed the Immigration and Public Works Act (1870) through the General Assembly. Over time the bill set up a system by which the New Zealand government would borrow large sums of money from Britain. The money provided government-subsidized immigration and money for infrastructures such as railroads, telegraph lines, roads, public buildings, and port facilities. This act and the subsequent amendment rejuvenated a war-weary colony. In the context of New Zealand's long history of electrification, the legislation and practices concerning the telegraph provided the foundation for a nationalized electric power regulation.

During the mid-nineteenth century, Colonial Office and Public Works policies designed to regulate the introduction of electric communication and power increasingly granted authority to local political bodies. New Zealand's first piece of legislation concerning electricity was the Electric Telegraph Act of 1865, which gave the colony's General and Provincial governments control over telegraphic communication. The act formed the basis of New Zealand's laws concerning electrification and made the regulation, construction, and maintenance of electric communication equipment a central government function. After Vogel's Public Works Act (1870), telegraphic legislation and construction remained under government control. Mine owner and urban industrialist proposals for electric power gained traction in New Zealand because electric power promised to modernize the country's cities and manufacturing potential. Thus, Parliament saw a need for new legislation to define where and how companies and regional authorities could install electric lighting and power supply technologies. In response, Parliament passed the Electric Lines Act of 1884. This law consolidated the preceding laws relating to telegraph and telephone lines and created restrictions involving electric lighting for public places. The act stated:

> The Governor may from time to time construct and maintain electric lines for lighting purposes in or connected with public offices or buildings under the control of the Governor in any part of the colony... Any local authority having power by law to construct public works within the district under its jurisdiction may construct and maintain electric lines for lighting purposes within such district as a public work, in like manner, and with the like powers, authorities, and liabilities as may by law be exercised in respect of, or as are attached to, the construction of such public works, but subject to the provisions of this Part of this Act.[23]

These clauses began establishing electric power as a public work in New Zealand alongside the telegraph, public water supply, and railway transport. The next and most significant amendments to this law were the Municipal Corporations Act of 1886 and its 1887 revisions. This act granted local authorities the right to use water resources to generate electricity and sell it for lighting and motive power. Under the Public Works Department, these legislative actions guided New Zealand's national transition to electric power.

As with many other industrial nations, during the 1890s, demand for electric power grew in New Zealand. Lighting accounted for much of that demand, but the utility of electric power also applied to public transportation, department store elevators, and factory machinery. The success of numerous hydroelectric schemes in New Zealand, such as the town of Reefton (1887), Phoenix Mine (1887), Wellington City Corporation (1888), Stratford (1898), and Parihaka Pa (1899), frequently featured in New Zealand papers.[24]

> The little West Coast town of Reefton has the electric light for private use. And yet Auckland and the other larger towns are contented to jog along contentedly behind Reefton.[25]

[23]"Electric Lines Act," *Statutes of New Zealand* 32, 1884, 147–148.

[24]"Electric Light," *Star*, December 1, 1888, 3, https://paperspast.natlib.govt.nz/newspapers/TS18881201.2.39.3; "The Electric Light Proposals," *Northern Advocate*, May 21, 1898, 3, https://paperspast.natlib.govt.nz/newspapers/NA18980521.2.22.20; Martin, *People, Politics, and Power Stations*, 31–32.

[25]The Fretful Porcupine," *Observer*, May 13, 1893, https://paperspast.natlib.govt.nz/newspapers/TO18930513.2.7

> New discoveries of payable stone continue to be made in the Phoenix mine [sic], enhancing the value of the mine considerably. The electric apparatus has been working with the greatest ease and regularity.[26]

These schemes relied on municipal or provincial funding. Hydropower proved to be the most successful and widespread of the available options. The success of hydroelectric projects had to do with geographic availability, but water manipulation also lived up to hydroimperialistic ideas about making the country's natural resources profitable. Still, some cities located far from water sources adopted steam or suction gas systems.[27]

While the demand for electric power grew, the national government increasingly found electricity regulation an effective means of maintaining and expanding its influence. As many have noted, governments distance themselves from infrastructures to control and organize populations through technological domains that seem separate.[28] For instance, during the 1890s the New Zealand Parliament granted municipalities increased fiscal freedom to borrow money or restructure school boards by passing acts like the Christchurch City Borrowing Act (1899) or Wellington Education Board Transfer Act (1897). Simultaneously, the national government moved to fund, build, and regulate electric power and communication networks. By managing infrastructures like electric power systems in New Zealand, Parliament played an active role in municipal boundaries and regional industries.[29] In 1896, with the passage of the Electrical-Motive-Power Act, the pattern of electrification shifted toward national over regional or municipal power boards.

> Notwithstanding anything to the contrary contained in any other Act, it is hereby declared that, from and after the coming into operation of this Act, it shall not be lawful for any local authority to grant to any person any right or concession for the purpose of either generating or using electricity as a motive-power without in each instance the previous consent of the Governor...

The act stipulated that private enterprises could no longer gain control over any New Zealand waterway to produce hydroelectric power, nor could any local body grant permission to generate electric power without government consent. It allowed the Public Works Department to ultimately decide how to begin developing a large-scale hydroelectric power station.[30]

---

[26] "Mining Intelligence," *Lake Wakatip Mail*, April 30, 1886, 5.

[27] Reilly, *Connecting the Country*, 24–25.

[28] Technological infrastructures allow for an "apparatus of governmentality." Michel Foucault, *The Birth of Biopolitics: Lectures at the Collége de France, 1978–1979* (New York: Picador, 2010), 70; Brian Larkin, "The Politic and Poetics of Infrastructure," *Annual Review of Anthropology* 42 (2013): 328.

[29] Reilly, *Connecting the Country*, 25.

[30] "Electric Motive Power Act," 17th October 1896, 60, no. 47, 149.

4.4 The Canterbury Region and Lake Coleridge Power Station

The Public Works Department and other government offices worked on electrification projects around New Zealand at the turn of the twentieth century, but Christchurch maintained the most support. Before moving on to its electrification, it is necessary to mention the region's colonial foundations, namely the removal of the indigenous population. Although the Public Works Department continued to operate using colonial bureaucracy and rhetoric from the 1890s well into the twentieth century, it bears stressing how these political and technological developments hinge on the colonization of lands and peoples. Prior to European colonization, the area was home to the Ngāi Tahu *iwi*, or nation. Following the Treaty of Waitangi (1840), which legislatively brought New Zealand and the Māori under the auspices of Great Britain and after the settlement of the region by the Church of England under the Canterbury Association (1850), colonial authorities and settlers increasingly assimilated, bought out, or forced out the Māori of the region.

By the 1870s, the city became a major agricultural hub and port city central to the colonial economy. During this period until 1900, the city and region saw a massive population boom due to sheep farming, wheat farming, and the meatpacking industry. The population comprised primarily of settlers from England, Scotland, and Ireland. Reliable census data on Māori in the region and in Christchurch itself does not exist.[31] People from the region mentioned the presence of Māori in urban spaces like Christchurch, Dunedin, and Greymouth but they and their culture remained suppressed.[32] Despite this, it is important to note that this region's settlement and urban successes depended on the removal and integration of Māori. Even before the formalization of the Lake Coleridge Dam or the founding of the Dominion of New Zealand, the imagined future of New Zealand was one without the Māori. As termed by Mark Thurner, communities like the Māori can usefully be described as unimagined communities or those that would "disturb the implied trajectory of unitary national ascent."[33] Too often Christchurch is remembered for being an innocent settlement in contrast with the North Island cities involved in the violence of the New Zealand Wars (1845–1872). The thriving pastoral economy, trading ports, and eventually national electric power stations depended on the removal of the Māori. Despite the physical and rhetorical attempts to unimagine them, the removal of Māori and their assimilation, embrace, and resistance to settler culture is

[31] "Census of New Zealand," April 3, 1881, *Statistics New Zealand*, Digitized Collection, https://www3.stats.govt.nz/historic_publications/1881-census/1881-results-census.html?_ga=2.250866822.1081525362.1597344575-1777297613.1597344575

[32] Atholl Anderson, Judith Binney, Aroha Harris, *Tangata Whenua: A History* (Wellington: Bridget Williams Books, 2015): 287.

[33] Mark Thurner, "Unimagined Communities: Megadames, Monumental Modernity, and Development Refugees," in ed. Rob Nixon, *Slow Violence and the Environmentalism of the Poor* (Cambridge: Harvard University Press, 2011): 150.

exceedingly complex because it is not just a story of victimization. Throughout the twentieth century, Māori communities utilized electric systems in New Zealand, worked on the Public Works teams that hoisted powerlines, and designed their own electric power infrastructures.[34]

As Christchurch and the surrounding pastoral townships thrived during the 1870s, the citizens increasingly saw themselves as an industrial community. In 1883, a passenger train carried 200 passengers from Christchurch to the electric lighting display at Kaiapoi Woollen Mills, illuminated by two arcs and sixty incandescent lamps.[35] The spectacle captivated many members of the Christchurch City Council. Over the next decade, the City Council pushed for public electric lighting and power supply, especially with the highly successful electric mining and city utilities reports.[36] In addition to the general interest in electric power, they advocated for hydropower, which had proven to be an effective solution for increasing settlement and attracting investors around New Zealand. And since the Electro-Motive Power Acts had not yet been passed, the city remained responsible for the power station. Numerous suitable water sources, such as the Waimakariri River, Rakaia River, Lake Coleridge, and Opihi River, surrounded Christchurch. However, comprehensive surveys of the hydroelectric scheme area did not begin until 1899 when the city council contracted Arthur Dudley Dobson (1841–1934), the city surveyor and engineer, to study the potential of the Waimakariri River.[37] This survey marks the beginning of the planning that resulted in constructing the Lake Coleridge Power Station.

> There is no doubt in my mind that if we had that scheme we should be able to carry out our water supply scheme, complete the drainage system, and light out streets by electricity. I don't think there would be the slightest difficulty at all about it.[38]

Even before the decision to build a dam in the region was made, excitement among local politicians and residents mounted because the project represented progress for the city of Christchurch and New Zealand itself. Dobson's first survey concluded that the river's width and shingle bed was not suitable for a dam.[39] It was "Out of the question," except for a single gorge. Dobson preferred to divert water from higher up

[34] Brian White and Isabelle Chambefort, "Geothermal Development History of the Taupo Volcanic Zone," *Geothermics* 59 (2016): 162.

[35] "Electric Lighting in the Colonies," *New Zealand Herald*, June 13, 1885, 1 (Supplement), https://paperspast.natlib.govt.nz/newspapers/NZH18850613.2.58; Rosemary Britten, *Lake Coleridge: The Power, the People, The Land* (Christchurch: Hazard Press, 2000): 100–101.

[36] "Untitled," *Cromwell Argus*, June 29, 1886, 2, https://paperspast.natlib.govt.nz/newspapers/CROMARG18860629.2.7; "Untited," *New Zealand Times*, August 15, 1889, 4, https://paperspast.natlib.govt.nz/newspapers/NZTIM18890815.2.18

[37] "The New City Surveyor," *The Press*, December 12, 1900, 2, https://paperspast.natlib.govt.nz/newspapers/CHP19001212.2.8

[38] "Untitled," *Poverty Bay Herald*, January 9, 1899, 2, https://paperspast.natlib.govt.nz/newspapers/PBH18990109.2.33H

[39] D. M. Calder, "Plant Ecology of Subalpine Shingle River-Beds in Canterbury, New Zealand," *Journal of Ecology* 49, no. 3 (1961): 581–594.

the river through channels to a lower elevation power station. The report convinced members of the Christchurch City Council to gather information and cost estimates for such a power station. The Swiss firm Escher Wyss of Zurich, who later played a role in supplying material for the dam at Lake Coleridge, offered to supply materials for £3500 with an additional expense for the engineers and supervisors.[40]

During the early twentieth century Escher Wyss, also known as Escher Wyss AG, developed turbines for numerous hydroelectric power plants including Brusi-Campocologna Hydro Electric Plant (1904) in Switzerland, Necaxa Falls Power Station (1905) in Mexico, and the Vemork Hydro Electric Plant (1911) in Norway.[41] Although distant from the electrifying and industrial centers of Europe and North America, the New Zealand electrical engineering community maintained business with companies like Escher Wyss, Brush Electric Lighting, and Siemens.

Many on the Christchurch City Council were keen to begin construction on the scheme. They hoped to do so starting in October 1900, Canterbury's Jubilee year, because it was "naturally an occasion for erecting some tangible central memorial commemorative of the solid progress accomplished by Canterbury."[42] For the city council, the construction of a large-scale public work accomplished more than supplying electricity. It meant that New Zealand achieved the status of a modern, self-sufficient state.

The local press, especially the *Star* and *Lyttleton Times*, highlighted the electrical potential of Christchurch by explaining Dobson's arguments. For instance, one widely circulated article stated, "There is, said Mr. Dobson, no city of any size south of the Line which has such natural advantages for electrical working as has Christchurch." Regional politicians echoed this argument and stressed that electric power could reshape Christchurch, and New Zealand, into a modern manufacturing center, a prospect they hoped would excite and gain the local population's support. One of the most pronounced advocates for such a dam was Thomas Edward Taylor (1862–1911), who claimed the river could make Christchurch the New Zealand manufacturing capitol. For politicians like Taylor, electric power's introduction highlighted a half-century of "almost magical development" of the city of Christchurch, and indeed New Zealand itself. Later in his article, Taylor titled an entire section, "The Colonies are All Behind," in which he claimed, "New Zealanders are sometimes rash enough to boast of their reforming zeal and general progressiveness

[40] "A Big Scheme," *Lyttleton Times*, June 11, 1900, 3, https://paperspast.natlib.govt.nz/newspapers/LT19000611.2.19

[41] "Development of the Necaxa, Mexico, Water Mile," *Electrical World and* Engineer, October 28th, 1905, 729–735; Frank C. Perkins, "A 50,000-Volt transmission System: The Largest Hydroelectric Plant in Switzerland," *Scientific American Supp.* January 9, 1909, 24–26; Einar A. Brofos, "Norwegian 250,000 Hydroelectric Development," *Electrical World*, December 9, 1909, 1411–1414.

[42] T. E. Taylor, "Harness the Waimakariri," *The Press*, October 27, 1900, 3, https://paperspast.natlib.govt.nz/newspapers/CHP19001027.2.6

in comparison with the Old Country."[43] Despite the introduction of electric power systems in New Zealand, Taylor felt the country lagged behind Great Britain, the United States, and other European countries in constructing electric power infrastructures. This is only partially true, especially among the British settler colonies. Complaints of falling technologically behind were common almost everywhere in the late-nineteenth century.[44] He admitted that Christchurch had "no Niagara" but had, according to Dobson, enough power to replace existing coal-powered plants. Advocates like Taylor and Dobson believed that hydroelectric power could be produced more cheaply than natural gas or other sources.[45] In 1900, wrapped up in the excitement of the prospect of the introduction of electric power, Taylor and a group of businessmen founded the New Zealand Electrical Construction Company. The company was intended to supply the city with equipment for an electric tramway system but ultimately failed to meet financial and legal requirements.[46]

Though it is easy to lose the string of colonialism and electrification in these exchanges, it is essential to connect these decisions back to the Crown's original arguments about the purpose of electrification in New Zealand– to create a modern model colony. The ensuing debates over the location of dams featured familiar colonial economic concerns such as profitability and self-sufficiency. Locations needed to demonstrate the potential to generate power for industries that would profit New Zealand and Empire, while also easing potential demands on Great Britain. As had happened frequently with the provincial, regional governments, the Christchurch City Council could not effectively administer plans, or pay, to build a large hydroelectric facility. In 1901, they contracted Robert Hay (d. 1928), a civil engineer from Dunedin, to investigate the river further.[47] Generally, he agreed with the earlier survey completed by Dobson. Despite this site's promise, the proposed scheme caused a significant violation of water rights in Selwyn County, located northwest of the city's precinct. Residents feared the construction might cause blockages and washouts with the potential to devastate the region's agricultural productivity. Ultimately, Hay recommended against the construction of the scheme.

---

[43] "A Big Scheme," *Star*, June 11, 1900, 4, https://paperspast.natlib.govt.nz/newspapers/TS19000611.2.62

[44] Renfrew Christie, *Electricity, Industry, and Class in South Africa* (Albany: State University of New York Press, 1984), 5; Hughes, *Networks of Power*, 262–264; Rennie, *Power to the People*, 233–234; David Nye, *American Illuminations: Urban Lighting 1800–1920* (Cambridge: The MIT Press, 2018), 78.

[45] "Electricity as a Motive Power," *Timaru Herald*, October 23, 1900, 4, https://paperspast.natlib.govt.nz/newspapers/THD19001023.2.33

[46] "Town and Country," *Lyttleton Times*, January 18, 1904, 4, https://paperspast.natlib.govt.nz/newspapers/LT19040118.2.22

[47] "The Waimakariri Scheme," *Lyttleton Times*, May 14, 1901, 2 https://paperspast.natlib.govt.nz/newspapers/LT19010514.2.5; "Loss to Dunedin," *New Zealand Herald*, November 29, 1928, 8, https://paperspast.natlib.govt.nz/newspapers/NZH19281129.2.14.2

The Christchurch City Council and the Selwyn County Council never reached an agreement, which foreshadowed increasing tensions between national projects and local interests. Large scale state electrification projects in New Zealand and elsewhere during the early twentieth century frequently favored state over local interests.[48] The frustrations generated by these localized land and water disputes increasingly catalyzed the movement to centralize the implementation of electric power projects, giving responsibility solely to the Public Works Department. Although many remained convinced of the Waimakariri River's utility, some were willing to seek alternative options. In 1901, Taylor put together another survey of the region, including Dobson, city council members, and journalists, of Lake Coleridge and the Rakaia Gorge.[49] Like Taylor, some electric power stakeholders thought that Lake Coleridge would still prove just as challenging to build but that the outcome would pay for the hardship and continued to advocate for the Waimakariri River scheme.

Taylor's excitement kept the scheme alive despite the hesitation of the Christchurch City Council. Electrification project momentum in New Zealand benefitted from the promise of national and imperial prestige, not solely from local interests. After touring European power stations, Taylor encouraged the city to hire a Swiss engineer, Colonel Théodore Turretini (1845–1916), but the city council ignored his proposal. Turretini had achieved international repute by working with Thomas Edison and William Siemens and circulating in electrical workshops in Germany, France, and the United States.[50] The debate continued into 1902, especially in the aftermath of the City of Christchurch Electric Power and Loan Empowering Bill's proposition. William Whitehouse Collins (1853–1923) introduced the bill into Parliament to grant the city the ability to control and seek out financial assistance with electric light and electric power. The bill sparked debates over financial responsibility for the Waimakariri scheme. Taxpayers in Christchurch worried that they would pay for everything while the entire region benefitted.

Additionally, local townships and utility councils, such as Selwyn County and the Waimakariri-Ashley Water Supply Board, feared a loss of irrigation supply. During the late nineteenth and early twentieth century, the pastoral regions throughout Canterbury depended heavily on the agriculture and wool industry. Any

---

[48] Ronald R. Kline, *Consumers in the Country: Technology and Social Change in Rural America* (Baltimore: Johns Hopkins University Press, 2000); Robert D. Lifset, *Power on the Hudson: Storm King Mountain and the Emergence of Modern American Environmentalism* (Pittsburgh: University of Pittsburgh Press, 2014); Ada Ordor and Yinka Omorogbe, eds. *Ending Africa's Energy Deficit and the Law: Achieving Sustainable Energy for All in Africa* (Oxford: Oxford University Press, 2018).

[49] "Waimakariri Power Scheme," *Lyttleton Times*, August 22, 1901, 7, https://paperspast.natlib.govt.nz/newspapers/LT19010822.2.77

[50] Serge Paquier, "Turrettini, Théodore," in *Dictionnaire Historique de la Suisse*, accessed April 1, 2023, https://hls-dhs-dss.ch/fr/articles/003896/2014-02-25/

misdirection or decrease of their water resources threatened that prosperity.[51] Eventually, they pressured the City Council to add a clause to the bill assuring that water would not be taken from the river within a certain distance of the gorge.[52] On November 17, 1902 the bill successfully passed through Parliament authorizing the city to raise a loan of no more than £250,000 to contract, construct, and maintain a power station on the Waimakariri River.[53] However, many still protested the scheme, mainly when engineers reported the Lake Coleridge proposal was less dangerous.[54] The city council continued to hesitate on finalizing plans for the dam, when the passage of the Water Power Act (1903) complicated their decision further.

Parliament's passage of legislation to control the country's vast water resources represents one of the most vital links between colonialism and hydro-electrification in New Zealand. By directing water usage for electric power, politicians assured the national government's power over settler and Māori energy interests. At the same time, the Water Power Act proved an inconvenience for many settlers and Māori farmers dependent on rivers for irrigation. The removal of water rights from the Māori proved especially harsh because of water's central role to Māori cosmogony and community life. Yet another British tool for creating an ideal New Zealand did not include Māori culture.[55] The Water Power Act stated:

> The sole right to use water in lakes, falls, and rivers or streams for the purpose of generating or storing electricity or other power shall vest in His Majesty.[56]

By making water the primary source of New Zealand's electric power production, a resource exclusively controlled by the national government, this act effectively placed electric power under their control too. For the first time, almost all electric power facilities were influenced by the national government. Seeking to maximize the country's hydroelectric potential, the New Zealand Government contracted L. M. Hancock, an American electrical engineer, and P.S. Hay, a civil engineering superintendent from the Public Works Department, to report on potential sites. As illustrated in *The Electrician,* they surveyed the entire country. They concluded

[51] David Greasley and Les Oxley, "The Pastoral Boom, the rural land market, and long swings in New Zealand Economic Growth, 1873–1939, *Economic History Review* 62, no. 2 (2009): 324–349.

[52] "City of Christchurch Electric Power and Loan Empowering Bill," in *Parliamentary Debates: House of Representatives* (Wellington: John Mackay, 1902), 250.

[53] "Christchurch City Council," *Lyttleton Times*, November 22, 1902, 9, https://paperspast.natlib.govt.nz/newspapers/LT19021122.2.86

[54] "Harnessing the Waimakariri," *Lyttleton Times*, August 29, 1902, 5, https://paperspast.natlib.govt.nz/newspapers/LT19020829.2.66; "Interview with Mr. T. E. Taylor," *The Press*, September 18,1902, 5, https://paperspast.natlib.govt.nz/newspapers/CHP19020918.2.24; "River Eyre Protective Works," *The Press*, November 13, 1902, 5, https://paperspast.natlib.govt.nz/newspapers/CHP19021113.2.19

[55] D. Williams, "Ko Aotearoa Tenei: Law and Policy Affecting Māori Culture and Identity," *International Journal of Cultural Property, 20* (2013): 311–331.

[56] *Water-Power Act 1903* (New Zealand) 3 EDW VII 1903 No. 26, 54–55, http://www.nzlii.org/nz/legis/hist_act/wa19033ev1903n26248/

the potential yield of electric power and the revenue from selling that power justified Parliament's eventual investment in these projects around the island.[57]

In Canterbury, Hancock reported, much to the dismay of citizens of Christchurch, that the Waimakariri River "was best left to its own sweet will."[58] Like Taylor, council members and advocates for the scheme complained that as an expensive foreign consultant, Hancock could not give the region fair treatment. However, Hannock did suggest that Lake Coleridge showed promise, even though the hyperbolic Taylor said Hancock's report "was not worth the paper it was written on."[59] In 1904, the "Hancock Report" made it clear that Lake Coleridge, not the Waimakariri River, ought to be the national government's priority and begrudgingly, the city council relented. Hancock calculated that Lake Coleridge would yield 94,677 horsepower, whereas Waimakariri only promised 20,440. Furthermore, he argued that the area around Lake Coleridge had room for expansion and more favorable terrain.[60]

Despite its intense enthusiasm for electricity, the Public Works Department had other priorities that made it necessary for boosters to fight for their projects. Public Works Department funding in the first years of the twentieth century focused primarily on transportation, connecting the islands by increasing people and materials' flow. Electric power did not receive as much attention because of the previous commitment to securing the islands by road and rail. Many public works officials believed that trade and transportation would produce unity among the island's citizens and municipalities, particularly the North Island Main Trunk (1908).[61] The proposed water diversions and scale of a dam at Lake Coleridge, which was projected to cost some £1,200,000, discouraged the Public Works Department from seriously considering the plan, still preferring transportation investment. Ultimately electric power stakeholders succeeded by promoting the manufacturing potential of electric power.

Christchurch citizens and politicians believed that increased electric supply would help the town, and New Zealand more broadly, to become a model state.

> If the Government installed its power schemes, the profits would go into the consolidated revenue, and would be used to the advantage of the whole Dominion... The proper use of water-power would prove to be a source of enormous profit in the near future, and that it would probably rival the railways in regard to the production of wealth.[62]

---

[57] "Australasia - Municipal, Foreign, and General Notes," *The Electrician*, January 13, 1905, 520.

[58] "The Water Power Report," *New Zealand Herald*, November 1, 1904, 4; https://paperspast.natlib.govt.nz/newspapers/NZH19041101.2.22

[59] "Christchurch Monopolies," *Lyttleton Times*, May 19, 1905, 3; https://paperspast.natlib.govt.nz/newspapers/LT19050519.2.13

[60] "Electrical Energy: New Zealand's Wonderful Possibilities," *The Press*, October 31, 1904, 10; https://paperspast.natlib.govt.nz/newspapers/CHP19041031.2.68

[61] Bill Pierre, *The North Island Main* Trunk (Wellington: Reed, 1981), 289–290.

[62] "Power for Christchurch," *Star*, January 13, 1910, 1; https://paperspast.natlib.govt.nz/newspapers/TS19100113.2.5

With expansion to serve Christchurch's manufacturing sector and rising electric power usage for lighting, the City Council argued that either the town or the national government should fund a major hydroelectric facility to keep New Zealand's (and Christchurch's) economy expanding. In 1909, the Christchurch City Council decided that the town's existing steam plants could not meet the expanding demand for electric power. The New Zealand Government had not broadened its regulation of electric power since 1903, when the Water Power Act had been signed. The council reconsidered the Waimakariri River scheme and quickly dismissed it because of the same water rights concerns raised earlier.[63] Considering all possible alternatives, they even unsuccessfully petitioned to build a city-run power station on Lake Coleridge.

After the petition failed the mayor, Charles Allison (1846–1920), began urging the government, specifically the Minister of Public Works, Roderick McKenzie (1852–1934), to fund a power station at Lake Coleridge. He claimed that "without electricity, Christchurch will lose its status as a manufacturing city."[64] With Christchurch as a center for distribution, the Canterbury region had long been the South Island's major agricultural producer of seed, wheat, wool, and meat. Since the introduction of refrigeration technology, companies such as the Canterbury Frozen Meat and Dairy Produce Export Company served as the foundation of the Christchurch economy. Additionally, the town hosted numerous port-related businesses, metal processing factories, and engineering firms. At the turn of the century, Christchurch politicians competed with other cities, such as Wellington and Auckland to expand its importance and attract a large labor pool. Increasingly, electricity was needed to keep and attract businesses to Christchurch. In response to Allison, McKenzie felt that the government should supply electric power to the city but believed the expense prohibitive.

Profit and the potential of manufacture continued to drive Christchurch's and Parliament's interest in constructing a state-funded and regulated power station. Finally, in 1910, Lawrence Birks (1874–1924), an Australian electrical engineer and lead engineer of the Tourist Department in Rotorua submitted a report on the Lake Coleridge's hydroelectric potential Hutt River schemes to the New Zealand government.[65] Another report written by R. W. Holmes, chief engineer of the Public Works Department, supported Birks and came to similar conclusions, much as Hancock's report had done. After years of resisting, Taylor agreed to drop his promotion of the Waimakariri River scheme in favor of Lake Coleridge.[66] Even though they initially feared the high cost, the Government passed the Aid to Water

[63] "The Waimakariri Scheme" *Lyttleton Times*, February 5, 1909, 5; https://paperspast.natlib.govt.nz/newspapers/LT19090205.2.20

[64] "State Water Power," *Lyttleton Times*, August 24, 1910, 7; https://paperspast.natlib.govt.nz/newspapers/LT19100824.2.42

[65] John E. Martin, "Birks, Lawrence," *Te Ara: Encyclopedia of New Zealand,* accessed June 11, 2020, https://teara.govt.nz/en/biographies/3b34/birks-lawrence

[66] "The Lake Coleridge Hydroelectric Scheme," *The Press*, December 3, 1910, 10, https://paperspast.natlib.govt.nz/newspapers/CHP19101203.2.60

Power Works Act (1910) so that money could be raised to credit the Public Works Fund for hydroelectric power and irrigation.[67] Mackenzie did not offer an explicit explanation for the change in Public Works policy. However, local papers argued that the success of other major hydro projects in America, Canada and Switzerland had supported impressive industrial results and suggested the same would be true for New Zealand.[68] Thus Christchurch became the primary focus in New Zealand for implementing state-funded electrification.

Despite the implementation of policies that removed legislative barriers to large-scale construction, building the power station continued to challenge the commitment to electrification but it allowed for the increased government acquisition of land. The Public Works Act (1908), an amendment to the 1864 bill, made it possible for the New Zealand Government to purchase the Lake Coleridge Power Station land. From the 1860s onwards, this legislation has proven problematic because it has granted the government ill-defined powers in land acquisition for roads and electric infrastructures that have disproportionately resulted in Māori land being taken as they had to be compensated at a lower rate.[69] Or in some cases the Crown, and later Dominion government, just laid claim to lands entitled to Māori. For example, on November 25th, 1914, the same day that the Prime Minister of New Zealand, William F. Massey (1856–1925) delivered a speech commemorating the opening of the Lake Coleridge Power Station, the Waiau Native Reserve submitted a petition requesting that the government return 1000 acres on the west bank of the Waiau River about 70 miles north of Christchurch. Due to clerical oversight, neglect, and possibly contempt, the government had taken the land in the 1860s, disregarding treaty protocols.[70]

On November 4, 1910, the official Public Works survey began, and construction projects followed with major building starting the following year. Before this venture, the area around Lake Coleridge had been open farmland. After 1910, the area played home to more settlers, machinery, and construction than in the initial European settlement of the region. Before the building could begin, the ground had to be tested for strength and waterproofing, a holding tank or surge chamber was needed to reroute water. Builders constructed two tram lines to transport water up the Rakaia River and remove soil. In continuing with the Public Works Department's transportation mission, Frederick W. Furkert (1876–1949) surveyed and built roads between the station and Coalgate, the nearest railway head, so that tractors, "Furkert's Motor Cars," could easily move material to the station. Furkert joined the Public Works Department in 1894 and served as the engineer-in-chief of the

---

[67] *Aid to Water Power Works Act 1910* (New Zealand) 1 GEO V No. 25, 70, http://www.nzlii.org/nz/legis/hist_act/atwwa19101gv1910n25352/

[68] "Water Power," *Bush Advocate* November 9, 1910, 4, https://paperspast.natlib.govt.nz/newspapers/BA19101109.2.8

[69] David V. Williams, *'Te Kooti tango whenua': The Native Land Court 1864–1909* (Wellington: Huia, 1999), 170.

[70] "Election Campaign: Premier in Christchurch," *Grey River Argues,* November 26, 1914, 7, https://paperspast.natlib.govt.nz/newspapers/GRA19141126.2.39

PWD for twelve years from 1920 until he retired in 1933. As a graduate of the Wellington Technical School and an associate member of the British Institute of Civil Engineers, Furkert reminds us of how New Zealand's young electrical infrastructure remained connected to international engineering networks. The traction engines and horse-drawn wagons moved material between 1910–1914. They transported 1750 ironbark piles for lengthy 66 kV transmission line, 500 wooden poles for local distribution in Christchurch, 130 miles of copper 7/14 Standard Wire Gauge (SWG), 60 miles of copper 7/12 SWG, 420 miles of 7/.135 aluminum wire, 14,080 yard of insulated cable for the 11 kV underground lines, 12 General Electric transformers for the powerhouse and substation, and three Escher Wyss Francis-type 1500 kW generators.[71] The workers using a combination of rail, tractor, and animal power transported some 12,00 tons of material in total (Fig. 4.2).[72]

Fig. 4.2 Traction engines at Coalgate railway station. (The Press (Newspaper). Negatives. Ref: 1/1–008122-G. Alexander Turnbull Library, Wellington, New Zealand)

[71]W. Wilson, "Hydro-Electric Power in New Zealand," *Journal of the Institute of Electrical Engineers*, January 7, 1916, 283; Reilly, *Connecting the Country*, 38.

[72]Martin, *People, Politics, and Power Stations*, 48.

Throughout the dam's construction, the project required more funding and political maneuvering than expected. Some New Zealanders, like the electrical engineer, Frederick Black, published a searing critique of the people that blindly followed the Public Works Department:

> Nobody knows what the project will ultimately cost, nobody knows what will be done with them, and nobody- or almost nobody cares. The main thing is that nearly two and a quarter million sovereigns are down on the list to be spent. The Government say and the public believe, without knowing why they believe, that the pouring of this golden stream into New Zealand rivers will create industrial conditions in the land at present undreamt of, that a new era will dawn wherein electricity – blessed, mysterious force – will permeate all our lives, and that, by virtue of the great water gift of the gods we shall soon tower above all the nations in the Pacific.[73]

Black's critique represents the minority opinion of the government officials and citizenry of Christchurch. The cost of electric power remained a secondary concern and suggests installing the electrical infrastructures: control of water sources, powerlines, and power stations. Even though Black himself would eventually be a consulting engineer on the project, his initial objections help capture the complexity involved in completing this project. In 1910, the plans for the scheme included designs with different theoretical horsepower that included generators, power houses, tunnel, penstock piping, water redirection, and storage tanks. The plans promised as little as 7477 hp. (£74,925) or as much as 34,700 hp. (£246,225).

Although earlier portrayals of the environment as perfect for a large power station excited the city council and Public Works Department, the miserable working conditions, unknown geological barriers, and discontented workers proved troubling.[74] Despite these realities, the political ambition and resolve of the Public Works Department and the City Council of Christchurch spurred the project. The Public Works Department only maintained that a central governing body could handle such a work, so the department became increasingly involved in the dam's construction. The supervising engineer, F. T. M. Kissel, began constructing a "town" for the laborers and future workers for the power station, even moving his own family there.[75] The town first consisted of eight cottages built to manage the tough conditions, namely fierce wind and rain, in tents. Kissel replaced G. S. Bogel who had been one of the first engineers on the 1910 survey.[76] Throughout the construction process, Kissel and his crew were frequently stalled by miscommunication and delay in material delivery on the Public Works Office's part in Christchurch coming

---

[73] Frederick Black, "Water Power Scheme: Government Proposals Criticized," *Star*, October 18, 1910, 4, https://paperspast.natlib.govt.nz/newspapers/TS19101018.2.62

[74] "Strike at Lake Coleridge," *West Coast Times*, September 24, 1912, 3, https://paperspast.natlib.govt.nz/newspapers/WCT19120924.2.20

[75] "Lake Coleridge," *Star*, November 4, 1912, 4, https://paperspast.natlib.govt.nz/newspapers/TS19121104.2.58

[76] "Electric Power," *Lyttleton Times*, November 15, 1910, 9, https://paperspast.natlib.govt.nz/newspapers/LT19101115.2.89; "Hydro-electricity," *Lyttleton Times*, January 17, 1911, 7, https://paperspast.natlib.govt.nz/newspapers/LT19110117.2.66

from Great Britain, Switzerland, and the United States. Word spread that work was available, and laborers flocked to Lake Coleridge. Besides the prospect of a new manufacturing center in Christchurch, the dam's scale brought a sizable population to the region, some 4000 between 1906–1911, to the area and boosted the economy. They made progress on the tunnels, buildings, and pipelines. Work frequently stalled due to harsh weather and miserable work conditions. One of the workers, James Ryan, claimed that after working on public works projects around New Zealand, "these were the worst conditions he had ever known." Besides the working conditions, Ryan and other workers interviewed for this article express their discontent with the living conditions. Part of this discontent stemmed from the unequal treatment of laborers and other Public Works Department members who got to live more comfortably. According to them, the rate for food and board was outrageously high.[77] In addition to these personal grievances, other workers took issue with the safety of the equipment and construction speed. As a result, the Canterbury General Laborers' Union formed and renegotiated wages and contracts between contracted companies and the Public Works Department. The environment, ever ignored by the architects of these sociotechnical imaginaries, challenged the commitment of the Public Works Department and Christchurch officials. Too often, historians take the widespread geographic availability of water for granted and suggest that hydroelectric power must have been the most logical, apparent, or straightforward course of action for stakeholders interested in electric power systems.

As had been the case with many colonial projects before, such as gold mining, herding, and lumber extraction, New Zealand's terrain proved troublesome, so electric power advocates continued to assert that centralized authority and modern technology would fix the problems. After all it had worked before, when the Colonial Office had directed finances toward expanding mines to the West Coast of the South Island or to exploit the tourist industry potential of the hot springs on the North Island. In 1913, under Lawrence Birks' direction, the Public Works Department took over the Lake Coleridge Scheme to make regulation easier by removing an administrative step. He appointed Evan Parry (1865–1938) the lead electrical engineer. Parry was one of Lord Kelvin's former students at Glasgow, another reminder of how this local electrical project remained connected to global electrical science and engineering networks of the period. After the takeover, many supporters believed construction could continue unimpeded, saying things like all work "is likely to go better now."[78] Meaning that now that the national government is at the helm, smaller oversights will disappear. They did not. The geological difficulties alone tested the resolve of everyone involved. The power station was the first to be constructed on a glacial moraine. This loose gravel made construction difficult

[77] "Miners Complain," *New Zealand Times*, July 12, 1912, 8, https://paperspast.natlib.govt.nz/newspapers/NZTIM19120702.2.87

[78] "Lake Coleridge Notes," *Press*, February 21, 1913, 9, https://paperspast.natlib.govt.nz/newspapers/CHP19130221.2.83

because of the deep foundation that had to be concrete, a temporary solution that plagued the power station throughout the twentieth century.[79]

In early 1914 as construction neared completion, negotiations concerning the cost of electricity demonstrated that the Public Works office and city council forgave prior difficulties because of the economic potential of hydropower.[80] All that remained was to finish the tunnel and pipelines, construct the powerhouse, and connect the switchboard. This is not to suggest that the construction progressed without issue. There are numerous stories about the difficult terrain, labor issues, and missed timelines. What is important here was the consistent adherence to the economic potential no matter the local detriment. By March 1914, posts and wires had already been erected through most of the 85-mile distance to Christchurch, where citizens eagerly awaited increased access to lighting and manufacturing jobs. By July, the six major sections of the dam – the tunnel, the pipeline, the powerhouse, the transmission lines, the Addington substation, and the Christchurch City Council's plant – had been completed.[81] The power station generated 66,000 volts transformed at Addington, a suburb of Christchurch, to 11,400 volts for distribution. The Public Works Department and the City Council negotiated and agreed that the city would supply 4500 kW. Escher Wyss, the Swiss firm mentioned earlier, supplied the three 1500 kW generators. The council agreed to pay £8 per kilowatt for the first 300 kW and £5 for every kilowatt above 300 kW in quarterly installments.[82] Years later, Birks used this cost comparison as an example of the virtue of hydroelectricity when he classified it as "poor man's light."[83]

> Under the present rates electricity in the house has been found decidedly cheap, and when the Lake Coleridge scheme comes into force the expense will be much less.[84]

Indeed citizens in Christchurch excitedly welcomed electric power into the city. One argued that "we have at last banished darkness from our thoroughfares."[85] The city set up new street lighting networks, transportation, and domestic electric power for

[79] "Glaciation at Lake Coleridge," *Star*, September 4, 1913, 1, https://paperspast.natlib.govt.nz/newspapers/LT19130904.2.108; R. Speight, "Ice Wasting and Glacier Retreat," *Journal of Geomorphology* 3 (1940): 131–141.

[80] "Lake Coleridge Explored," *Sun*, April 4, 1914, 14, https://paperspast.natlib.govt.nz/newspapers/SUNCH19140404.2.104

[81] "The City's Electricity, Progress of the Great Scheme from the Lake to the Citizen's Home," *Lyttleton Times*, July 11, 1914, 15, https://paperspast.natlib.govt.nz/newspapers/LT19140711.2.14 8

[82] "Lake Coleridge Current," *Press*, October 21, 1914, 2, https://paperspast.natlib.govt.nz/newspapers/CHP19141021.2.5

[83] "The White Coal: A Visit to Lake Coleridge," *Manawatu Times*, November 29, 1917, 6, https://paperspast.natlib.govt.nz/newspapers/MT19171129.2.31

[84] "Cheap Current: The Lake Coleridge Power," *Star*, July 10, 1913, 4, https://paperspast.natlib.govt.nz/newspapers/TS19130710.2.77

[85] "Untitled," *Sun*, May 24, 2015, 6, https://paperspast.natlib.govt.nz/newspapers/SUNCH19150524.2.38

lighting, heating, and small appliances. Regional industries, especially the lucrative frozen meat and dairy industry, quickly bought into the power supply.[86]

Consumers and the politicians that advocated for the dam's affordability drove interest in the completion of the Lake Coleridge dam. Furthermore, the citizens of Christchurch had already begun to see electric power applications in the city, such an electric-powered trash incinerator (1903) and an electric tram line. Coal-fired steam generators powered these systems by Westinghouse engines.[87] They believed that the cheaper energy promised by the council and engineers would turn Christchurch into a modern manufacturing city. In almost every positive article about the Lake Coleridge Power Station, authors emphasized the cost benefits of electric power, namely that it would be cheaper than natural gas and get less expensive as more subscribers took advantage. Even though the city council eventually agreed to these rates, they feared that local companies might circumvent the city and purchase power directly from the Government. To address these concerns, the City Council and the Minister of Public Works, William Fraser (1840–1923), arranged a meeting. On March fifth, 1914, at the meeting, Fraser stated:

> The Government had sunk a lot of money in the Lake Coleridge Scheme and was anxious to see it a commercial success, but it would not make both a wholesale and retail profit.[88]

However, he also acknowledged that while he did not wish to compete with the City Council, local bodies could not control the Government if in the future they elected to negotiate with other parties. World War I and the resulting material shortages marred the early years of the dam's operation and reminded New Zealander's of their military connection to a much larger Empire. One paper even claimed, "Wars and rumors of wars have interfered with the progress of the Lake Coleridge Scheme."[89] Still, the station's opening excited the settler community about the potential economic growth and visions of self-sufficiency attached to the introduction of electric power.

> The City of Christchurch has never been repaid the preliminary expenses which were incurred; and seeing that that body really inaugurated the idea of hydroelectric power in New Zealand, and as the scheme has proved a blessing and a gold-producer, I think the National Government should, at any rate, repay the City Council of Christchurch.[90]

Despite the remaining tensions over the initial investments in surveying and material between the City of Christchurch and the Public Works Department, Lake Coleridge proved the Public Works Department and national government's effectiveness in the

---

[86] "Frozen Meat Industry," *The Press,* April 13, 1915, 6, https://paperspast.natlib.govt.nz/newspapers/CHP19150413.2.48

[87] John C. Martin, *People, Politics, and Power Stations*, 28–29.

[88] "Lake Coleridge: The Supply of Power is the Government a Competitor," *Lyttleton Times*, March 5, 1914, 3, https://paperspast.natlib.govt.nz/newspapers/LT19140305.2.14

[89] "Light and Power: Lake Coleridge Scheme," *Sun*, September 11, 1914, 8, https://paperspast.natlib.govt.nz/newspapers/SUNCH19140911.2.55A

[90] Henry Thomas Joynt Thacker (1820–1939) member of Parliament and Mayor of Christchurch, "State Supply of Electrical Energy Bill," *Parliamentary Debates* 181, October 24 (1917): 408.

construction and operation of a hydroelectric scheme. Such a success was necessary given increasing frustration across the island over increasing national-government incursion on regional water rights, especially at the Waihi Gold Mine, which was having difficulty building its own hydroelectric dam.[91] Soon after Christchurch, Lake Coleridge began supplying power to nearby Lyttleton Harbor and Tai Tapu. Excitement swept the region as more people waited, sometimes impatiently, to be connected to the new electric power supply. After 1915, the city and surrounding boroughs bought into the power scheme, and the Public Works Annual Report stated:

> The plant has worked smoothly and efficiently, without serious interruption, and the country will learn with satisfaction that this important commercial venture on the part of the Government shows every promise of becoming a financial as well as an engineering success.[92]

4.5 Conclusion

> The important fact is that with the practical completion of the Lake Coleridge installation New Zealand is fairly entered upon an undertaking that is going to benefit Canterbury enormously in the immediate future, and that may revolutionise the industries of the Dominion with the next quarter century.[93]

After completing the dam, stakeholders heralded the power station as a turning point for the Dominion. In a speech during the opening ceremony, Prime Minister Rhodes (1861–1956), referenced new hydroelectric projects, the Waihi and Whangarei schemes, proclaimed, "that in time to come the whole of the power required in the Dominion would be generated by the current generated in the streams and rivers of the country."[94] Furthermore, they celebrated New Zealand's "endless supply," which had, thanks to the political maneuvering a decade earlier, come under the control of the State.

> The late Mr. Seddon, with his far-seeing statesmanship made the abundant water power of New Zealand a State monopoly, and when the turbines begin to spin at Lake Coleridge the people of this country will appreciate the significance of that step to a land that possesses practically unlimited supplies of the "white coal."[95]

---

[91] Reilly, *Connecting the Country*, 35.

[92] William Fraser, "Public Works Statement," *Appendix to the Journals of the House of Representatives*, 1915, Session I, D-01, xi-xii.

[93] "Lake Coleridge," *Lyttleton Times*, November 26, 1914, 6, https://paperspast.natlib.govt.nz/newspapers/LT19141126.2.31

[94] "City's Electricity," *Star*, November 26, 1914, 8, https://paperspast.natlib.govt.nz/newspapers/TS19141126.2.86

[95] Thomas Hughes, *Networks of Power: Electrification in Western Society, 1880–1930* (Baltimore: John Hopkins University Press, 1983): 292.

The orchestration of the hydroelectric works at Lake Coleridge reinforced British settler colonialism and underlines the practices embedded in New Zealand's national electric power system. Much like other late-nineteenth-century technological infrastructures such as the telegraphs and railroads, electric power systems justified the increasing number of calls for centralized governance. Perhaps most among the British settler colonies, New Zealand strived to achieve self-sufficiency at the end of the nineteenth century. The same policies guided the Dominion government. A large-scale nationally-funded electric power supply promised to increase the productive capabilities of Christchurch and subsidiary communities. Lake Coleridge was not an outcome of New Zealand's newly gained independence, rather it is the product of more than a half-century of colonial policies aimed at creating a self-sustaining and productive British population in the Pacific. Hydropower didn't rise as the obvious power source. Electrification stakeholders consistently aligned with those familiar colonial policies to ensure that project's success and consequently provided the model under which the Public Works Department continued to develop New Zealand's electric power infrastructure.

As with so many other stories of electrification, some narratives can lead to the conclusion that specific methods of power production were inevitable. As the story of Lake Coleridge suggests, these systems developed because they benefitted political systems that relied on exploitation for expansion and production. The establishment of New Zealand's first national electric power system, which began with the Lake Coleridge Power Station in Canterbury, epitomizes hydroimperialism in New Zealand because its construction hinged upon the repurposing of indigenous land, the establishment of national electric-power regulations, and support from a state agency, the Public Works Department. The centralization of electric power production behind a central authority was one of the many infrastructure developments that strengthened the cause for New Zealand's new national government. Rather than being seen as New Zealand's first step toward a modern electrical power system, Lake Coleridge was an adaptation of British colonialism, a sociotechnical imaginary, that provided the physical infrastructure to justify a national government. Even at the end of the twentieth century, when the state-run Electricity Corporation of New Zealand (ECNZ) disbanded and reconsolidated into private power boards, the physical infrastructure remained and enforced centralization, large-scale production, and exploitation of land. New Zealand's electrical infrastructure, like so many around the world, remains grounded in that colonial vision.

Another way to consider the longer impact of these colonial policies in New Zealand's electric infrastructure and challenge presumptions about the inevitability of specific electric power systems is to consider how some Māori communities have created electric power infrastructures outside those older policies, or directly confront those systems. Despite the cessation of large-scale government acquisition of land in the early twentieth century and the success of land trusts, or organizations that develop reclaimed Māori land, Public Works authorities sought Māori lands because the land could only be sold through the government, which artificially lowered the valuation.[96] Even as Māori regained control of land through the trusts, it remained difficult for individuals to hold property. The expansion of the electric power grid during the mid-century was one of the many public works initiatives that excluded smaller Māori properties. Beginning with Lake Coleridge, the Public Works Department argued for the centralization of electric power production, enlarging the country's power production capabilities, and use of electric power to fuel New Zealand's productivity and expand the government's control over the country's energy infrastructure. Yet with few exceptions, Māori communities did not receive electric power until the 1950s.[97]

During the 1980s, several Māori Trusts, such as the Tuaropaki Land Trust, began to reimagine and decolonize electric power production for their community. Over the next three decades, they reclaimed their lands in the Mokai Geothermal Fields. They began to design geothermal electric power systems, the Tuaropaki Power Company (1983), to provide electricity to their community. They used the geothermal wells and the profits from the electric power to create Māori-run industries to help fund education, technology, and legal initiatives in their community. For example, the Tuaropaki Trust constructed the Miraka (2010), a dairy processing facility powered by steam and electric power from the Tuaropaki plants. With the partnership of other trusts, they also founded Gourment Mokai (2002), a multi-hectare geothermal glasshouse. Their third facility, the Ngaire George Sustainability Center (2012), practices vermicomposting, or worm composting. The center is supplied by geothermal wastewater, compost waste, and dairy waste. The facility provides plants, fertilizer, and other agricultural products for trusts around New Zealand. Many of these geothermal technologies were and remain remarkable technologies. At the same time, they show historians of electricity another electrical imaginary at work, one that Māori communities developed to access electric power systems previously designed to supply a New Zealand in which they were not supposed exist. Although electric power stakeholders in Christchurch and the Public Works Department imagined and built a specific type of infrastructure, in which electrification produced and powered a modern, industrial, and economically productive New Zealand for settlers, not all people in New Zealand adhered to such a pattern of electrification.

Bibliography

Anderson, Atholl, Judith Binney, and Aroha Harris. 2015. *Tangata Whenua: A History*. Wellington: Bridget Williams Books.

Boast, Richard. 2008. *Buying and Selling the Land: Government and the Māori Land in the North Island, 1865–1921*. Wellington: Victoria University Press.

[96]Richard Boast, *Buying and Selling the Land: Government and the Māori Land in the North Island, 1865–1921* (Wellington: Victoria University Press, 2008): 191–192.

[97]Reilly, *Connecting the Country*, 123–126.

Broich, J. 2007. Engineering the Empire: British Water Supply Systems and Colonial Societies. *Journal of British Studies* 46 (2): 346–365.
Chatterjee, Animesh. 2018. 'New Wine in New Bottles': Class Politics and the 'Uneven Electrification' of Colonial India. *History of Retailing and Consumption* 4 (1): 94–108.
Chatterjee, Elizabeth. 2020. The Asian Anthropocene: Electricity and Fossil Developmentalism. *Journal of Asian Studies* 79 (1): 3–24.
Chikowero, Moses. 2007. Subalternating Currents: Electrification and Power Politics in Bulawayo, Colonial Zimbabwe, 1834–1939. *Journal of African Studies* 33 (2): 287–306.
Gooday, Graeme. 2008. *Domesticating Electricity: Expertise, Uncertainty and Gender, 1880–1914*. London: Pickering & Chatto.
Greasley, David, and Les Oxley. 2009. The Pastoral Boom, the rural land market, and long swings in New Zealand Economic Growth, 1873–1939. *Economic History Review* 62 (2): 324–349.
Gucciardo, Dorotea. 2011. *The Powered Generation: Canadians, Electricity, and Everyday Life*, Dissertation, University of Western Ontario.
Hasenöhrl, Ute. 2018. Rural electrification in the British Empire. *History of Retailing and Consumption* 4 (1): 10–27.
Hasenöhrl, Ute, and Jan-Henrik Meyer. 2020. The Energy Challenge in Historical Perspective. *Technology and Culture* 61 (1): 295–306.
Hoag, Africa H. 2008. Turning Water into Power: Debates Over the Development of Tanzania's Rufiji River Basin. *Technology and Culture* 49 (3): 624–651.
Hughes, Thomas P. 1983. *Networks of Power: Electrification in Western Society, 1880–1930*. Baltimore: The Johns Hopkins University Press.
Hungerford, H., and S. Smiley. 2016. Comparing Colonial Water Provision in British and French Africa. *Journal of Historical Geography* 52: 74–83.
Kapoor, Nathan. 2019. Who Has Seen the Wind: Imagining Wind Power for the Generation of Electricity in Victorian Britain. *Technology and Culture* 60 (2): 467–493.
Kline, Ronald R. 2000. *Consumers in the Country: Technology and Social Change in Rural America*. Baltimore: Johns Hopkins University Press.
Larkin, B. 2013. The Politics and Poetics of Infrastructure. *Annual Review of Anthropology* 42: 327–343.
Lewis, M. 2007. The personal equation: Political economy and social technology on India's canals, 1850–1930. *Modern Asian Studies* 41 (5): 967–994.
Lifset, Robert D. 2014. *Power on the Hudson: Storm King Mountain and the Emergence of Modern American Environmentalism*. Pittsburgh: University of Pittsburgh Press.
Loomba, Ania. 2015. *Colonialism/Postcolonialism*. 3rd ed. New York: Routledge.
Macfarlane, Daniel, and Peter Kitay. 2016. Hydraulic Imperialism: Hydroelectric Development and Treaty 9 in the Abitibi Region. *American Review of Canadian Studies* 46 (3): 380–397.
Martin, John H. 1991. *People, Power and Power Stations: Electric Power Generation in New Zealand 1880–1990*. Wellington: Bridget Williams Books and Electricity Corporation of New Zealand.
Nixon, Rob. 2011. *Slow Violence and the Environmentalism of the Poor*. Cambridge: Harvard University Press.
Offer, Avner. 1993. The British Empire, 1870–1914: A Waste of Money? *Economic History Review* 46 (2): 215–238.
Pawson, Eric, and Neil C. Quigley. 1982. The Circulation of Information and Frontier Development: Canterbury 1850–1890. *New Zealand Geographer* 38 (2): 65–76.
Petchey, Peter. 2019. New Zealand's Technological Participation in the International Goldfields: The Example of the Stamper Battery. *Technology and Culture* 60 (2): 494–522.
Pritchard, Sara B. 2012. From hydroimperialism to hydrocapitalism: "French" hydraulics in France, North Africa, and beyond. *Social Studies of Science* 42 (4): 591–615.
Railsback, L. Bruce. 2017. Rain, Riches, and Empire: The Relationship between Nations Ruling Distant Lands, Nations of Great Wealth, and Regions of Regular Moderate Atmospheric Precipitation. *Weather, Climate, and Society* 9 (3): 455–469.

Reilly, Helen. 2008. *Connecting the Country: New Zealand's National Grid 1886–2007*. Wellington: Steele Roberts.

Rennie, Neil. 1989. *Power to the People: 100 Years of Public Electricity Supply in New Zealand*. Electricity Supply Association of New Zealand.

Rydell, Robert, and Nancy Gwinn, eds. 1994. *Fair Representations: World Fairs and the Modern World*. Amsterdam: Free University Press.

Tvedt, T. 2011. Hydrology and Empire: The Nile, Water Imperialism and the Partition of Africa. *Journal of Imperial and Commonwealth History* 39 (2): 173–194.

White, B.R., and I. Chambefort. 2016. Geothermal development history of the Taupo Volcanic Zone. *Geothermics* 59 (B): 148–167.

Williams, David V. 1999. *'Te Kooti tango whenua': The Native Land Court 1864–1909*. Wellington: Huia.

Nathan Kapoor is an Assistant Professor of History at Illinois State University, Normal, IL. He specializes in the history of electric power technologies in the British Empire. He received his PhD in the history of science, technology, and medicine from the University of Oklahoma. His current book project examines the relationship between electrification and colonialism in New Zealand between the nineteenth and twenty-first centuries. His other research projects concern the history of wind and geothermal power systems.

Chapter 5
Formation and Transformations of the Cuban Electric Company/Unión Eléctrica, 1920s–1980s

William J. Hausman, John L. Neufeld, and Rui Pereira

Abstract The formation and transformation of Cuba's electrical sector followed a pattern similar to that in many Latin American countries. Towns initially were electrified by companies established by local capitalists, frequently with the support of foreign direct investors. By the end of the 1920s, the apex of the world-wide utility holding company movement, the Cuban properties had been consolidated by the globally expanding American & Foreign Power Company (A&FP) into a single company, Compañia Cubana de Electricidad (Cuban Electric Company). While Cuban Electric expanded service over the next 30 years, this was accompanied with periodic turmoil related to the company's labor and pricing policies. A&FP consistently expressed optimism in its annual reports in the 1950s and was imagining continued expansion of Cuban Electric on the eve of the Cuban revolution. The ultimate success of the revolution brought nationalization of the company in 1960, abruptly ending Cuban Electric's position as a key component of a U.S. multinational enterprise. While electricity production in Cuba, both before and after nationalization, did not appear to be substantially different than that in several other Latin American countries, there is evidence that there was a statistically significant decrease in the growth of electricity production in Cuba after nationalization.

Keywords Cuban electrification · American & Foreign Power Co. · Multinational enterprise · Nationalization · Fidel Castro · Electric bond & share · Foreign Claims Settlement Act · Fulgencio Batista

---

W. J. Hausman (✉)
Department of Economics, Emeritus, William & Mary, Williamsburg, VA, USA
e-mail: wjhaus@wm.edu

J. L. Neufeld
Department of Economics, Emeritus, University of North Carolina, Greensboro, NC, USA

R. Pereira
Department of Economics, William & Mary, Richmond, VA, USA
e-mail: rmpereira@wm.edu

W. B. Carlson, E. M. Conway (eds.), *Electrical Conquest*, Archimedes 67,
https://doi.org/10.1007/978-3-031-44591-0_5

5.1 Transitions Anticipated and Unanticipated

The trajectory of Cuban electrification in the twentieth century was in many ways typical of developing countries around the world, particularly in Latin America.[1] While many of the transitions to electrification were imagined and implemented by talented actors, others were not, and in fact many were unanticipated. The transitions explicitly imagined by actors in the industry occurred during the early to middle years of the twentieth century, as large networks developed, and electrification spread around the world. Unanticipated transitions occurred later in the century, beginning after the Second World War and extending into the 1970s.

Cuban towns initially were electrified around the turn of the century by companies founded by local capitalists, but frequently owned and controlled by foreign investors from the United States, Canada, Great Britain, and Germany. By the mid-1920s, during the heyday of the world-wide utility holding company movement, U.S. investors came to dominate Cuba's electrical sector. By the end of the decade, the Cuban properties had been consolidated into a single U.S.-owned company, Compañia Cubana de Electricidad (Cuban Electric Company), a subsidiary of the globally expanding American & Foreign Power Company (A&FP). Once near-national consolidation had been achieved, Cuban Electric proceeded to grow, a condition that engendered sharp reactions on the part of labor, consumers, and all levels of government. The unanticipated depression of the 1930s dealt a serious but temporary blow to the prospects of the company, the ongoing electrification of Cuba, and the course of Cuban economic development. At several tense points in the 1930s, labor unions briefly took control of the management of the company, imagining that they could do a better job of operating the company than the foreign owners. Despite the economic recovery following the depression, tensions continued periodically to plague the company, frequently interrupting progress.

The decade of the 1950s was one of modest progress for the company, with a spurt in investment late in the decade, just as Cuba was undergoing yet another change in government, this one quite tumultuous. When revolutionaries ultimately seized control of the government of Cuba in 1959, the fate of the foreign-owned electric utility hung in the balance. While seizure of the property was not inevitable and had not been explicitly imagined by the leaders of the revolution (who rather were focused on agricultural reform), foreign relations with the United States deteriorated quickly, and nationalization of the property was executed by legal decree in 1960. The new government consequently was thrust suddenly into the electric utility business. Neither Fidel Castro (a lawyer), who took the title Prime Minister, nor Ernesto (Che) Guevera (a doctor), who became Finance Minister, Minister of Industries, and President of the National Bank, had any deep knowledge of economics, nor experience in running a business. They were able, however, to rely on expertise, and heavy subsidies, provided by the Soviet Union, which stepped in as the United States became increasingly antagonistic. The newly

[1]Hausman contributed the text of this chapter, Neufeld the graphs, and Pereira the quantitative analysis.

restructured electric utility, renamed Unión Eléctrica, took advantage of both the expertise and the subsidies.[2]

The first transformation of Cuban electric utilities – from a hodgepodge of local utilities with both domestic and foreign ownership structures, to a nationwide, consolidated, foreign-owned utility embedded in an international network – was imagined, planned, and carried out by U.S.-based utility executives associated with the Electric Bond & Share Company (EB&S). Visionaries such as Sidney Z. Mitchell, President of the company, sought to create a giant holding company that sat at the top of a pyramid comprised of diverse sub-holding companies (such as A&FP) and numerous operating electric utilities (such as Cuban Electric).[3] EB&S, however, was much more aggressive than other U. S.-registered utility holding companies in its pursuit of foreign acquisitions.[4]

There are specific characteristics of electric utilities operating in foreign countries that provide potential challenges for the foreign owners. Electric utilities are extraordinarily capital intensive. Relatively large sums of money have to be invested before any electricity – and hence revenue – can be generated. Initial investments and the frequent but necessary expansions of systems as technology improves cannot be funded solely out of retained earnings. Hence, there is a need for outside investors, especially in relatively capital-poor developing countries. But once in place, the ownership of these investments are difficult to transfer.[5] Everywhere, the perception of electricity as a commodity (or service) gradually shifted from luxury item, limited mostly to the central business districts of large cities, to a necessity of life.[6] Thus, prices and access became issues about

[2] Oil was the primary fuel for Cuban electric utilities, and Cuba imported nearly 90% of its fuel needs from the Soviet Union. The oil was heavily subsidized. The collapse of the Soviet Union caused an immediate and deep recession in Cuba and led to a new national energy policy. Michael Panfil, Daniel Whittle, and Korey Silverman-Roati, *The Cuban Electric Grid* (New York: Environmental Defense Fund, 2017), 8, https://edf.org/electricity-in-Cuba

[3] Sidney Alexander Mitchell, *Sidney Z. Mitchell and the Electrical Industry* (New York: Farrar, Straus & Cudahy, 1960). Mitchell died in 1944. Several of his obituaries assert that before the Wall Street Crash of 1929 Mitchell was the richest man in the world. See, for example, *Chicago Tribune*, Feb. 19, 1944, 12.

[4] On the various types of utility holding companies in the mid-1920s, see United States Federal Trade Commission, *Control of Power Companies* (69th Congress, 2nd sess., Doc. No. 213, Washington, USGPO, 1927).

[5] This situation has been described as "asset specificity" or the "obsolescing bargain." The theory is that in the early stages of foreign investment, investors are able to bargain for favorable terms from the host country. However, once the investment is in place, it becomes an immobile sunk cost, shifting bargaining power to the host country. Two foundational theoretical works are Raymond Vernon, *Sovereignty at Bay: The Multinational Spread of US Enterprises* (New York: Basic Books, 1971), and Oliver E. Williamson, "The Economics of Organization: The Transaction Cost Approach," *American Journal of Sociology* 87 (Nov. 1981): 548–77.

[6] As the final report of the task force investigating the massive 2003 blackout affecting Canada and the United States noted, "Modern society has come to depend on reliable electricity as an essential resource for national security; health and welfare; communications; finance; transportation; food and water supply; heating, cooling, and lighting; computers and electronics; commercial enterprise; and even entertainment and leisure – in short, nearly all aspects of modern life." U.S.-Canada Power System Outage Task Force, *Final Report*, April 2004, 5, https://www3.epa.gov/region1/npdes/merrimackstation/pdfs/ar/AR-1165.pdf

which sensitive consumers and governments began to agitate and imagine alternative outcomes, with much lower prices. Likewise, also because of the capital intensity, labor found itself in a relatively strong bargaining position and over the years made its grievances well known. In the United States, regulation could ameliorate some of these conditions. But when the electric utility is foreign owned, the intensity of the agitation is amplified. Hence, it is not surprising that the Cuban Electric Company was subjected to nearly constant complaints from city authorities and consumers, and frequently had difficult labor problems to confront. These conditions did not lead to the radical solution that occurred in Cuba in every country, but it did lead to the gradual demise of foreign ownership of electric utilities in most Latin American countries from the late 1950s into the 1970s. This second great transformation was much more painful to foreign investors than the first.

This chapter traces these two transformations, one distinctly imagined and implemented by utility executives, the second more a reaction created by consumers, labor, and governments that imagined an alternative structure for the industry, one centered on domestic or government (public) ownership. Finally, we examine production and electricity utilization data under the two regimes, before and after nationalization. There is evidence that the annual growth rate of electricity generation in Cuba slowed after nationalization. Nevertheless, the nationalized electric utility managed to extend availability of electricity from 50% of the population in 1959 to 95% in 1989.[7] When annual electricity generation from the 1920s to the 1980s is compared to that in other Latin American countries, Cuba is situated near the middle of the group both before and after nationalization. A more recent analysis of hourly utilization of installed generating capacity (a measure of efficiency) for 2012–15 shows Cuba somewhat on the low side but not an outlier.

5.2 Early Electrification in Cuba

Within a decade after Thomas Edison opened the Pearl Street station in New York (1882), domestic and foreign investors were establishing local electric utilities in Cuba. Many of these early stations were associated with existing railways, gas lighting companies, or urban tramways, enterprises that already had some access to capital. By the turn of the century, Havana, Cárdenas, Camagüey, Cienfuegos, Pinar del Rio, Santa Clara, and Santiago de Cuba had at least rudimentary electric service.[8] These developments occurred despite disruptions caused by the Cuban War of Independence (1895–98) and Spanish-American War (April – August 1898).[9]

[7] Panfil, et al., *Cuban Electric Grid*, 8.

[8] Jorge R. Piñón, "Cuba's Energy Challenge: Fueling the Engine of Future Economic Growth," Institute for Cuban & Cuban-American Studies, Occasional Paper Series, March 2004.

[9] On the difficulties faced during the war by the Spanish-American Light and Power Company, a U.S. company incorporated in 1883 with plants in Havana and Matanzas, see *New York Times*, Dec. 8, 1896.

Shortly following the U.S. victory in the Spanish-American War and subsequent occupation, the Platt Amendment (1901) and Cuban-American Treaty of Relations (1903) recognized Cuban independence, asserted the right of the United States to guarantee that independence, but permitted the U.S., under certain circumstances, to interfere with Cuban affairs.[10] After these political events, and because of geographical proximity, business relations between Cuba and the United States grew closer. The United States became the major market for Cuban sugar, and Cuba became a major recipient of U.S. capital investment.

Foreign investment in Cuban electric companies increased in the first decade of the twentieth century, along with consolidations and frequent ownership changes. By 1925, total generating capacity of central station electric power in Cuba was 105,000 kilowatts. In addition, sugar refineries had installed over 15,000 kilowatts of generating capacity to power their operations as well as providing electric lighting to surrounding towns.[11] By 1929, the Cuban Electric Company, a subsidiary of the American & Foreign Power Company, would control roughly 90% of central station capacity in Cuba and 75% of total generating capacity.[12]

5.3 The Rise of the American & Foreign Power Company

A&FP was a subsidiary holding company of EB&S, a top-level U.S.-based holding company. Utility holding companies generated no electricity themselves; they held controlling interests in the operating companies that did generate electricity and in sub-holding companies. Holding companies earned revenue from dividends paid by their subsidiaries and from fees charged operating companies for finance, engineering, and managerial services.

The General Electric Co., one of the two major electrical equipment manufacturers in the U.S., created EB&S in 1905 and gave it a portfolio of domestic and international stocks and bonds that General Electric previously had accepted in payment for electrical equipment. EB&S created five second-tier sub-holding companies over the course of the next two decades, including A&FP in 1923. The securities EB&S had received from General Electric were distributed to these companies in return for a controlling share of their stock. The sub-holding compa-

[10]The 1903 treaty was replaced in 1934 with a treaty that rescinded the special relationship, accessed January 6, 2023, http://www.latinamericanstudies.org/us-cuba/treaty-5-29-34.htm

[11]*McGraw Central Station Directory and Data Book, 1925* (New York: McGraw-Hill Co., c. 1925), 954–61.

[12]*McGraw Central Station Directory, 1929* (New York: McGraw-Hill Catalog and Directory Company, Inc., c.1929), 784–90. The vast majority of non-central station capacity was owned by sugar mills.

nies then acquired local utilities, consolidating them where possible, and modernizing them.[13]

World War I was the catalyst for EB&S's interest in purchasing and managing foreign electric utilities. The U.S. government was concerned about the war's effects on the Panama Canal and suggested that EB&S purchase the electric utilities in Panama City and Colón, which was accomplished in 1917. In 1920, EB&S acquired utilities that had been seized from Germans by the Guatemalan government. EB&S acquired its first Cuban property, Cia. Electrica de Alumbrado y Traccion de Santiago, in 1922.[14] By the end of the following year, EB&S owned electric utilities in Camagüey, Cienfuegos, Cárdenas and Santa Clara, Cuba.[15]

On December 18, 1923, EB&S announced the incorporation of the American & Foreign Power Company in Maine. The purpose of the new company was to "enter the public utility field of Cuba, Central America, and South America."[16] Total capital was to be a minimum of $50 million and EB&S intended to maintain a controlling interest in the company.[17] Other securities were to be marketed to the public, including in New York, London, Paris, and Amsterdam.[18] The next day, the *New York Times* reported that subscriptions for the preferred stock had exceeded the amount available.[19] E.B. Lee, a spokesperson for EB&S, exuded optimism over the new company: "The present is considered a most opportune time for the formation of an enterprise such as the American and Foreign Power Company, Inc., as many foreign exchanges are depreciated, making it possible for the new company, with its large available capital, to acquire properties at low price. It is understood that financial returns obtainable from public utility companies in foreign countries are much larger than in the United States. In fact, the Electric Bond and

[13]The sub-holding companies could issue bonds and non-voting preferred stock to raise capital, which then could be used to purchase sufficient shares of voting common stock in operating companies to gain control. Rates of return on the common stocks held by the promoters could be substantial. The Federal Trade Commission estimated rates of return in 1924 and 1925 ranging from 19% to 55%. U. S. Federal Trade Commission, *Control of Power Companies*, xxiii–xxiv.

[14]See William J. Hausman and John L. Neufeld, "U.S. Foreign Direct Investment in Electric Utilities in the 1920s," in *The Free-Standing Company in the World Economy, 1830–1996*, eds. Mira Wilkins and Harm Schröter (Oxford: Oxford University Press, 1998), 372–73, and William J. Hausman and John L. Neufeld, "The Rise and Fall of the American & Foreign Power Company: A Lesson From the Past," *The Electricity Journal* 10 (Jan./Feb. 1997): 49.

[15]*New York Times*, Dec. 5, 1923, 32.

[16]*New York Times*, Dec. 19, 1923, 29.

[17]This would be equivalent in 2019 dollars to between $3 and $4 billion, using either labor cost or relative income to adjust. Samuel H. Williamson, "Seven Ways to Compute the Relative Value of a U.S. Dollar Amount, 1774 to present," MeasuringWorth, 2023, accessed Jan. 7, 2023, https://www.measuringworth.com/calculators/uscompare/relativevalue.php

[18]*New York Times*, Dec. 19, 1923, 29.

[19]*New York Times*, Dec. 20, 1923, 25. The 400,000 shares of preferred were priced at $96 a share, thus raising $38.4 million for the company. Each preferred share came with a share of the common stock. EB&S retained the rest of the common. Both classes of preferred stock carried a cumulative dividend of 7%.

Share Company expects these foreign utility investments to yield an average return perhaps twice as much as is possible in this country."[20] While implicitly recognizing exchange rate risk, Lee completely missed the possibility of political risk, both of which would later bedevil the company.

5.4 The American & Foreign Power Company Consolidates Its Position in Cuba

In 1924, its various Cuban properties contributed 74% of A&FP's total operating revenues. By late summer 1925, the company (along with EB&S) was purchasing shares on the open market of the Havana Electric Railway, Light and Power Company, whose capitalization at the time exceeded that of A&FP.[21] In December, A&FP took control of the electric railway.[22] EB&S transferred its shares in the Havana electricity plants to A&FP, giving A&FP control over nearly all central station utilities in Cuba.[23] By the end of 1927, A&FP also controlled local electric utilities in Panama, Guatemala, Ecuador, Colombia, Brazil, and Venezuela. Its consolidated assets totaled $379 million.[24]

A&FP continued to expand its holdings in 1928, adding fourteen utilities, mostly in South America, increasing its total assets to $543 million. In May 1928, all Cuban holdings were consolidated into one operating company, Compañia Cubana de Electricidad (Cuban Electric Company). A&FP stated in its 1928 annual report that this "simplified and improved operating conditions by eliminating many small companies with the expense incident to maintaining them. It has improved the service and the power generation efficiency and has enabled the operating company to reduce rates. It has simplified and strengthened the capital structure."[25] The Cuban subsidiary contributed 57% of the gross revenue of A&FP in 1928, more than twice the next largest contributor, its Brazilian subsidiaries.[26] In 1929, after acquisition of properties elsewhere (Chile, Mexico, and Shanghai, China), the Cuban subsidiary's contribution to A&FP's total revenue fell to 29%, still more than subsidiaries in any

[20] *New York Times*, Dec. 19, 1923, 29.

[21] *New York Times*, Aug. 8, 1925, 13.

[22] The takeover was effected by creation of a new company, the Havana Electric and Utilities Company. The president of Havana Electric Railway, Light and Power Company, Frank Steinhart, was allowed to remain as president of the new company. *New York Times*, Dec. 19, 1925, 30.

[23] In 1926, the street railway properties were split, with A&FP retaining ownership of the electric plants but not the railways themselves. *New York Times*, Aug. 12, 1926, 25; Aug. 15, 1926, E10; Oct. 30, 1926, 29; Nov. 26, 1927, 19; Jan. 19, 1928, 47.

[24] *Moody's Manual of Investments, Public Utility Securities* (New York: Moody's Investor Services, 1928), 1653.

[25] American & Foreign Power Company Inc., Annual Report for 1928, May 23, 1929, 6.

[26] *New York Times*, Mar. 15, 1929, 40; June 11, 1929, 44.

other country. A&FP claimed over $750 million in assets on its consolidated balance sheet for 1929.[27]

At this point, the emerging world-wide Depression began to adversely affect A&FP. Gross revenues of the Cuban Electric Company declined by 1% in 1930, while net revenues rose slightly. Because of exchange rate depreciation in several countries, overall gross dollar revenue of A&FP was down about 3% over the course of the year. The company, nevertheless, continued to acquire properties in 1930, including several small ones in Cuba and several in India. Total assets of the company grew to $987 million.[28]

The years immediately following would prove to be very difficult for Cuban Electric and its parent, A&FP. In 1931, total assets of A&FP reached almost $1 billion but then fell to roughly $750 million in 1932, where they stabilized for the rest of the decade.[29] Operating revenues fell by roughly 18% in dollar terms for both Cuban Electric and A&FP overall. The Depression hit Cuba very hard. Between 1929 and 1933 exports fell by 70% and imports by 80%. Between 1928 and 1933, net national product in current prices fell by 26%.[30] These economic difficulties generated considerable political and labor unrest.

A&FP had a very close relationship with the leader of Cuba at the time, General Gerardo Machado, and the Cuban Electric Company was considered by many to be virtually a division of the government of Cuba, even though it clearly was foreign owned. One of the properties EB&S acquired in 1923 was previously owned by Machado, who had risen to prominence in the Cuban War of Independence. He became mayor of Santa Clara and owned the Compañia Cubana de Electricidad in that city. When EB&S bought the company in 1923, Machado retained the title of vice-president. A&FP supported Machado and invested $500,000 in his successful campaign for president in 1924. Machado generally favored foreign investment and suppression of labor unions, and clearly looked after the company's interests. The company's close ties to Machado, however, may have contributed to consumer and labor unrest. There were periodic local boycotts, and municipalities passed

[27] American & Foreign Power Company Inc., Annual Report, 1929, 7, 26; *New York Times*, May 28, 1930, 43. EB&S maintained a special relationship with A&FP. It was the only sub-holding company in the system in which EB&S had retained majority ownership. In March 1929, EB&S reorganized, creating a new company with the same name. In the process, EB&S increased its investment account instantaneously by $400 million, all of which was attributed to an increase in the "market value" of A&FP. U.S. Federal Trade Commission, *Utility Corporations* (U. S. Senate, 70th Congress, 1st session, Document 92, parts 23 and 24, Washington: USGPO, 1930), 410.

[28] American & Foreign Power Company Inc., Annual Report, 1930. The Cuban peso was pegged to the U.S. dollar at parity. The company also raised $50 million by selling gold debentures to a bank syndicate and sold a $20 million debenture of the Cuban Electric Company to its parent company, EB&S.

[29] Consolidated assets in nominal dollars did not approach $1 billion again until 1954. American & Foreign Power Co., Annual Reports.

[30] B.R. Mitchell, *International Historical Statistics: The Americas, 1750–1988*, 2nd edition (New York: Stockton Press, 1993), 189, 279, 424–28, 750.

numerous resolutions demanding lower lighting rates. Machado vetoed forty-six of these municipal resolutions during his term.[31]

Troubles, both in Cuba and for A&FP, escalated during 1931–34. In the summer and fall of 1931 there was a nationwide campaign in Cuba against high electric rates, including demands that the company lower average rates from 16 cents to 10 cents per kilowatt hour, reduce the minimum flat rate, and reduce industrial rates proportionally. Numerous mass meetings were held, boycotts threatened and initiated, and one municipality proposed converting the local electric plant into a cooperative.[32] At the protest's peak, the company may have lost as many as 40,000 customers due to boycotts.[33] As noted above, operating revenues in Cuba fell by 18% in 1931, but there was otherwise no mention of the Cuban protests in the company's annual report, and it appeared to be weathering the storm.[34]

The boycott of 1931 gradually subsided, and 1932 was relatively quiet, at least in Cuba. Operating revenues of the Cuban subsidiary, however, fell by another 21% in 1932, and A&FP was forced to suspend dividend payments on its preferred stock.[35] The price of sugar was 59% lower in 1932 than it had been in 1929, putting tremendous strain on the nation's sugar industry and its workers.[36]

According to Thomas O'Brien, by the beginning of 1933 "the national economy disintegrated [and] Cuban society rapidly advanced into a state of undeclared civil war."[37] In March 1933, a bomb exploded at the Cuban Electric Company's headquarters in Cienfuegos. No casualties were reported, but it did considerable damage. Patrolmen in different parts of that city and in Havana reportedly came under fire.[38] Organized labor was active in protests.[39] In July, bus drivers went on strike. Soldiers fired on demonstrators in Havana on August 1, killing several people. Across the country, more workers joined the strike, including tobacco workers and journalists in Pinar del Rio, and typographers, journalists, and dock workers in Havana. The military took control of Havana on August 9 and Machado, who had resigned on

---

[31] Thomas F. O'Brien, "The Revolutionary Mission: American Enterprise in Cuba," *American Historical Review* 98 (June 1993): 774–76.

[32] *New York Times*, July 5, 1931, 36.

[33] O'Brien, "Revolutionary Mission," 776.

[34] Electric Bond & Share wrote down its investment in A&FP in December 1931 by the same amount, $400 million, it had written up the investment in 1929. U. S. Federal Trade Commission, *Utility Corporations* (United States Senate, 70th Congress, 1st session, Document 92, part 66, Washington: USGPO, 1934), 1266, 1274.

[35] It had never paid dividends on its common stock.

[36] "The Foreign Power System," New York: American & Foreign Power Company Inc., c. 1953, 41.

[37] Thomas F. O'Brien, *The Revolutionary Mission: American Enterprise in Latin America, 1900–1945* (New York: Cambridge University Press, 1996), 231.

[38] *New York Times*, March 18, 1933, 30.

[39] Ralph Lee Woodward, Jr., "Urban Labor and Communism: Cuba," *Caribbean Studies* 3 (Oct. 1963): 17–50, examines the role of labor unions, communism, and Cuban governments from colonial times through the early years of the Castro regime.

August 11, fled the country.[40] By the end of the year, several American-owned sugar mills announced that they were suspending operations due to unreasonable union demands.[41] The *Wall Street Journal* reported that there had been little damage to the properties of the Cuban Electric Company, and that it was one of the few organizations whose employees did not join the strike. However, the company had suffered a loss of revenue due to "the revolution."[42]

After a few months of collective leadership, Ramón Grau San Martin was named President of Cuba. Protests and boycotts against the power company continued. With the agreement of A&FP, a commission to study the rate issue was created. It recommended rate cuts averaging 25%. On Dec. 6, 1933, President Grau issued a decree ordering the company to reduce its electric and gas rates by 45% across the country, without reducing wages or reducing personnel. The company protested vehemently. Other American-owned firms also faced demands to lower rates; for example, the Cuban Telephone Company, a subsidiary of the International Telephone and Telegraph Company, was ordered to reduce rates for personal and business lines by 40% and 44%, respectively.[43]

On December 16, 1933, employees and managers of the Cuban Electric Company announced that a strike scheduled to have begun that day would be postponed for 30 days. The company had agreed to thirty of forty-one demands, including those for a 48-hour work week and 30 days of paid vacation, but wanted more time to consider the remaining demands. The company also announced that it was appealing the 45% rate reduction to the Supreme Court.[44] The company failed to accept enough of the employee's demands, and negotiations broke down. The strike was to begin on January 14, 1934. The secretary of the Light and Power Workers' Union announced that "It will be a strike of blood and fire."[45] There were reports of physical

[40] "Cubans General Strike to Overthrow President, 1933," Global Nonviolent Action Database, accessed Jan. 9, 2023, http://nvdatabase.swarthmore.edu/content/cubans-general-strike-overthrow-president-1933

[41] *New York Times*, Dec. 7, 1933, 10; *Washington Post*, Dec. 7, 1933, 9.

[42] *Wall Street Journal*, August 15, 1933, 9.

[43] *New York Times*, Dec. 7, 1933, 10; *Washington Post*, Dec. 7, 1933, 9. The U.S. Ambassador to Cuba, Sumner Welles, immediately met with the President, at Grau's request, to discuss the political situation. At this meeting Welles objected to the rate decrees, arguing that "the lack of study and the confiscatory nature of many of the decrees he had issued affecting legitimate and vested interests in Cuba must necessarily impair confidence." According to Welles, Grau backed off and indicated he would accept modifications if the Supreme Court so ordered. Welles also speculated that Fulgencio Batista, with the support of army officers, was prepared to remove Grau from office. Grau, he wrote, was nothing but a "figurehead." Telegram of Welles to Acting Secretary of State, Dec. 7, 1933, Foreign Relations of the United States, Diplomatic Papers, 1933, The American Republics, Vol. V.

[44] *New York Times*, Dec. 16, 1933, 11; *Washington Post*, Dec. 16, 1933, 2. The articles also noted that a demonstration of 35,000 people had been held in Havana that day to express gratitude to the Grau administration for its recent decree mandating use of at least 50% Cuban labor in foreign-owned firms.

[45] *New York Times*, Jan. 14, 1934, 14.

intimidation and violence against more moderate employees of the company, and the company's attorney was beaten by employees in the main office building.[46] When the strike indeed began, the Cuban government seized control of the company, and service was reduced but not entirely suspended. The President of the Cuban Electric Company, S. J. Mahoney, immediately protested to the U.S. State Department. President Roosevelt's personal envoy in Havana did not consider the seizure permanent, but thought it to be a "provisional intervention."[47] Union members were put in day-to-day control of the company.[48]

At the same time that the government was assuming control of the electric company, army Sergeant Fulgencio Batista executed a successful coup, as U.S. Ambassador Welles had predicted.[49] Carlos Mendieta, a revolutionary war veteran and opponent of Machado, was made President of Cuba.[50] On January 20, it was announced that employees and officials of the Cuban Electric Company were trying to settle the labor dispute.[51] Several days later, Pres. Mendieta formed a commission to study the problems raised by the recent strike and the decree mandating the 45% rate reduction.[52] Jesus Correro, Luis Machado, a cousin of former President Machado, and Eduardo Chibas, Sr. were appointed to the commission.[53] Within days, managerial control was returned to the company. Electricity, gas, water, and streetcar service workers across the country immediately commenced a strike. Military guards were placed at power plants and main offices of Cuban Electric. It was reported that the company had accepted all of the long-standing demands of the union, but employees had balked at replacement of the general manager.

President Mendieta issued a stern proclamation asking for cooperation "to free Cuba from a state of chaos and ruin."[54] Simultaneously, the government suspended all authorizations for public demonstrations. American officials were reportedly pleased by the actions of the Mendieta government, but tobacco workers and bus drivers remained on strike, and student protests were ongoing.[55] Chaos continued

---

[46] *Ibid.*

[47] *New York Times*, Jan. 16, 1934, 19; *Wall Street Journal*, Jan. 16, 1934, 2.

[48] O'Brien, "Revolutionary Mission," 780.

[49] Telegram of Welles to Acting Secretary of State, Dec. 7, 1933. Welles had left his post on Dec. 13, 1933.

[50] Batista became Army Chief of Staff, which gave him control of the Cuban Army.

[51] *Wall Street Journal*, Jan. 20, 1934, 3.

[52] *New York Times*, Jan. 26, 1934, 13.

[53] *Washington Post*, Jan. 29, 1934, 3.

[54] *New York Times*, Feb. 4, 1934, 21.

[55] In response to the U.S.-friendly coup, Franklin D. Roosevelt on March 9, 1934, issued an executive order creating the Second Export-Import Bank of Washington (the first bank having never made any loans). Its initial purpose was to facilitate trade with Cuba. The first loan made by the bank was to finance Cuba's purchase of silver ingots from the U.S. This may have been to bolster the value of the Cuban peso, which was linked one-to-one to the dollar. The two banks were later merged into the Export-Import Bank. http://www.exim.gov/about/history-exim/historical-timeline/legislative-history (accessed March 23, 2021).

with intermittent bombings and shootings. Electric power was only partially restored, and the former general manager of the company, and strike leaders, were placed under arrest.[56] The labor dispute finally ended in May when the company capitulated to most of the union demands.[57] Another general strike occurred in 1935. It was begun by high school students and teachers, spread to University of Havana students and faculty, and then was joined by labor unions. This strike was brutally repressed. The university was closed (for 3 years) and many leaders of the strike were assassinated or abducted and executed. The reaction to this repression was so intense that Carlos Mendieta faced dissent within his government, eventually lost support, including Batista's, and was forced to resign in December. Batista remained the power behind a series of puppet Presidents until he assumed the office himself in 1940.[58]

In its 1933 annual report, A&FP for the first time acknowledged the difficulties of its Cuban subsidiary, lamenting the decline in revenues (total revenues by 12.4% and net revenues by 27%), the rate reduction decree, the strike, the temporary government takeover, and the difficulties the government and municipalities were having in paying their bills. Nonetheless, the company remained hopeful for the future, praising the reduction in the tariff on Cuban sugar of 25% and the establishment of a quota for Cuban sugar imports into the United States.[59] Operating revenues of the Cuban Electric Company continued to decline in 1934, finally reaching a trough. They were 18% lower in 1934 than they had been in 1933. Overall, operating revenues declined by 52% from 1929 to 1934.[60]

The Cuban economy improved slowly after 1934. Net national product rose by 23% from 1933 to 1940.[61] Gross operating revenues of the Cuban Electric Company rose from $10.6 million to $12.0 million over the same period, a 13% increase. In 1940, the Cuban subsidiary contributed 20% of the operating income of A&FP, still more than its subsidiaries in any other country.[62]

[56] *Washington Post*, Feb. 5, 1934, 2.

[57] *Wall Street Journal*, June 1, 1934, 9.

[58] "Cubans General Strike to Overthrow President, 1935," Global Nonviolent Action Database, accessed Jan. 9, 2023, https://nvdatabase.swarthmore.edu/content/cubans-general-strike-overthrow-president-1935

[59] American & Foreign Power Company Inc., Annual Report, 1933.

[60] American & Foreign Power Company Inc., Annual Report, 1934.

[61] Net national product figures for Cuba were not available for 1934. Mitchell, *Historical Statistics*, 750.

[62] American & Foreign Power Co., Annual Reports, 1933–1940.

5.5 Aftermath of the Crash and Depression: Company Reorganizations

Holding companies had long been accused by progressives of using their excessive power to victimize both utility customers and outside investors. The Depression brought the collapse of several prominent utility holding companies, and most of those that survived suspended dividend payments.[63] EB&S survived but was weakened severely. With Franklin D. Roosevelt's New Deal, the critics of holding companies came into power, and they acted swiftly and punitively. The Public Utility Holding Company Act of 1935 (PUHCA) contained a "death sentence" that mandated the dissolution within 3 years of most holding company systems in the U.S. The newly-formed Securities and Exchange Commission (SEC) was made responsible for enforcement of PUHCA. All holding companies not exempted had to register with the SEC. Despite the three-year mandate, full implementation took decades because of litigation initiated by the companies, and the law's complex requirements.[64]

A&FP (but not its parent, EB&S) was granted an exemption from most of the law's severe requirements because none of its subsidiaries operated in the U.S. Although EB&S remained the largest shareholder of A&FP, the two companies had no officers, directors, or employees in common. In December 1939, A&FP resumed paying dividends.[65] In January 1943, the international division of Ebasco Services, then a subsidiary of EB&S, was transferred to Ebasco International, a subsidiary of A&FP.[66]

Eventually, EB&S divested all its subsidiaries, retaining only its service company, Ebasco Services, Inc. and a controlling interest in A&FP. A&FP was required by the SEC to simplify its financial structure, which then consisted of debentures, serial notes, and bank notes, three classes of preferred stock with dividend arrearages of $410 million, common stock, and option warrants. Difficulty obtaining financing delayed A&FP's financial simplification until 1949, when EB&S transferred debt it held of Cuban Electric to A&FP and restructured AF&P's debt to the parent company.[67] A&FP was then able to obtain bank financing and reorganized its financial structure in February 1952.

---

[63]The collapse of Samuel Insull's Chicago-based empire was probably the best known and most widely publicized. For example, see Forrest McDonald, *Insull* (Chicago, University of Chicago Press, 1962).

[64]John L. Neufeld, *Selling Power: Economics, Policy, and Electric Utilities Before 1940* (Chicago: University of Chicago Press, 2016), 147–52. The SEC won its case against EB&S in the Supreme Court on Mar. 28, 1938. The court held that the registration provisions of the act were constitutional. This decision essentially ended litigation on the constitutionality of the act. *New York Times*, Mar. 29, 1938, 9. U.S. Securities and Exchange Commission, *Tenth Annual Report* (Philadelphia, by the Commission, 1944), 84.

[65]American & Foreign Power Company, Annual Report, 1939, 8–9. The company subsequently registered as a public utility holding company with the SEC. Kantor v. American & Foreign Power Co., U.S. Court of Appeals for the First Circuit – 197 F.2d 307 (1st cir. 1952), 2.

[66]*New York Times*, Oct. 23, 1942, 33; Jan. 26, 1943, 27. The SEC did not grant EB&S's request for exemption from provisions of PUHCA until 1959. *New York Times*, June 25, 1959, 39.

[67]U.S. Securities and Exchange Commission (*Sixteenth Annual Report*, Washington: USGPO, 1950), 68–70. *New York Times*, Oct. 27, 1944, 27; *Wall Street Journal*, Dec. 20, 1949, 16. A&FP had written off its Shanghai property during World War II.

5.6 Progress and Setbacks in the 1950s

From 1940 to 1952, the operating revenues of the Cuban Electric Co. rose sharply from $12 million to $47.3 million.[68] In 1952, Cuban Electric contributed 25% of the total operating revenues of A&FP, which had risen from $61.4 million to $187.2 million.[69] Total assets on the consolidated balance sheet of A&FP rose from $755 million to $914 million. In its 1952 annual report, A&FP announced a planned increase in generating capacity in Cuba of 45%, later raised to 63%, by the end of 1955. In addition to two $12 million loans from the Export-Import Bank, the company was successful raising funds in Cuba through a consortium of local banks and security sales to the public, about which it was very proud.[70] While frequently expressing concern about the heavy reliance on sugar exports, A&FP nevertheless felt that the Cuban economy had untapped potential. Consolidated operating revenues of A&FP in the 1950s remained flat, fluctuating around $200 million per year, while consolidated assets of the company rose from $954 million in 1953 to $1.2 billion in 1959.[71] Cuban Electric raised the annual dividends paid to A&FP during the 1950s from $.50 per share in 1952 to $1.50 per share in 1956–58. A&FP owned 88% of the outstanding shares of common stock.[72] A&FP's annual reports exuded optimism during this period, but there remained periodic unrest in Cuba, unacknowledged publicly by the company, and the situation changed dramatically by the end of the decade.

[68] Adjusting for changes in the U.S. consumer price index, operating revenues slightly more than doubled (a 108% increase). Williamson, MeasuringWorth.

[69] In 1951, the International Bank for Reconstruction and Development (World Bank), in collaboration with the government of Cuba, issued a report by a technical and economic mission that had been conducted in 1950. The report noted the relatively high rates for industrial electric power. For non-industrial uses, the report cited the fact that the system owned by the Cuban Electric Co. was operating close to capacity with outdated and inefficient equipment, that expansion was necessary, and that "the true financial position of the Company is hard to assess" (p. 325). The report went on to state that "By standards of other countries, however, rates are high and, as already noted, some big industrial consumers who could buy the Company's power are finding it cheaper to generate their own. There are widespread public complaints, too, about the cost of electric power" (p. 326). The report suggested that Cuban Electric focus on expansion of non-industrial power. International Bank for Reconstruction and Development, "Report on Cuba," Washington, D.C., 1951, 170, 323–27.

[70] In 1957 Cuban Electric held a bond sales contest for employees. The grand prize was an 18-day trip for two to Europe, which was won by César Rodŕiquez, who sold $112,500 in first mortgage bonds (roughly one million in 2019 dollars). Nine other employees won trips to Washington, New York, Mexico City, and Miami. *American & Foreign Power Panorama* (by the company, September 1957).

[71] A&FP eventually was compensated for the expropriation of most of its Argentine properties, but the proceeds had to be invested in Argentina. American & Foreign Power Company, Annual Reports, 1952–1959.

[72] *Moody's Public Utility Manual* (New York: Moody's Investor Services, Aug. 1965), 1548.

5.7 Fidel Castro and the Expropriation of the Cuban Electric Company

After a lengthy, sporadic revolutionary campaign, Fidel Castro took control of the Cuban government on January 1, 1959.[73] The full implications for foreign investors were not immediately clear.[74] Castro clearly had been advocating radical reforms, but opposition to the Batista regime seemed central to his concerns. Some accounts held that after ousting Batista, Castro intended to rule for several years and then hold free elections. Once taking over, however, Castro acted quickly to consolidate his power, including naming himself Prime Minister in February. It became clear that there would be no elections.[75]

A Castro spokesperson in the U.S., Constantine Kangles, U.S. general counsel for the Republic of Cuba, reassured American investors that Castro would use U.S. programs, such as the New Deal and Fair Deal, as templates, and that he was primarily concerned with improving life in rural villages. He asserted that "the Cuban regime has no intention of nationalizing property, foreign or domestic." Kangles cited Manuel Urrutia, Castro's choice for Provisional President, as the source of these positive statements.[76] Speaking to the press on the evening of February 3, 1959, Castro outlined "sweeping plans" for the transformation of the Cuban economy but made no mention of nationalizing industries.[77] Other news from

[73]The U.S. recognized the Provisional Government of Cuba on Jan. 7, 1959. On the same day, Castro's 9-year-old son, Fidel, Jr., who had been attending public school in Queens, left New York for Havana. *New York Times*, Jan. 7, 1959, 12, 16. The U.S. business community in Cuba emphatically recommended immediate recognition. They recognized that Castro was in control, that the "government was much better than they had hoped for," and that early recognition would assist in curbing Communist "strength." Telegram from the Embassy in Cuba to Department of State, Jan. 6, 1959. Foreign Relations of the United States, 1958–1960, Cuba, Vol. VI.

[74]Several early accounts in the *New York Times* were optimistic, noting that the end of the revolution "would bring an improvement in economic and business conditions." Grant Hylander, an A&FP vice president, noted that there had been no major damage to electrical properties and that "he did not expect the property to be nationalized." *New York Times*, Jan. 3, 1959, 3.

[75]Lily Rothman, "How Fidel Castro Went from Revolutionary to Ruler," Time.com, Nov. 26, 2016, accessed Jan. 9, 2023, http://time.com/3666177/fidel-castro-cuba-history/; also see in the left-leaning magazine *Jacobin*, Samuel Farber, "Cuba's Challenge," accessed Jan. 9, 2023, https://www.jacobinmag.com/2015/06/cuban-revolution-fidel-che-raul-castro/. In a visit to the United States in April, Castro denied, in a speech to the National Press Club, that his brother Raúl was a Communist and asserted that there were no Communists in his government. R. Hart Phillips, "Reds' Alleged Role in Castro's Regime Alarming Havana," *New York Times*, April 24, 1959, 1.

[76]*New York Times*, Jan. 4, 1959, 8. Urrutia (Manuel Urrutia Lleó) lasted six months before being accused of treason by Castro. After years of house arrest and asylum in the Venezuelan and Mexican Embassies in Havana, he received a safe conduct pass to the U.S. and became a leader of vehemently anti-Castro organizations and a professor. See his obituary in the *New York Times*, July 6, 1981, D7.

[77]*New York Times*, Feb. 4, 1919, 10. The same article noted that a group of 600 Cuban Electric Company workers had begun a hunger strike and were camping in front of the Presidential Palace. They were recent hires and were not permitted to return to their jobs by employees who had seized the offices and plants of the company when the Batista regime fell.

the new regime was not so reassuring. The *New York Times* reported that there had been "widespread reports" that Castro favored nationalization of all utilities, including the properties of A&FP and International Telephone and Telegraph (IT&T).[78] In late February, the government announced the creation of a commission to investigate the operations and rates of both Cuban Telephone (an IT&T subsidiary) and Cuban Electric. The commission was headed by the Minister of Communications, Enrique Oltuski, an American-trained engineer. Oltuski asserted that the study would be thorough and that he expected to be able to lower rates, but that there would be no government intervention in the operation of the companies.[79]

The commission released its report on Cuban Electric in late June, but A&FP claimed not to have been given a copy until August, after a rate reduction already had been ordered.[80] The report criticized the company for its perceived inefficiencies and was particularly harsh regarding what it claimed were its "excess profits" and "overvaluation" of assets. It noted the critical importance of access to electricity as a component of modern life and asserted that strict regulation or state ownership was necessary in order to reduce inefficiencies.[81] The Agrarian Reform Law of May 17, 1959 had shocked the sugar interests (with its prohibition of foreign ownership) and ranchers (with its dismantling of large land holdings), but mining, petroleum, hotel, and utility interests "seemed a little bit more hopeful than they did even a few months ago." The regime was perceived to be more honest than previous ones, and "economists. . .feel that Dr. Castro is slowly learning that, if he is to succeed, he must be much more considerate of foreign investors, particularly Americans."[82] That was the last positive bit of news the owners of the Cuban Electric Co. received.

On August 20, 1959, Cuban Electric was ordered to cut its basic rate from 9 cents to 6 cents per kilowatt-hour, and its overall rates by 30.5%, retroactive to Aug. 1. Castro and his cabinet cited the commission report, highlighting excessive engineering service fees, excessive fuel costs, and inadequate administrative procedures. The company also was ordered to continue its expansion program.[83] The President of A&FP, Henry Sargent, immediately returned to New York from

[78]Gene Smith, "Utilities in Cuba Await Their Fate," *New York Times*, Jan. 18, 1959, F1.

[79]*New York Times*, Feb. 24, 1959, 12; *The Militant*, March 2, 1959, 1. Two investigators from the Commission, both engineers, visited Cuban Electric managers in April, requesting detailed information on rates and kilowatt-hour sales. The company's response was deemed inadequate. "Relations with Rate Investigating Committee," typescript, April 29, 1959 (from the files of John Plunket, who was President of Cuban Electric when it was nationalized; files provided by Paul Plunket, John Plunket's nephew).

[80]American & Foreign Power Co., Annual Report, 1959, 18.

[81]Russell M. Tremayne, "The Cuban Electric Company: A Study of Expropriation" (M.A. Thesis, Boise State University, 1982), 8–9. The company had been subject to a maximum rate of 9.3 cents per kilowatt-hour for the first 100 kilowatts of use since 1933, but by the 1950s this would have been non-binding. Tremayne, "Cuban Electric," 12.

[82]Gene Smith, "Castro Believed More Amenable," *New York Times*, Aug. 9, 1959, F1. Castro's degree was in law from the University of Havana in 1950.

[83]*New York Times*, Aug. 21, 1959, 5; *Wall Street Journal*, Aug. 21, 1959, 2. A&FP, in its 1959 Annual Report, indicated that the overall rate reduction amounted to 22%.

Argentina, where Argentine properties taken over by the government were being appraised, to discuss the situation.[84] The timing of the announcement was unfortunate for the company. It occurred a day after an issue of $15 million of A&FP 6% convertible debentures due in 1984 had been oversubscribed in the bond market but not yet delivered by a consortium of investment banks led by First Boston. The price of A&FP's stock immediately fell 17% to a level 25% below the exercise price provided by the debentures, sharply reducing their value to investors. The next day, "in a financial move without parallel in recent memory," the issue (which came to be known as the "Castro convertibles") was cancelled. Henry Sargent vowed to go to Cuba to discuss the situation, hoping to bring back the bond issue in the near future.[85] This never happened.

In September, Roberto Acosta, head of the government's rate investigation commission, asserted that the government had no intention of "confiscating or destroying" the Cuban Electric Company, wishing only to regulate it. The commission recommended reducing the book value of the "inflated assets" of the company ($292 million in Dec. 1958) by $86 million. An additional $46 million of "accumulated reserves" was also removed from the value of assets, leaving a figure of $160 million, on which the company was to be allowed a 7% return, which would have sharply reduced revenue. The company strenuously objected to the reductions and requested that it be allowed to send "a group of experts to work with the investigating committee." Acosta welcomed the idea and stated that "we are willing to reconsider our decision if the company can prove to us that we are wrong."[86] In October, Communications Minister Oltuski reiterated that the company had not yet demonstrated a sufficient reason to justify suspension of the decree reducing rates but claimed that he remained ready to discuss the matter further.[87] Nothing came from this. In November, A&FP cut its dividend by half, to 12.5 cents per share and opened negotiations for a $10 million loan from the Cuban government. The loan was to partially replace the $15 million bond issue that had been cancelled when the rate reduction was announced. It was to be used for expansion of the electrical system in Cuba. Cuban Electric executives remained "optimistic about prospects for negotiating a revision of the rate cuts." They had submitted a 108-page answer to the order cutting rates and felt that they had "a good case."[88]

[84] *New York Times*, Aug. 26, 1959, 47.

[85] *New York Times*, Aug. 26, 1959, 39; Aug. 27, 1959, 37.

[86] *New York Times*, Sept. 12, 1959, 25. The A&FP Annual Report, 1958, p. 28, states that the capitalization of the Cuban subsidiary was $272 million in 1958. The 1960 Annual Report, p. 7, noted that when Cuban Electric was expropriated, there were approximately $280 million in outstanding securities.

[87] *New York Times*, Oct. 14, 1959, 16.

[88] *Wall Street Journal*, Nov. 2, 1959, 12; Nov. 6, 1959, 25; *New York Times*, Mar. 10, 1960, 43.

Cuban Electric had stopped making dividend payments, most of which went to A&FP, in September 1959. The company defaulted on an $8 million bank loan due in November, and suspended payments on its Export-Import Bank construction loan.[89] In March 1960, A&FP announced that its net earnings for 1959 had declined by 31%, due mostly to the reduction of electricity rates in Cuba. Negotiations between A&FP and the government regarding rates had reached an impasse by early 1960 and it was clear that A&FP was preparing for a potentially more disturbing intervention.[90]

Relations between the U.S. and Cuba deteriorated through the first half of 1960, as Cuba moved closer politically to the Soviet Union and the U.S. responded. On July 6, 1960, Cuba passed Law No. 851 (the Nationalization Law), a direct response to President Eisenhower's decision to cut Cuba's 1960 sugar quota, essentially shutting off the U.S. market completely for the rest of the year, leaving Cuba with huge inventory of unsold sugar that had little prospect of being purchased by other countries.[91] This law set up the legal means for the seizure of American-owned properties.[92] On August 6, 1960, Cuba nationalized the Cuban Electric Company, the Cuban Telephone Company, three American-owned oil refineries, and thirty-six American-owned sugar mills "in retaliation for the 'economic aggression' of the United States." The properties were immediately seized by the military and civilian militia.[93] Relations continued to deteriorate. A U.S. embargo on shipments of material to Cuba in October 1960 led to the nationalization of remaining American-owned businesses in Cuba.[94]

---

[89] American & Foreign Power, Annual Report, 1959, 32.

[90] Tremayne, "Cuban Electric," 30–32.

[91] *New York Times*, July 7, 1960, 1.

[92] Tad Szulc, "Havana Is Ready to Seize More American Property," *New York Times*, July 7, 1960, 1. For the text of the law, see "Cuba Nationalization Law," *American Journal of International Law* 55 (July 1961): 822–24. The law contained a provision for compensating the owners of seized properties with government bonds. Payments were tied to sugar purchases made by the United States in excess of 3 million tons annually, a completely unrealistic provision. Alan Dye and Richard Sicotte, "The U.S. Sugar Program and the Cuban Revolution," *Journal of Economic History* 64 (Sept. 2004): 673–704, explore the relationship between U.S. sugar quota policies in the 1950s and the coming of the revolution.

[93] *New York Times*, Aug. 8, 1960, 1. Compensation was set according to the Nationalization Law.

[94] R. Hart Phillips, "Cubans Expect Quick Seizure of American-Owned Property," *New York Times*, Oct. 20, 1960, 9.

5.8 The Aftermath of Nationalization

In its 1960 Annual Report, A&FP acknowledged the loss of its Cuban properties, for which it was not expecting to be compensated.[95] It stated that the amount of securities of Cuban Electric outstanding was approximately $280 million, 72% of which were held by U.S. interests: 60% by A&FP and 12% by the Export-Import Bank, other U.S. banks, and mutual investment funds.[96] In January 1961, two New York banks charged off loans they had made to Cuban Electric: Manufacturers Trust for $1.5 million and Irving Trust for $750,000.[97]

In 1962, the U.S. Internal Revenue Service (IRS) ruled that U.S. citizens and corporations could claim tax deductions for certain losses of property seized in Cuba. The Commerce Department had estimated total losses to have a book value of roughly $1 billion, later raised to nearly $2 billion by the Foreign Claims Settlement Commission.[98] The Cuban government's compensation offers were recognized as completely unrealistic. A&FP wrote off its entire investment in Cuban Electric (listed for tax purposes as $168 million) in 1960, which wiped out the company's tax liabilities for the next half decade. The IRS cautioned, however, that should any of these losses eventually be recovered, such compensation would be considered taxable income when received.[99]

A&FP (and its parent EB&S) would struggle after the loss of its largest property and gradually withdraw from the business of operating foreign electric utilities.[100] In its 1961 Annual Report, the company noted that it was then receiving over one-third of its income from a diversified portfolio of non-utility direct investments and foreign government bonds and that it intended to increase this percentage. The SEC had recognized the changing nature of the company by exempting EB&S

---

[95] Several years later, in 1962, the executives of Cuban Electric discussed, and received a legal opinion on, the hypothetical possibility that the properties would be returned by the Provisional Government, a notion they rejected. John Plunket, President of Cuban Electric, argued that they "should give no consideration to accepting the return of the properties," which "were taken under a law and a procedure which were perfectly valid under the Constitution in effect at the time. The only thing left to discuss is the compensation for the taking." "Memorandum to Mr. D. G. Lewis," Nov. 23, 1962 (John Plunket file). A legal opinion provided in December argued that the seizure was, under U.S. and international law, illegal without just compensation. Reid & Priest, "Memorandum," Dec. 26, 1962 (John Plunket file).

[96] American & Foreign Power, Annual Report, 1960, 7.

[97] *Wall Street Journal*, Jan. 26, 1961, 4.

[98] https://www.justice.gov/fcsc/claims-against-cuba, accessed Jan. 9, 2023.

[99] *Wall Street Journal*, Nov. 7, 1962, 2.

[100] In addition to the confiscation of its Cuban properties, A&FP had been forced to sell its Mexican properties to the government in 1959. It previously had lost its Argentine properties and was engaging in extended negotiations over compensation. American & Foreign Power Co., Annual Report, 1959.

from PUHCA in August 1960.[101] A&FP, still a subsidiary of EB&S, became a registered investment company in February 1961 and continued to shed, or be shorn of, its utility properties. In April 1967, EB&S and A&FP merged, adopting the name Ebasco Industries, which in 1969 was acquired by Boise Cascade, a paper manufacturer and distributor of office products.[102] Ebasco Industries at that time was described as "a closed-end investment company with subsidiaries engaged in engineering design, construction, and consultation in the utility, chemical and construction fields throughout the world."[103] There was no mention of foreign utilities, but the company still owned several South American utilities and other properties and investments it had purchased with compensation it received for other expropriations in South American countries. In 1973, Boise Cascade sold Ebasco Services, the engineering division of Ebasco Industries, to Halliburton.[104] Boise Cascade purchased OfficeMax, an office supply company, for $1.3 billion in 2003, taking the OfficeMax name for the new company.[105] Finally, in 2013 another office supply company, Office Depot, merged with OfficeMax, the new company keeping the Office Depot name, but also retaining the OfficeMax brand.[106] This corporate genealogy would be trivial were it not for the fact that at the time of this merger, Office Depot held the single largest claim against the government of Cuba for expropriations that occurred after Fidel Castro seized power in 1959.[107] At $267,568,413.62 (plus 6% interest from August 1960 – approximately $7.4 billion as of 2017), this claim represents nearly three times the next largest claim and is 14%

[101] U.S. Securities and Exchange Commission, *27th Annual Report, fiscal year ended June 30, 1961*, 87th Congress, 2 ns Session, House Document No. 269 (Washington: USGPO, 1962), 100–01.

[102] *Wall Street Journal*, Feb. 28, 1967, 4; Mar. 1, 1967, 25. *New York Times*, July 15, 1968, 43.

[103] *Wall Street Journal*, July 24, 1969, 13.

[104] After the U.S. Justice Department filed an antitrust suit, Halliburton sold the division to Enserch (previously Lone Star Gas) in 1976. Raytheon then purchased the division in 1993. From there the company begins to fade into obscurity. It was mentioned as a "legacy" company of Washington Group International in an article on the World Trade Center terrorist attack, in which the company lost 18 employees. Washington Group was later purchased by URS Corp, itself purchased in 2014 by AECOM, a Fortune 500 company. While Ebasco Services is not specifically identified as a subsidiary on the company's web page, its power business unit does claim Ebasco's heritage; accessed March 23, 2021, http://siteselection.com/ssinsider/special/sp010917b.ed3.htm; https://en.wikipedia.org/wiki/URS_Corporation; https://en.wikipedia.org/wiki/AECOM; accessed March 23, 2021, https://www.aecom.com/blog/meeting-worlds-need-power-innovation/

[105] OfficeMax sold the paper, forest products, and timberland assets of the company in 2004 to a newly-formed limited partnership that retained the Boise Cascade name. It is now fundamentally a building products company. *Mergent Industrial Manual, 2015, Vol 1* (New York: Mergent, 2015), 399.

[106] As of June 2020, the company reorganized and currently is a holding company using the named ODP Corp., accessed Jan. 9, 2023, https://money.cnn.com/quote/profile/profile.html?symb=ODP

[107] By 1976, Boise Cascade had shed all of its Latin American investments not related to forest products, leaving the Cuba claim as the only remaining asset from its merger with Ebasco Industries. *New York Times*, Nov. 17, 1972, 67; *Wall Street Journal*, April 25, 1973, 2; May 13, 1976, 12.

of all claims certified by the Foreign Claims Settlement Commission.[108] The claim had passed successively from Cuban Electric to A&FP, Ebasco Industries, Boise Cascade, OfficeMax, currently resting with Office Depot (now ODP Corp.). In December 2014, as the Obama administration moved to normalize relations with Cuba, the issue of these more than half-century old claims (and corresponding counter claims), seemed relevant. However, beginning in 2017, the Trump administration re-imposed much tighter restrictions, and the claims again became an historical curiosity. The issue possibly may be revived under the Biden administration, with the new administration signaling in 2021 that it would relax travel and remittance restrictions.[109]

5.9 Claims Under the Foreign Claims Settlement Act

In 1964, "the U.S. Congress directed the Foreign Claims Settlement Commission ... to determine the validity and amount of claims of U.S. nationals against Cuba based on losses resulting from the nationalization, expropriation, intervention, or other taking of properties between January 1, 1959 (the triumph of the Cuban revolution) and October 16, 1964 (the date of the program's authorization)."[110] The Cuban Electric Company clearly was covered by this action and a claim was duly filed.[111] In addition, Ebasco Industries filed a separate claim (later assigned to Boise Cascade).[112]

---

[108] Foreign Claims Settlement Commission (FCSC), Claim No. CU-2578; Decision No. CU-4122, accessed Jan. 9, 2023,https://www.justice.gov/fcsc/cuba/documents/1501-3000/2578.pdf. Of 8821 claims filed, 5913 were certified for a total of $1.9 billion, accessed Jan. 9, 2023, https://www.justice.gov/fcsc/claims-against-cuba Office Depot also inherited an $11.7 million claim for Cuban Electric mortgage bonds held by Ebasco Industries. FCSC, Claim No. CU-3548; Decision No. CU-3866 accessed Jan. 9, 2023, https://www.justice.gov/fcsc/cuba/documents/3001-4500/3548.pdf

[109] Accessed Jan. 9, 2023, https://home.treasury.gov/system/files/126/cuba_fact_sheet_20190906.pdf. Relations with Cuba will remain a contested political issue. See Nicholas Kristof, "The Embargo on Cuba Failed. Let's Move On," *New York Times*, Jan. 23, 2019. On Biden administration policies see Los Angeles Times, Feb. 12, 2021, accessed Jn. 9, 2023, https://www.latimes.com/politics/story/2021-02-12/biden-resume-remittances-travel-to-cuba

[110] Richard E. Feinberg, "Reconciling U.S. Property Claims in Cuba: Transforming Trauma into Opportunity," Brookings Institution Latin America Initiative, December 2015, 16–17. For information on the Foreign Claims Settlement Commission see https://www.justice.gov/fcsc, accessed Jan. 9, 2023.

[111] Foreign Claims Settlement Commission (FCSC), Claim No. CU-2578; Decision No. CU-4122, accessed Jan. 9, 2023, https://www.justice.gov/fcsc/cuba/documents/1501-3000/2578.pdf

[112] Foreign Claims Settlement Commission, Claim No. CU-3548; Decision No. CU-3866, accessed Jan. 9, 2023, https://www.justice.gov/fcsc/cuba/documents/3001-4500/3548.pdf

Ebasco's claim on behalf of A&FP against the Cuban government included the value of land ($114,000), mortgage bonds issued by the Cuban Electric Co. ($11.3 million, plus interest), and contractual rights to receive bonds to be issued by Cuban Electric to replace collateral ($29.9 million), for a total claim of $42.7 million. The Commission found the claims for land and the mortgage bonds to be valid but denied the claims for the contractual rights to receive bonds in the future.[113] The total certified claim was $11.75 million.

The original claim on behalf of the Cuban Electric Co. was for $323.6 million. It included the loss of utility plants ($285.3 million), other current assets ($30.2 million), other investments, losses on equipment sales, liability to suppliers, deferred debits, retirement plan funding, and resettlement of employees (which together summed to $8.1 million). The Commission agreed on the value of the utility plants, but denied many of the other claims, including that for memberships in tennis, country, and yacht clubs, and most claims relating to employee resettlement and retirement funding. The total loss was set at $319.4 million, from which was deducted debts owed to agencies of the government of Cuba ($41.1 million, including that owed to the Cuban bank, Financiera Nacional de Cuba). The Commission also deducted the $11.75 million already certified as a loss to Ebasco, making the total certified loss $266.5 million. After objections to the denials from Boise Cascade's lawyers, a new hearing was held, and a final decision was issued on August 19, 1970. The Commission denied most of the objections, but added $1.05 million to the original claim, making the final certified loss $267.6 million (roughly $24 million less that the book value declared by the company in Dec. 1958).[114]

[113]The decision was issued on Sept. 11, 1969. The Commission held that the holder of the collateral for the bonds, the Export-Import Bank, would be the proper claimant, but that as an agency of the U.S. government the Bank was not eligible to make a claim under the Act. FCSC, Decision No. CU-3866, 6.

[114]FCSC, Claim No. 2578. From Aug. 6, 1960, until the time of settlement of the claim, 6% annual interest was to be added to the claim. No funds have yet been paid out. Boise Cascade also inherited from Ebasco the largest claim in the China Claims Program due to the loss of A&FP's Shanghai Power Co. Its certified claim was $54 million. The China Program was completed when China agreed to a payment of $80.5 million to the U.S. government (roughly the amount of Chinese assets frozen in the U.S.). There were 381 certified claims totaling $197 million. Payments amounted to $1000 plus 39.03% of the claim. It does not appear that interest was paid. This forced Boise Cascade to initiate a campaign to find private Shanghai residents who owned shares of the company in 1949. *New York Times*, Jan. 28, 1979, A22; Fox Butterfield, "Boise Seeks Lost Shareholders," *New York Times*, Dec. 26, 1979, D1; FCSC, Completed Programs – China, accessed Jan. 9, 2023, https://www.justice.gov/fcsc/completed-programs-china. A spokesperson for Boise Cascade acknowledged receiving partial compensation for the loss of A&FP's China property but expressed skepticism that people would be buying Boise Cascade stock because of the outstanding Cuba claim. In addition, given the potential tax consequences, a company lawyer "said it was unclear whether the claim is, in fact, 'a potential liability or a potential asset.'" *New York Times*, Oct. 6, 1991, F15.

5.10 Assessing the Impact of Nationalization

The immediate impact of the revolution and nationalization on the production of electricity in Cuba was not dramatic. Output increased successively in every year from 1958 to 1963, with the exception of 1962, and total output of electricity from 1958 to 1963 increased by 18%.[115] While it is impossible to know what the trajectory of Cuban electrification would have been without the existence of A&FP and subsequent nationalization (a classic counter factual), some context is provided by analyzing the growth of electricity production before and after nationalization and by comparing electricity production in Cuba to that in several other Latin American countries. Figure 5.1 shows annual electric utility generation (excluding sugar mills and other industrial enterprises). Irrespective of obstacles faced by the utility over the years, there was steady growth throughout with an obvious but brief acceleration in the late 1950s (due to an A&FP investment boom), but the impact of nationalization is not evident in this figure. Figure 5.2 presents annualized cumulative growth rates in electricity generation (including sugar mills and other industrial enterprises) in Cuba and the U.S. The trajectories are similar until the Second World War, and the impact of the depression is obvious. There was divergence during the war years, followed by a return to steady post-war growth in

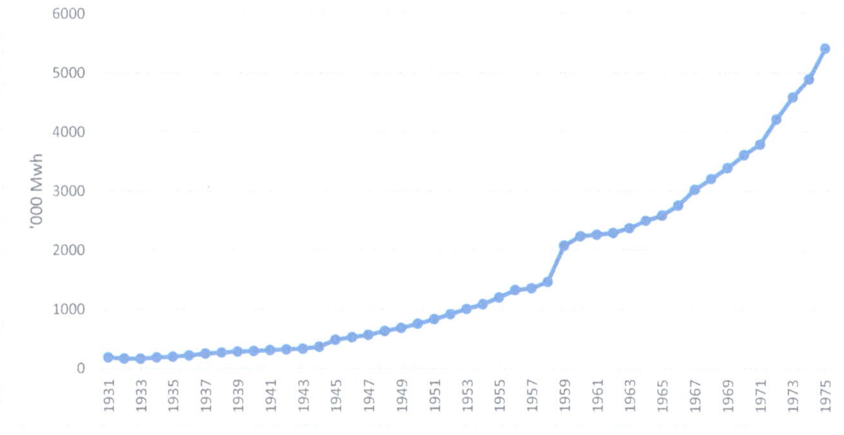

Fig. 5.1 Annual Electric Utility Generation in Cuba, 1931–75. (Source: Susan Schroeder, *Cuba: A Handbook of Historical Statistics*, Boston: G. K. Hall, 1982, 296.)

[115] Susan Schroeder, *Cuba: A Handbook of Historical Statistics* (Boston: G.K. Hall, 1982), 296. Sugar mills, many of which were foreign owned, were nationalized around the same time that Cuban Electric was nationalized. G.B. Hagelberg and Jose Alvarez, "Command and Countermand: Cuba's Sugar Industry under Fidel Castro," Association for the Study of the Cuban Economy, Nov. 30, 2006, accessed Jan. 9, 2023, https://www.ascecuba.org/asce_proceedings/command-and-countermand-cubas-sugar-industry-under-fidel-castro/. The second, sixth, seventh, and tenth largest claims certified by the FCSC were sugar companies, accessed Jan. 9, 2023, https://www.justice.gov/fcsc/final-opinions-and-orders-5

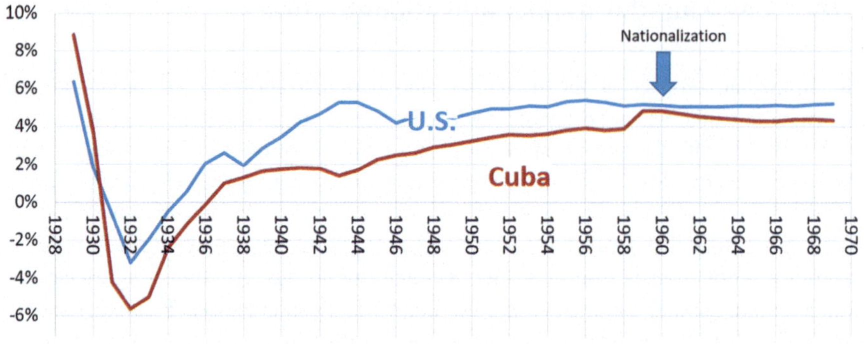

Fig. 5.2 Annualized Cumulative Growth Rates in Per Capita Generation, 1928–1969. (Source: Schroeder, *Cuba: A Handbook*, 296.)

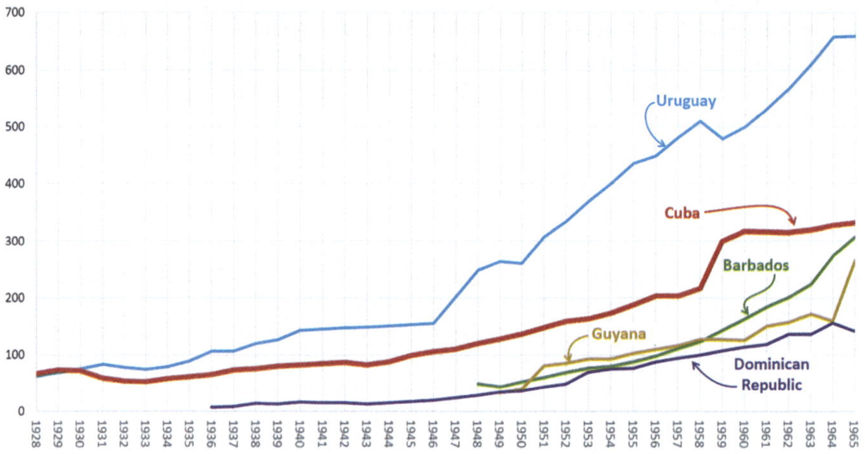

Fig. 5.3 Annual Per Capita Generation (kWh), 1928–65. (Source: B.R. Mitchell, *International Historical Statistics: The Americas, 1750–1988*, 2nd edition (New York: Stockton Press, 1993))

both countries until divergence again in the late 1950s. The investment boom in Cuba in the late 1950s is evident in this data, as is the gradual but not catastrophic decline in growth in Cuba after nationalization. Figures 5.3 and 5.4 put electricity production in Cuba in perspective by comparing it to production in four similarly situated countries from the late 1920s to 1965 and to six other countries from the early 1960s to 1988.[116]

[116]The electric utility in Uruguay was government owned throughout. The utility in Paraguay was taken over by the government in 1948, that in the Dominican Republic in 1955, that in Guyana in the 1960s, and that in Jamaica in 1974. Prior to the takeovers, utilities in these countries mostly were foreign owned. The utility in Barbados was foreign owned throughout. William J. Hausman, Peter Hertner, and Mira Wilkins, *Global Electrification: Multinational Enterprise and International Finance in the History of Light and Power, 1878–2007* (New York: Cambridge University Press, 2008), 32–33, Appendix B.

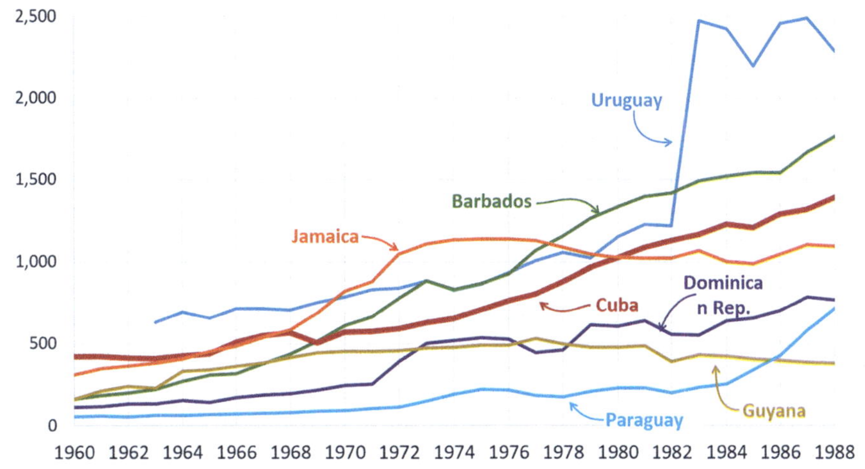

Fig. 5.4 Annual Per Capita Generation (kWh), 1960–88 (includes industrial generation). (Source: Mitchell, *International Historical Statistics*.)

Cuba does not appear to be distinctive either before or after nationalization. Figure 5.5 uses recent (2012–15) production and installed capacity data to calculate utilization rates, or how efficiently, on an average daily basis, the electric utility sector is using its installed capacity. Cuba falls toward the lower end of this scale, but not abnormally so.[117] The Cuban electric utility sector did not become a massive outlier after nationalization.

Because statistically significant breaks in the growth rate of electricity production may not be evident in the visual analysis, we performed an econometric analysis of the time series data for Cuban electric utility generation. Table 5.1 presents decadal summary statistics for the growth rate of electricity generation in Cuba by decade between 1931 and 1975. Electricity production by public utilities in Cuba grew by an average of 8.1% per year between 1931 and 1975. The greatest growth in electricity production in Cuba occurred in the 1950s, during which decade A&FP was expanding its investments in Cuba. Electricity production grew at an average annual rate of 12.1% per year in the 1950s. The largest single year of growth in electricity production, perhaps surprisingly, was the revolutionary year 1959, during which time electricity generation grew by 41.7%. In contrast, growth in electricity production slowed substantially during the 1960s, growing at an average rate of 5.1% per year. Growth in electricity production picked up briefly in the 1970s but did not reach the rates observed during the 1940s and 1950s. Based on a series of statistical

[117] Paraguay stands out as being particularly efficient, but it is unique in that it exports nearly 75% of it its production, which is entirely hydroelectric. *2015 Energy Statistics Yearbook*, Tables 30, 33.

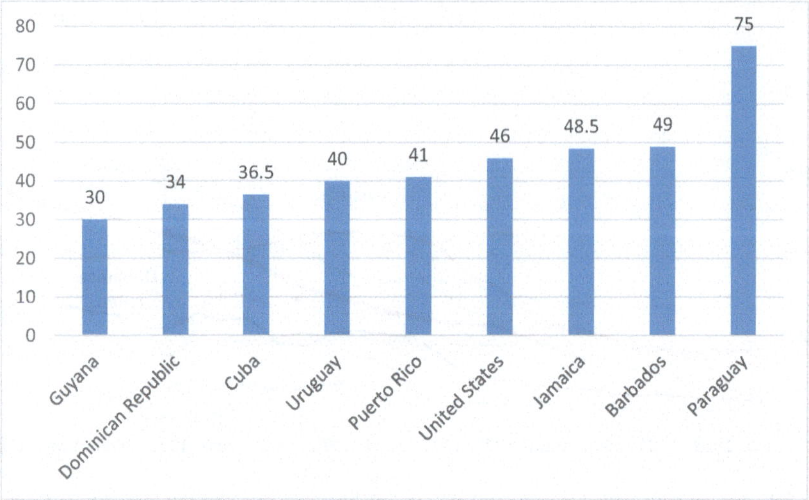

Fig. 5.5 Hourly Utilization of Installed Electric Capacity (kwh/kw) 2012–15. (Source: *2015 Energy Statistics Yearbook*, Table 31, https://unstats.un.org/unsd/energystats/pubs/yearbook/documents/2015eyb.pdf)

Table 5.1 Statistics for the Growth Rate of Cuban Electricity Generation, 1931–1975

	Years	Mean	Median	Standard Deviation	Minimum	Maximum
Whole sample	44	8.09	7.52	7.62	−9.74	41.70
1930s	8	5.39	6.58	7.71	−9.74	13.22
1940s	10	9.27	7.66	8.39	3.44	31.81
1950s	10	12.05	9.91	10.70	2.27	41.70
1960s	10	5.07	5.52	2.66	1.12	9.57
1970s	6	8.13	7.78	2.48	4.99	11.16

(Source: Schroeder, *Cuba: A Handbook.*)

tests, we conclude that the nationalization of the Cuban Electric Company in August of 1960 coincided with a statistically significant decline in the growth rate of electricity production in Cuba.[118]

[118] Unit root test results indicate that electricity generation in Cuba was stationary and slowly mean reverting around a permanent structural break in 1960, while the level of electricity generation grew according to a stochastic trend. The Lumsdaine-Papell and Bai Perron tests provide evidence that the growth rate of electricity generation in Cuba as stationary around breaks in 1946 and 1960. A brief explanation of unit root tests can be found at https://en.wikipedia.org/wiki/Unit_root_test, accessed Jan. 9, 2023.

5.11 Conclusion

The formation and transformation of Cuba's electrical sector followed a pattern that would have been familiar in many other developing countries around the world, especially in Latin America. Towns initially were electrified by companies established by local capitalists, frequently with the substantial support of foreign direct investors. By the end of the 1920s, the apex of the world-wide utility holding company movement, the Cuban properties had been consolidated into a single company, Compañia Cubana de Electricidad (Cuban Electric Company), by the globally expanding American & Foreign Power Company. In 1930 the Cuban Electric Company contributed more revenue to A&FP than utilities in all other countries in which the company owned properties (including Brazil, Mexico, Venezuela, Argentina, Chile, Colombia, Panama, Guatemala, Costa Rica, Ecuador, India, and China). Over the next 30 years Cuban Electric expanded service, touted repeatedly in the annual reports of A&FP. This expansion occurred, however, along with periodic turmoil related to the company's labor and pricing policies. In 1950 the World Bank organized an economic and technical mission to Cuba. Much of the commentary was quite critical of the electric power sector. The report cited the "high cost of public power" and urged industrial users, particularly sugar mills, to invest in isolated plants in order to "minimize stoppages due to hurricane damage and possible labor unrest." The report further argued that the company probably was "overcapitalized in relation to its real value."[119] A&FP consistently expressed optimism in its annual reports in the 1950s and was imagining expansion in Cuba on the eve of the revolution, including a large but ill-timed investment spurt at the end of the decade.

Beginning in the late 1940s and extending into the 1970s, the major trend in the electrical sector in Latin America was for foreign direct investments to be withdrawn, taken over by local interests, or nationalized by governments.[120] The process was sudden and dramatic in Cuba. The ultimate success of the Cuban revolution in 1959 brought a shocking nationalization of the company in 1960, thus abruptly ending Cuban Electric's position as a key element of a multinational enterprise. While electricity production in Cuba, both before and after nationalization, did not appear to be substantially different than that in several other Latin American countries, there is evidence that there was a statistically significant decrease in the growth of electricity production with nationalization, a situation most likely not imagined by the new government of Cuba. The Cuban revolution, of course, led to radical changes in the entire structure of the Cuban economy, so the slowdown in

---

[119]International Bank for Reconstruction and Development, "Report on Cuba," Washington, D.C., 1951.

[120]The landscape changed again in the late 1970s and 80s. Government-owned electric utilities were increasingly subjected to severe criticism. Governments began liberalizing policies, privatizing and restructuring utilities, and opening up again to foreign investment. Hausman, et al., *Global Electrification,* chapter 7.

electricity production may have been largely a reflection of general economic stagnation, and may well have had nothing to do with the business capabilities of Unión Eléctrica.

Bibliography

American & Foreign Power Company Inc. Annual Report. Various years

Dye, Alan, and Richard Sicotte. 2004. The U.S. Sugar Program and the Cuban Revolution. *Journal of Economic History* 64: 673–704.

Hagelberg, G.B., and Jose Alvarez. 2006. *Command and Countermand: Cuba's Sugar Industry under Fidel Castro*. Association for the Study of the Cuban Economy, Nov. 30

Hausman, William J., Peter Hertner, and Mira Wilkins. 2008. *Global Electrification: Multinational Enterprise and International Finance in the History of Light and Power, 1878–2007*. New York: Cambridge University Press.

Hausman, William J., and John L. Neufeld. 1997. The Rise and Fall of the American & Foreign Power Company: A Lesson From the Past. *The Electricity Journal* 10, 46: –53.

———. 1998. U.S. Foreign Direct Investment in Electric Utilities in the 1920s. In *The Free-Standing Company in the World Economy, 1830–1996*, ed. Mira Wilkins and Harm Schröter, 361–390. Oxford: Oxford University Press.

International Bank for Reconstruction and Development. 1951. *Report on Cuba*. Washington, D.C.

McDonald, Forrest. 1962. *Insull*. Chicago: University of Chicago Press.

McGraw Central Station Directory. 1929. McGraw-Hill Catalog and Directory Company, Inc., New York.

McGraw Central Station Directory and Data Book. 1925. McGraw-Hill Co., New York.

Mitchell, B.R. 1993. *International Historical Statistics: The Americas, 1750–1988*. 2nd ed. New York: Stockton Press.

Mitchell, Sidney Alexander. 1960. *Sidney Z. Mitchell and the Electrical Industry*. New York: Farrar, Straus & Cudahy.

Moody's Manual of Investments, Public Utility Securities. 1928. Moody's Investor Services, New York.

Moody's Public Utility Manual. 1965. Moody's Investor Services, New York.

Neufeld, John L. 2016. *Selling Power: Economics, Policy, and Electric Utilities Before 1940*. Chicago: University of Chicago Press.

O'Brien, Thomas F. 1993. The Revolutionary Mission: American Enterprise in Cuba. *American Historical Review* 98: 765–785.

———. 1996. *The Revolutionary Mission: American Enterprise in Latin America, 1900–1945*. New York: Cambridge University Press.

Panfil, Michael, Daniel Whittle, and Korey Silverman-Roati. 2017. *The Cuban Electric Grid*. New York: Environmental Defense Fund.

Piñón, Jorge R. 2004. *Cuba's Energy Challenge: Fueling the Engine of Future Economic Growth*. Institute for Cuban & Cuban-American Studies, Occasional Paper Series, March

Schroeder, Susan. 1982. *Cuba: A Handbook of Historical Statistics*. Boston: G.K. Hall.

Tremayne, Russell M. 1982. *The Cuban Electric Company: A Study of Expropriation*. M.A. Thesis, Boise State University

U.S.-Canada Power System Outage Task Force. 2004. Final Report. April. https://www3.epa.gov/region1/npdes/merrimackstation/pdfs/ar/AR-1165.pdf.

U. S. Federal Trade Commission. 1927. *Control of Power Companies*. 69th Congress, 2nd sess., Doc. No. 213. Washington: USGPO.

U.S. Federal Trade Commission. 1930. *Utility Corporations*. U. S. Senate, 70th Congress, 1st session, Document 92, parts 23 and 24. Washington: USGPO.

U. S. Federal Trade Commission. 1934. *Utility Corporations*. United States Senate, 70th Congress, 1st session, Document 92, part 66. Washington: USGPO.
U.S. Securities and Exchange Commission. 1944. Tenth Annual Report. Philadelphia, by the Commission.
Vernon, Raymond. 1971. *Sovereignty at Bay: The Multinational Spread of US Enterprises*. New York: Basic Books.
Williamson, Oliver E. 1981. The Economics of Organization: The Transaction Cost Approach. *American Journal of Sociology* 87: 548–577.
Woodward, Jr, and Ralph Lee. 1963. Urban Labor and Communism: Cuba. *Caribbean Studies* 3: 17–50.

William J. Hausman is Chancellor Professor of Economics, Emeritus, William & Mary. He has written extensively on the history of electrification in the United States and worldwide. He co-authored, with Mira Wilkins and Peter Hertner, *Global Electrification: Multinational Enterprise and International Finance in the History of Light and Power, 1878-2007* (Cambridge University Press, 2008).

John L. Neufeld is Professor of Economics, Emeritus, University of North Carolina, Greensboro. He has published numerous papers on the history of electric utilities in the United States, and was a member of the North Carolina Energy Policy Council. He is author of *Selling Power: Economics, Policy, and Electric Utilities Before 1940* (University of Chicago Press, 2016).

Rui Pereira is Lecturer of Economics, William & Mary. He is interested in applied policy analysis. His specialty is applied microeconomics, environmental and energy economics and applied computational, mathematical and statistical methods in economics. His research has appeared in the *Public Finance Review, Energy Economics,* and *Applied Economics Letters.*

Chapter 6
Between Material Dependencies, Natural Commons and Politics of Electrical Transitions: State as Networks of Power in Greece, 1940–2010

Stathis Arapostathis and Yannis Fotopoulos

Abstract This chapter examines the electrical system of Greece in the context of political priorities, and the propagation of particular sociotechnical visions by political actors and technical experts. We contend that state-led energy transitions shaped both the structure of the electrical network and their role and agency in state formation. Two distinct transitions have been identified, which are associated with the transformation of the state of modern Greece. We argue that the frames of decision making were shaped by narratives, visions, and expectations, which in turn legitimized particular public policies and technical choices. These choices ultimately defined the configuration and technical aspects of electrical networks. Within the framework and under the auspices of technocratic ideologies, nature has been subjected to governance, with successive governments striving to convert natural commons into national resources. Within the framework of the co-production idiom, we integrate the political, social, and technical aspects of the electrification process in Greece. We aim to analyze the competitive dynamics and interrelationships between the oil, lignite, and hydropower industries in the production of electricity since the Second World War. We argue that an examination of the politics surrounding the electricity grid provides insight into the formation of the electrical state. This involves an analysis of how the state was shaped through technological infrastructures, visions, and ideologies.

Keywords Natural resources · Greek state · Politics · Technological infrastructures · Electrical transitions · Lignite · Renewables · Hydrocarbons

S. Arapostathis (✉)
National and Kapodistrian University of Athens, Athens, Greece
e-mail: arapost@phs.uoa.gr

Y. Fotopoulos
National and Kapodistrian University of Athens, Peristeri, Greece
e-mail: iofot9@phs.uoa.gr

W. B. Carlson, E. M. Conway (eds.), *Electrical Conquest*, Archimedes 67,
https://doi.org/10.1007/978-3-031-44591-0_6

6.1 Introduction

Monday, Sept. 14, 2020 was hailed as a landmark day in the history of Greece's energy sector, since a milestone 51% of the electricity demand was finally covered by renewable energy sources, while 10 years earlier, renewable sources like photovoltaics were barely used and there existed only 1250 MW of wind energy installed capacity (see Fig. 6.1).[1] For almost six decades, the predominant source of Greece's energy generation has been lignite, producing not only significant energy, but also significant greenhouse gas emissions. Between 1990 and 2017, lignite accounted for 34% of the country's greenhouse gas emissions, one of the highest in the European Union. Lignite is therefore the key culprit in Greece's standing as one of the nations doing the most climate damage within Europe.[2] In an effort to reduce negative climate impacts, last August, the electricity production (Fig. 6.1) by lignite decreased by almost 60% and fell to a single-digit share of the country's energy production.[3]

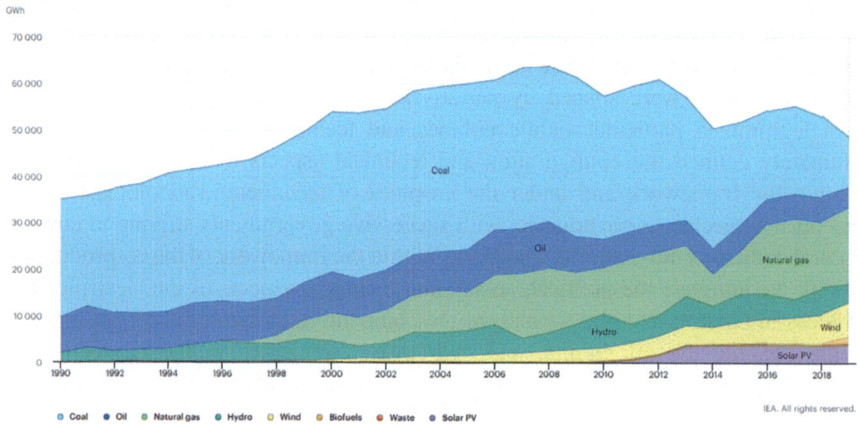

Fig. 6.1 Electricity generation by source, Greece 1990–2019. (Source: IEA, 2021, [Reproduced courtesy of IEA])

---

[1]D. Kadda, "Which investments "unlock" the package of 32 billion euros" (in Greek), *Capital.gr*, September 20, 2020, https://www.capital.gr/oikonomia/3482014/poies-ependuseis-xekleidonoun-to-paketo-ton-32-dis-euro; Ch. Floudopoulou, "PPC: The reasons that led to +315% from the low of 2019" (in Greek), *Capital.gr*, September 20, 2020, https://www.capital.gr/epixeiriseis/3481923/dei-oi-logoi-pou-odigisan-sto-315-apo-ta-xamila-tou-2019; WWF, "Even the market turns its back on lignite" (in Greek), February 14, 2019, https://www.wwf.gr/ta_nea_mas/?uNewsID=365925; WWF, "Chronicle of previously announced lignite damage in the Greek economy" (in Greek), *World Wide Fund for Nature (Greece)*, July 3, 2019, https://www.wwf.gr/ta_nea_mas/?uNewsID=824045

[2]F. Maurogiorgi, "The end of lignite-fired plants until 2028" (in Greek), *Energyin.gr*, September 24, 2019, www.energyin.gr/2019/09/24/τέλος-οι-λιγνιτικές-μονάδες-μέχρι-το-2028/

[3]Ch. Zafeirouli, "Public Power Company (PPC) . . . whipped off lignite" (in Greek), *Business Daily*, December 14, 2020, www.businessdaily.gr/oikonomia/30285_i-dei-exafanise-ligniti-epese-se-77-meridio-toy

Since the interwar period (1918–1940), the Greek energy sector has been characterized by its dependence on oil and coal. The coal, metallurgical coke (metcoke), anthracite, charcoal, and oil products that fulfilled a majority of Greece's energy needs, eventually accommodating 80% of total energy needs, were mainly imported from the United Kingdom such that Greece remained highly energy dependent.[4] After World War II (WWII, 1939–1945) and the Civil War (1946–1949), the dependence of Greece's energy sector on oil and coal increased further. The imports of oil and oil-based products were linked with investment conducted by local tycoons from the shipping industry in oil refineries. Trying to secure the market nationally they set business relations with major oil companies who aimed to control not only trade within Greece, but also with the Middle East and East Mediterranean. The overall energy status of the country showed very high dependence on oil, with 72% of primary energy consumption being oil (Fig. 6.2).[5] Greece rapidly increased primary energy production through the intensification of lignite extractions, but high net imports persisted and caused economic strain from 1975 to 2011.

The energy dependence of the country (ratio of net energy imports to total gross of domestic energy consumption) has changed at an average annual rate. The energy dependency of Greece rose to 67% while the oil dependency skyrocketed to 98% on average over the period 1990–2006 (Fig. 6.2).[6] Drawing on the analytical framework

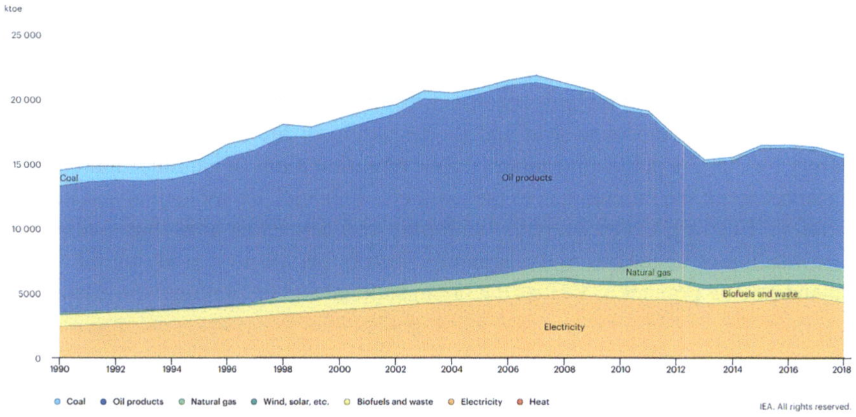

Fig. 6.2 An overview of the Greek energy sector with energy sources and their share in the consumption. (Source: IEA, 2021. [Reproduced courtesy of IEA])

---

[4]K. Kegel, "The investigation of the lignite problem of Greece. Summary of the report of prof. Kegel of the Freiberg / Sa Braunkohlen-Forschungs-Institut," *Τεχνικά Χρονικά* (April 1939): 313; S.N. Tsotsoros, *Energy and Development in the Postwar Period. The Power Public Corporation 1950–1992* (in Greek), (Athens: National institution of Research, 1995): 84–87; N. Pantelakis. *The electrification of Greece from the private initiative to the state monopoly, 1889–1956* (in Greek) (Athens: MIET, 1991).

[5]Tsotsoros, 1995, 57.

[6]Central Bank of Greece, Report of the Comander of Central Bank of Greece for 2008 (in Greek), *National Bank of Greece- Eurosytem* (Athens: April, 2008): 204, http://www.bankofgreece.gr/BogEkdoseis/ekthdkth2008.pdf

of co-production,[7] the present chapter studies the energy dependence and independence in relation to the electrical system along with political priorities, agendas and the promotion of specific sociotechnical imaginaries by political authorities and actors from the technical world, like engineers and utility managers.[8] We identify and study transitions of the electrical system, which means long-term transformations of more than 30 years, in order to integrate the making of the electrical system in broader energy transitions. In the Greek case, we have identified two distinctive transitions that emerged and linked to the state transformation. We claim that both transitions were state-led transitions, which means transitions that their governance and steering were dominated by a centralized organization, a state-owned company with the ownership of the relevant infrastructures. Concurrently this emphasis on transitions assists us to show the role of the electrical system in shaping the state priorities and to follow its structuration as a techno-political process. We show the importance of the establishment of narratives and visions in relation to technologies and the appropriate transformations of the network, and how they shaped new policy agendas and configured frames of decision making in public policy. In addition, we show how the state priorities and politics, understood nature as an important inseparable element of the state transformation over the long period covered in this chapter. Nature has become an object of governance, and the efforts of the different governments over time were towards the transformation of natural commons to national resources. In this process key figures- mostly technocrats and engineers- played an important role as entanglers[9] who either publicly supported and promoted visions and narratives of technical choices or they steered transitions of the electricity system towards specific directions. Those key actors participated in the politics of the electrical grid and the transition of the electrical system in the country by shaping public discourses, public representations of natural resources and narratives about the rational use of the country's financial, technological and natural resources. Existing historiography – with only very few exceptions- has studied the electrification of Greece from the economic history perspective, studying the financial and pricing policies as well as the organizational and institutional dimensions.[10]

[7] S. Jasanoff, ed., *States of Knowledge: The Co-production of Science and the Social Order* (London: Routledge, 2004).

[8] S. Jasanoff, and S.-H. Kim "Containing the Atom: Sociotechnical Imaginaries and Nuclear Power in the United States and South Korea," *Minerva* 47, no. 2, (2009): 119–146.

[9] E. van der Vleuten, "Radical change and deep transitions: Lessons from Europe's infrastructure transition 1815–2015," *Environmental Innovation and Societal Transitions*, 32 (2019): 22–32.

[10] Tsotsoros, 1995; Pantelakis, 1991. For approaches different from the economic historiography of electrification see A. Tympas, S. Arapostathis, K. Vlantoni, Y. Garyfallos, "Border-Crossing Electrons: Critical Energy Flows to and from Greece," in *The making of Europe's critical infrastructure*, eds. Per Hogselius, Anique Hommels, Arne Kaijser and Erik van der Vleuten (Palgrave, 2013): 157–181; S. Arapostathis and S. Karas, "Water management, expertise and Techno-politics in energy and agriculture in Greece, 1940–2014: the case of Acheloos River," *Hist.Technol.* 33 (2017): 179–204; S. Arapostathis and Y. Fotopoulos, "Transnational energy flows, capacity building and Greece's quest for energy autarky, 1914–2010," *Energy Policy* vol. 127 (April 2019): 39–50.

While those approaches have been important and have supplied the historical analysis with invaluable archival material, sources and comprehensive analysis, their focus on the social, political and material entanglements in the electrification process has been very minimal and incomplete. Under the analytical umbrella of the co-production idiom,[11] we bring together the political, social and technical dimensions of the electrification of Greece. Our approach shows that technical solutions were legitimized due to visions, policies and politics at the national level while they shaped political orders and participated in the national politics and the process of statehood.

The main focus of the chapter is the post-WWII years, yet our narrative draws on sources and linkages from the interwar period onward. We provide a well-organized section on the empowerment of major actors as well as the development of production and consumption percentages and the share of different energy sources and technologies in the Greek energy sector with an emphasis on power production. We reconstruct the competitive dynamic interrelation among the well-established oil, lignite and hydropower producers in the making of electricity system since WWII. We stress that the main electricity actors' emphasis was on the deployment of strategies for the demolition of oil's dominance in the electricity system through the support of lignite and hydroelectricity. Visions about the energy autarky of the country directed energy and public policies that resulted in the dominance of lignite in the 1980s. Lignite's use was legitimized in a context of the ideology of energy autarky and political populism that prioritized union politics in energy planning. In this context, while the politically and ideologically dominant discourse was that of autarky, security, and energy independence, it was transnational electrical and natural gas technological networks with Turkey, the Balkan countries, and Italy that increased the country' dependency on foreign energy sources and the technologies and systems relevant for their operation. Even when the integration of renewables – mostly wind technologies- in the energy sector was promoted under the quest for energy autarky and decarbonization, technological dependences have remained inescapable, since the technologies for these forms of energy production are produced abroad rather than in Greece. These 'hidden' dependencies may not appear in the forefront of the public representations and discourses, but they do powerfully shape the dynamics and politics of the electricity system and the energy sector and the introduction of niche innovations. We study the emergence of statehood by reconstructing the entanglements of electricity infrastructures, natural resources and technologies of energy production. We argue that by studying the politics of the electricity grid we unravel the making of the electrical state – the state as it was configured through technological infrastructures, visions and ideologies.

[11] Jasanoff, 2004.

6.2 Electrical Technologies as Politics by Other Means

6.2.1 From Interwar Sociotechnical Fragmentation to Postwar Technocratic Visions

The pre-WWII period was characterized by fragmentation, small-scale production units, and electricity networks that were established mainly for urban and suburban electric power demand. The factories predominantly used imported coal and petroleum. Such dependence on imported coal triggered the concerns of engineers, who started to explore potential exploitation of native natural resources such as lignite and water.[12] By WWI (1914–1918), no less than 20 cities in the Greek territory had established independent power factories. By 1939, there were 346 production units in the country, which provided electricity for only 438 municipalities, and corresponded to a mere 7.94% of the country's total electricity use. As a rule, electricity production was implemented by a number of private or municipal companies served the Greek territory.[13]

The installed power before WWII was at 65 MW and accounted for 52% of the total installed power (125 MW) of the country. Its power production in several cases was made with high-powered steam turbines. The fuels used were imported (coal-fired, oil-fired), while 8 small hydro-electric plants of 9.2 MW were established. Total electricity production in 1939 exceeded 310 MWh and overall consumption was estimated at 250.6 MWh.[14]

With the outbreak of World War II in Europe on 1st of September of 1939, the prospect of transporting primary fuels for energy production became unsafe and unstable and the Greek state was obliged to produce its own primary fuels for energy sources. The mining of a few well-known deposits of lignite began in Kalogreza (Athens) and Rafina in Attica, in Kymi (Evoia), and in Florina (Macedonia), due to the fact that the English coal supply was shrinking due to the war.[15] The post-WWII years brought major changes to Greece, impacting the industrial potential of the country, its infrastructures, and its technological systems. In the postwar energy sector, the imports from oil jumped to 86% of total consumption, while coal imports were reduced to 14%. In the post WWII period, the electrification of Greece was marked by the visions circulated during the War by local engineering elites, and actors affiliated with initiatives of foreign technical assistance from the United Nations and the Marshall Project.

In 1942, Theodoros Raftopoulos, technical advisor in the National Bank of Greece, argued that there existed two main sources necessary to achieve national energy autarky, hydropower and lignite. He insisted that while there had not yet been

[12]Pantelakis, 1991, 360–362.

[13]Pantelakis, 1991.

[14]Tsotsoros, 1995, 84–87.

[15]I. Papadimitriou, "The energy prospect of Greece" (in Greek), *ΕΝΕΡΓΕΙΑ 2002*, (Athens, 2002), 10–12.

a systematic attempt to exploit these sources, the post war period would be the most appropriate period to establish a program for the exploitation of hydropower.[16] Raftopoulos linked electricification with modernity and the industrialization of the country. He was a prominent member of a technocratic committee established by the Bank of Greece tasked with the study of finding a solution to Greece's energy needs. The dominant ideological inclination of this committee was a commitment to the technocratic ideals and rational organization of the Greek economy and society based on Taylorist ideas.[17] Energy was a privileged domain to promote such a modernist ideology through the development of a rationalist and technoeconomic evaluation of technical solutions. The committee was presided by K. Zavitsianos, director of the Bank of Greece and had acting direction of A. Diomidis, Head of the Council of the Bank of Greece. It comprised of leading upper- and middle-class elites, economists, engineers, agronomists, and industrial scientists of the period. The committee stressed the importance of the use of native sources in Greece as the only way to secure low-cost energy production. They identified lignite and hydropower as the only resources which merited investment that would lead to secure energy production within a nationwide grid, named the "National Energy Transmission Network". Extant surveys conducted by foreign experts had identified the rivers of Acheloos and of Aliakmonas as the most important sources of hydropower along with small rivers in the North Peloponnese. Ample lignite sources were identified elsewhere, in the area of Ptolemaida (Southwest Macedonia). The committee determined that the areas requiring high energy development and intensive energy consumption would be the metropolitan area of Athens, the zone on North Peloponnese that would comprise of Pyrgos, Patras, and Korinthos, and the metropolitan area of Thessaloniki in Central Macedonia. Those were the areas within the country that attracted the most industrial activity and thus the highest energy needs.[18]

Following the committee's conclusions, Raftopoulos identified as major priorities the establishment of hydropower stations in the rivers of Acheloos, Aliakmonas, and Ladonas. He stressed that those priorities emerged out of elaborate scientific research that questioned the then-established views that the hydropower potential of the Greek rivers was not appropriate for electricity production and built the grid based on that assumption. Furthermore, Raftopoulos suggested Greece harness the hydraulic potential of Lake Paralimni and the water fall of Anthidonos for hydropower production. Lake Paralimni is located in Central Greece, only 54 kilometers from Athens. The establishment of a hydropower station there would direct water through a pipeline from Lake Paralimni to the adjacent seafront. While there were drawbacks in this scheme, Raftopoulos insisted that it could be a critical infrastructure in order

---

[16] Th.I. Raftopoulos, "The energy economy and the sources of energy in Greece" (in Greek), *Οικονομικός Ταχυδρόμος (Economicos Taxydromos)*, April 13, 1942, 1.

[17] Tympas et al., 2013, 157–181.

[18] Th.I. Raftopoulos, *The national electricity network of Greece (in Greek)* (Athens: Argyris Papazisis, 1946), VI.

to cope with emergent 'peak load'[19] demands in the metropolitan area of Athens. He viewed the Paralimni hydropower station as a critical infrastructure that could contribute the amount of still-needed electricity that the current plants in Acheloos or Aliakmonas and the future plants on the borders of Epirus and Macedonia would not be able to provide.[20] The exploitation of the water of the Acheloos river went hand in hand with an extant grandiose scheme to reclaim Lake Mesologi in westerm Greece. The Acheloos river was conceptualized by Raftopoulos as a project offering benefits for both energy and agriculture.

In 1947, Romaidis, Laganas, and Mihalopoulos reported to the United Nations that Greece's only possible energy sources were hydropower, lignite, and wood. At the same time, they stressed that for neither hydropower nor lignite were there any analyses suggesting the exact potential of each of these resources to contribute to the energy needs of the country. Yet they argued that the production of electricity should be based on hydropower, while the production of thermal power should exploit Greece's lignite resources. They identified and tried to predict an increase from 1635kW/h (10^6) to 4132 kW/h of the future needs of the country over the next 20 years.[21] They suggested routes of transmission lines and distribution networks in the form of the "National Grid." The "National Grid" was distinguished by the several "electrical counties". The latter would comprise of high voltage networks of 66 KVolt, lower voltage transmission lines of 33 KVolts for the mid-range transmission within the different locales of the county, and low voltage distribution networks. At completion, they projected the Grid would be 1147 km while the length of the regional electrical counties networks would be around 678 km.[22]

In the pre-WWII period, as we have briefly shown above, the initial efforts and visions built by important entanglers were towards the exploitation of nature and the slow transformation of natural sources (mainly lakes and rivers and to a lesser extent lignite) towards natural commons, as integrated components of the future to-be-built national electricity grid.

6.3 Building the Electrical State on "White Coal"

As we have shown in the previous section, the exploitation of native natural resources was a key technocratic vision in the years after WWII that was shaped co-currently with design ideas for the grid. This emphasis had started during the

[19] Peak load describes a short period of time, as the demand for electricity is high and unpredictable, so peak electricity is more extensive. On the other hand, 'base load' refers to a stable and minimum demand for electricity over a 24-hour period.

[20] Ibid, 139.

[21] Romaidis et al., "Report on Energy Economy and Irrigation Projects" (in Greek) in Reports for the Exploitation of Resources in Greece, V.I Water and Energy Economy, *United Nations Relief and Rehabilitation Administration (UNRRA)* (Athens 1947): 186–187.

[22] Ibid.194.

interwar period when due to the expansion of electricity demand, waterfalls and rivers started to attract the interest of both local and foreign engineers. The Aliakmonas River, about 500 km from Athens, was considered by the American missionary Boucher as a possible backup source to the electrical system that provided electricity to the urban areas of Athens and Lavrio. Tripotamos and Vodas were considered for Thessaloniki and West Macedonia, according to the technical studies of the advisors from Stockholm (*Vattenbyggnadsbyran*). The domestic "white coal"[23] and the hydropower potential of the country was conceived of as a key source to build the backbone of a nation-wide network. Despite the fact that Aliakmonas was characterized by domestic experts as a utopian project due to the required expensive investment and low productivity yields it offered compared to thermal units, the general view was that, in order to maximize the efficiency of the thermal units, they needed to be backed by "white coal". Some seawater exploitation projects such as the Glafkos in Patra have been studied since the late nineteenth century, but due to the unsafe wartime environment and political upheaval, these studies were postponed. The Glafkos project was materialized only after the technical supervision of the German company *Allgemeine Elektricitäts-Gesellschaft AG* (AEG)[24] along with the help of a Greek engineer of National and Technical University, I. Kyriazis, during a period in which German techno-scientific capital had favorable conditions for establishing its connections to many relevant development plants and infrastructures.[25] The dictatorial regime of Metaxas (1936–1941) promoted a techno-political program under which infrastructures became of paramount importance. In Metaxas' view, the exploitation of nature was a prosperous activity not only for the electrification programme, but also for the establishment of heavy industry, thus further developing an industrialization programme, especially in the aluminum, ammonia, bauxite, and electrical industries. A key natural resource was the river Acheloos, one of the Greece's biggest rivers, located in the prefecture of Aitoloakarnania in central Greece.[26] The state, with its emphasis on the exploitation of natural commons, supported the scheme of the American enterprise Hugh R. Cooper and Co Inc. & Chemical Construction that made provision of three large scale dams (Kremasta, Prevetzan, Kriekoukion) in different locations along the river banks.[27] Even in Raftopoulos' early vision of an integrated national (mainland) network in 1942, hydroelectric power (see Fig. 6.3) was crucial for the expansion of

[23] D. Genidounia, "The utopias about Aliakmonas: Our technical problems" (in Greek), *Ελεύθερον Βήμα (Eleftheron Vima)*, September 27, 1930, 1; L.A. Kanellopoulos, "White Coal: The waterfalls of Tripotamos, a letter" (in Greek), *Ελεύθερον Βήμα*, June 17, 1937, 3; N. Broumis, "The White Coal: The national economy" (in Greek), *Ελευθερον Βήμα*, June 26, 1937, 3.

[24] A German producer of electrical equipment company founded in Berlin in 1883.

[25] M. Dimitriadou-Loumaki, *The German economic and political invasion in Greece the decade 1920–1930* (in Greek), Ph.D. Diss., University of Panteio, 2010.

[26] Arapostathis and Karas, 2017, 179–204.

[27] Economikos Taxydromos, "The Greek white coal: Contract for the exploitation of the power of Acheloos," February 19, 1940, 3; K. Laganas, "The energy economy of Greece," *Τεχνικά Χρονικά* (1–15 May, 1938): 429–444.

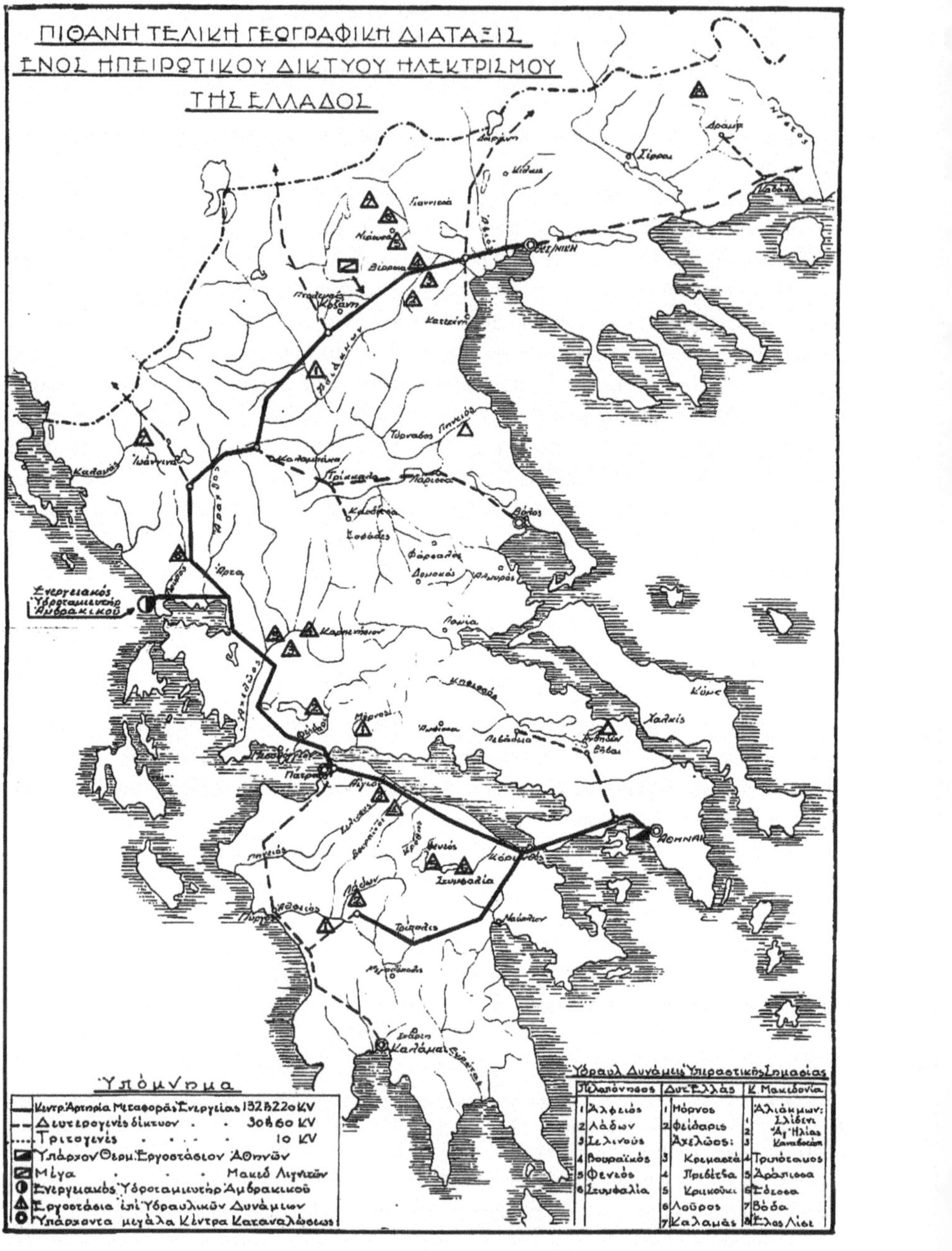

Fig. 6.3 Raftopoulos's early plans for the electricity grid. The route that he suggested passes through areas that are characterized by substantial hydropower potential, indicative of his interest in hydropower development. (Source: Raftopoulos, 1942: 32 [Reproduced courtesy of Technical Chamber of Greece])

the energy infrastructure; he allocated for hydropower the use of six rivers in Peloponnese (South Greece), seven in Macedonia (North-East Greece) and eight in Central-West Greece.

Until 1943, water exploitation was publicly promoted and represented as the most important resource for the energy economy: "The use of electricity in villages, agricultural crops and handicrafts will increase national incomes and raise the level of our culture, improving the social and economic conditions of the people's lives. Wide use of electricity presupposes that we can give the farmer a lot of electricity at a low price. This can only be achieved by exploiting domestic sources of energy. The energy sources of our place are, first of all, the waters, the waterfalls and then the lignite, the alcohol that comes from domestic products, the air and possibly other unexplored oil wells that some of the legends we've been hearing about lately".[28]

As Eythimis Tzamouranis, a leading journalist for *Eleftheron Vima*, on technological news from abroad (New York-USA), pointed out that, electricity production by the waterfalls and the exploitation of nature were seen as driving forces behind economic growth and the improvement of living standards:

> After the constant destruction of our forests, the waters unbridled, unhindered by the technical works of mountain hydrology, have become torrents and have caused incalculable disasters. There are lakes and swamps in the lowlands, outbreaks of malaria and other calamities that decimate the population ... Here is the water solution in the opinion of our technicians. Depending on the altitude, it is divided into three stages. In the mountains, technical works of mountain water and reforestation must be carried out in order to tame the torrents and to become calm streams that will bring wealth and happiness. Below where the waterfalls are formed, major technical works, dams and reservoirs will be built to regulate the fall for the normal operation of the power plants. These factories, the hydroelectric installations, will supply all of Greece with the electricity we need to acquire a truly national industry, to reform our agriculture and to raise the standard of living of our people. After that, the factories move and give us electricity, the water falls in the plains. Here's the third part of its mission and the work that needs to be done.[29]

The plans developed by local engineering experts provided the context for the social and political legitimization of technological solutions that would be designed and funded by the Marshall Plan that made the electrification of the country one of its priorities. A major area of the reconstruction in which US aid efforts had been concentrated over the final two years (1950–1952) of the aid, had been in the field of public utilities, including electricity production, the modernization of the water supply, and the sewerage systems of major cities. Greece consumed less electricity in comparison with the other countries of Western Europe. In many Greek cities, there was no electricity current. A survey of the Marshall Plan committee revealed that out of the 10,000 communities in Greece, only about 300 municipalities had

[28] E. Tzamouranis, "Water and our culture - The energy economy" (in Greek), *Ελεύθερον Βήμα* (August 15, 1943): 1.

[29] Ibid, 1.

electricity. Most of these were small villages whose electricity was produced by a small diesel engine that was operated by the local oil press or small sawmill.[30]

The Marshall Plan's long-term electrification plan for Greece envisioned factories powered by the flow of the waters of the major rivers as well as the exploitation of the domestic lignite resources. The plan envisaged the construction of three hydroelectric plants along the most important rivers in Greece and a thermal power plant in Aliveri, Evia. The lignite-fired power plant were to be fueled by lignite mines that were being prepared nearby.[31]

As we have already stressed, the emphasis on the exploitation of water resources for energy purposes continued to increase following the interwar period. The envisioned water policy was linked to an ideology of rational state organization that would be secured by engineering rationality. This was not something new in the culture of public policies in Greece. A blend of techno-rationalism and techno-nationalism had emerged since the interwar period when the use of water was first seriously studied. After WWII due to the American engineering advice of Electric Bond and Share Company (EBASCO) that was supported by the Marshall Aid, large-scale intervention and dam construction was integrated into the energy planning of the reconstructed energy sector.[32] The American representative and commercial agent atomic scientist Walker Lee Cisler (1899–1994) who introduced EBASCO to the energy reconstruction plan for Greece, was reassured that the energy program would maximize its capabilities under EBASCO's techno-economic studies. During a press conference he held with the Minister of Coordination and the President of the Coordinating Committee (Constantinos A. Doxiadis), he was asked about the role of hydroelectric in the reconstruction plan. Cisler stressed that Greece should maintain a "balanced" program in which hydroelectric and thermal units would have an equal share. Failure to do so would result in an imbalance which, in times of drought, would create irregularities in hydroelectric power and thus in the electricity system. In this context, lignite and hydroelectric power would be domestic electrification resources.[33] He also deemed lignite as the basis of the energy economy, meaning it was to be used more widely than electricity, covering the energy needs of transport, industry and more, as had been established as the "proper" paradigm in the United States and in Europe.

Under EBASCO's advice, Greece's energy policy was directed toward the exploitation of the native resources of water and lignite.[34] In EBASCO's report, the integrated electrical system was to be composed of hydroelectric plants, while the construction of thermal units depended on the study and exploration of new

[30] American Committee of Marshall Plan in Greece (ACMPG), *The Marshall Plan in Greece: the full account of the Marshall Plan aid to Greece: July 1948–January 1952 (in Greek)* (Athens:1952): 53.

[31] Ibid, 54.

[32] Arapostathis and Karas, 2017.

[33] To Vima, "The problem of hydroelectric power. Mr. Cisler recommends the use of lignite" (in Greek), *Το Βήμα (To Vima)*, August 30, 1949, 6.

[34] Embros "The report for the electrification was submitted for publication to the government by the Americans" (in Greek), *Εμπρός (Embros)*, February 28, 1950a, 6.

lignite fields. The company advised the exploitation of hydropower as a way to decrease the production cost by decreasing fuel imports. The report conceived of the rivers as national commons that should be considered as substantial natural resources for energy production within an integrated national grid.[35] In February of 1950, EBASCO submitted its proposals for the electricity network and the reconstruction of the power system. Four hydroelectric power stations and one thermoelectric one were proposed which would be based on the use of domestic lignite. This policy was implemented by the newly established Public Power Company (PPC), an organization socially and politically legitimized due to an emphasis on autarky that grounded in the intensification of electrification of provincial Greece and rural areas.[36] The expansion and further domestication of electricity[37] was developed in a deterministic argument that permeated the public discourses of the period by linking electricity to economic growth and progress.[38] At the end of the 1950s, the building of dams was conceived of as an ideal engineering solution. Any discussion of alternative technologies, most notably nuclear infrastructure, had been framed by contemporary journalists as an index of "hypermodernist privilege"[39] that would be expensive and far exceed the natural and budgetary capacities of the state. During the 1950s the Acheloos River in Central Greece emerged as an emblematic river with a series of dams on its banks until the early 1970s that made the river of a resource of national importance.[40]

In 1954, the inauguration of the first hydroelectric project of Louros, a project that was supervised by EBASCO and constructed by the French Omnium – Lyonnais (with the assistance of the Greek enterprise ETER) in Epirus (North Greece), became a hot spot for the technical elites and Royal family. In his speech, the President of the PPC, A. Dimitrakopoulos, highlighted the importance of this project, the first to be conceived of and delivered by Marshall Plan energy planning. In his view, hydroelectric power was vital to the industrialization of the suburban and rural areas of Greece, industrialization which would initiate their economic expansion. Dimitrakopoulos explained that the project was less important in the context of the country's energy planning but is of great importance to the local region as "It will serve mainly (power) production and then all other social needs. This will mean an increase in working and living conditions with the indistinguishable advantage of using domestic resources. Electricity is going to be plentiful, it is obvious".[41]

---

[35] EBASCO, *Electric Power Program of the Kingdom of Greece*, Electric Bond and Share Company (January, 1950), 2–1; To Vima, "The Embasco report on the electrification of our country" (in Greek), *Το Βήμα*, May 8, 1950, 3.

[36] Macedonia, "The electrification program is now entering a new phase" (in Greek), *Μακεδονία (Macedonia)*, August 15, 1957, 4.

[37] Ibid, 4.

[38] Embros, "Greece on the road to technological advancement" (in Greek), *Εμπρός*, December 14, 1957, 7.

[39] Empros, "Two symbols of national strength and wealth" (in Greek), *Εμπρός*, January 2, 1960, 6.

[40] Arapostathis and Karas, 2017.

[41] Technical Chronicles, "The inauguration of the hydroelectric project of Louros," *Τεχνικά Χρονικά (Technika Xronika)* 61, (July 1, 1954): 28–31.

It was the same period that the main electricity development program in northern Greece was designed and implemented to increase the productivity of Thessaloniki and other major cities of Macedonia, Thrace. The Agras program was named after the village by the river Vodas, near the Yugoslavian border and about 100 km northwest of Thessaloniki. The project was implemented with the construction of a hydroelectric plant at 40,000 kilowatts. A long power line, approximately 262 km passed through Macedonia, distributing electricity to the city centers of Thessaloniki, Serres, Drama, and Kavala, with substations that could distribute current to smaller cities nearby. These projects were scheduled for completion in December 1953. The plan also expanded the Vodas project with a construction of a large dam and another reservoir tank above Edessa. These construction projects and those of Agra and those of Ladonas in the Peloponnese were assigned to the Italian company Edison under the terms of the Greek-Italian Agreement on Economic Cooperation. The agreement provided that Greece would supply waste and raw materials to Italy, from which the Italian factories would manufacture machinery and spare parts.[42]

By the end of 1960s, exploitation of the rivers was understood to have driven national economic growth.[43] In this context, the dams and their related infrastructures have been presented to the public press as hypermodernist "symbols" of industrialization and as artifacts of abundance of electricity and of the country's development. The public press also characterizes the period 1950–1960 as the "Decade of Cosmogony" in order to highlight the importance of PPC projects in the fight for prosperity.[44] As we have shown those social representations of technological infrastructures and their modernists connotations were discursively and socially legitimized by the activities of foreign and local engineers and experts. The co-production of local and foreign expertise, was important in steering the implementation of crucial infrastructures (like the dams) in making nature economically exploitable and integrated into the expanding electricity grid. This practice was in line with the wider electrification industries in Europe and the US and acted as a hidden integration of Greece to the Western culture, practice and materialities.

6.4 In Pursuance of the Black Gold

The hydroelectric potential of Greece was emphasized as an optimal option during the energy reconstruction planning and the establishment of the National Grid, yet it was lignite that actually acquired the dominant position in the electricity system of

---

[42] ACMPG, 56.

[43] F. Kavvadia, "All rivers are exploited! A huge project to boost the national economy" (in Greek), *Ελευθερία (Eleftheria)*, September 18, 1966, 7.

[44] Embros, "1950–1960: A decade of cosmogony. The contribution of the PPC to the struggle for prosperity" (in Greek), *Εμπρός*, December 31, 1960, 18.

Greece.[45] Since the interwar period, there has been the continuous emphasis on the unexplored resources of Greece that emerged during the interwar period eventually became dominant after the war due to increasing American influence and the widespread view that lignite would make a significant contribution to the country's energy autarky.[46] Emphasis was placed on the reduction of imported coal and oil for electricity generation. In this regard, the American company EBASCO and American engineering practices have had a critical influence by stressing the advantages of solid over liquid primary energy sources.[47] While research on lignite sources, quantities, and sites had begun by the 1930s, nothing substantial was done by 1950. Yet the link between lignite and the industrialization of Greece and its post war reconstruction made headlines, dominating public discourses which were often shaped by foreign and local experts. American geologists, mineral engineers, and scientists who conducted research and surveys in West Macedonia in the Ptolemaida area, along with the Greek mineral scientist Ch. David, supported the view that the sources in the area of Ptolemaida were rich—the largest in the country.[48] The private company Greek-American Company of General Lignite Products acquired the rights for the extraction of the lignite with the 2092/39 Act of November 1939.[49] This company had already acquired the rights for the exploitation of lignite on May 30, 1930 following their successful bid at a public competition, beating 22 other mostly foreign competitors.[50] The predictions over the scale of the sources ranged from 143 M tonnes to 2 billion tonnes, garnering the attention of mineral and energy experts. The discourses about energy autarky legitimized the intense promotion of the use of lignite and its dramatic increase in the production of electricity over a period of the next forty years. While in 1938 lignite was responsible for only 8% of the power production, by the late 1970s, this percentage had increased to 80.9%. In the post-WWII period, the emphasis on the use of lignite was strong. In 1949, *EMBROS (ΕΜΠΡΟΣ)*, a leading journal of the period, published an article entitled "Lignite as factor for improving the standards of living",[51] which claimed lignite was poised to offset all imports of coal and oil. During the 1920s, lignite resources attracted the interest of reformist liberal Prime Minister Eleftherios Venizelos. He ordered researchers to identify the sites and extent of Greece's lignite resources.[52] This emphasis on using domestic lignite resources characterized the years after

[45] Tympas et al., 2013.

[46] Embros, "The program of hydroelectric projects is complied on the basis of which the use of lignite is required" (in Greek), *Εμπρός,* August 30, 1949, 6.

[47] Ibid, 6.

[48] Economikos Taxydromos, "Achelooss hydroelectric capabilities" (In Greek), *Οικονομικός Ταχυδρόμος,* February 26, 1940, 1.

[49] Embros, "From Kozani. The Exploitation of Lignite" (in Greek), *Εμπρός,* 18 October, 1949, 4.

[50] Embros, "100 million tons of lignite depocits, exist in the Kozani area" (in Greek), *Εμπρός,* 13 May, 1950b, 5.

[51] S. Drakopoulos, "The treasures of the Greek subsoil - Lignite as a factor in raising the standard of living" (in Greek), *Εμπρός,* November 20, 1949, 5.

[52] Ibid, 5.

WWII and the Greek Civil War. The American report submitted by EBASCO to the Greek government urged for the transition from oil-based energy production to lignite based electric production with the necessary construction of transmission networks.[53] Under American influence, research began to be conducted in the region surrounded Ptolemaida, as well as in Aliveri, and in Kymi on the island of Evoia. The geological research was conducted by the Research Station of Subsoil, which functioned under the leadership of geologists J. Papastamatiou and G. Aronis as a part of the Ministry of Coordination.[54] The research team under Papastamatiou that surveyed the area of Ptolemaida estimated the local lignite stocks at around 300 M tonnes. The research team under Aronis conducted surveys in Evoia using drilling machinery by EBASCO. The American consulting company had used Raymond Concrete Pile Company of Delaware for its expertise in drilling and for the provision of diamond core drilling machinery.[55] Mapping of the lignite stocks was deemed necessary for the technological transformation of existing oil-fired power stations.[56] The results of the research in conjunction with the emphasis of the American experts, as well as the prospects and visions of the use of native sources for power purposes, legitimized the expansion of the research into several regions of Greece from Serres and Mount Pagaion to the Peloponnese, the region around Pyrgos in the prefecture of Ilia, and the areas surrounding Corinth and Xylokastro.[57] In April 1951, the government reactivated the agreement with the Greek-American Lignite Company that was initially signed in 1939. This agreement made provision for the exploration of lignite sources in Ptolemaida in West Macedonia. The reactivation was pursued by the government in its attempt to further boost policies of extraction and exploitation of Greece's lignite resources.[58] In 1957, the public press *EMBROS* hosted an article entitled "Greece on the route to progress". It was there that the public investments for the electrification of the country – predominantly for lignite mining and dam construction- were acknowledged as important to secure Greece's integration into the "developing world":

> The era of underdevelopment ends now. The PPC with its constructions drives the Nation towards the highest levels of techno-economic development. The works implemented based on cutting edge scientific and technical knowledge that show how a nation can achieve abundance in electricity, the most important economic and cultural commodity of our century. Most importantly, the works do not produce electricity based on precious and rich sources like coal or oil but [based on] the neglected lignite and the unruly torrents.[59]

[53] Embros, 1950a, 6.

[54] Embros, "The mineral wealth of Greece. Scientific research is being conducted in all areas" (in Greek), *Εμπρός*, November 9, 1950c, 4.

[55] EBASCO, 1950, xiii.

[56] Embros, 1950c, 4.

[57] Ibid, 4.

[58] Embros, "The Implementation of the reconstruction work is accelerated" (in Greek), *Εμπρός*, April 27, 1951, 4.

[59] Embros, 1957, 4.

After the Marshall period, the exploitation of domestic lignite fields became a hot diplomatic theme for Greek governments, as the capital needed for the investment could not be provided by the domestic capital forces. Rejection for any further US economic assistance (Markezinis visit in Washington on May 7, 1953), forced Greek governments to turn to their capitalist allies in Europe—especially West Germany and France—in order to attract the much-needed capital for their industrialization and economic reconstruction program. This was made possible by a rekindling of their techno-diplomatic relations. The Erhard (the second Vice-Chancellor of W.Germany) - Markezinis agreement (on November 1, 1953) that resulted after much lobbying and political support by Bodosakis, a well networked Greek industrialist, initiated economic cooperation between the two countries on the basis of a $80 M (DM 336 M) subsidy. According to this agreement, German companies and their Greek associates would contribute to at least 10 mega-projects in different industrial sectors. The most prominent of these projects was the installation of a coal mining facility in Ptolemais along with a steam power plant to support the electrification program.[60] Despite American interests in the Greek mining industry, diplomacy prevailed and the US State Department consented to German contribution to the Greek economic program. Since the late 1950s, the extractive activities in the region of Ptolemaida and its energy-industrial plant were promoted as the infrastructures that secured the country's technological progress.[61]

During the first years of the Dictatorship in Greece (1967–1974), the willingness to continue intensive cooperation with Greece's European partners in the energy sector, such as the France and the West Germany, is notable. On the last day of the 1967 in the offices of PPC, the Minster of Coordination, Macarezos, the Minister of Industry and Energy, Kypraios, and the Ambassador of West Germany, Slitter, signed a contract for the provision of a lignite-fire power plant in Megalopoli in the Peloponnese. The AEG consortium was to supplement the machinery and build a 250.000 KW power plant, as well as provide the technical expertise and supervision of the overall implementation plan. In his speech, Colonel Kardamakis expressed his hope that "PPC will try to achieve the best performance of the project, on the one hand, on the basis of the superior technique of the Germans, and on the other, on the ability and quality of the AEG group. Technical Services (AEG) will assist and oversee the process".[62] This project was highlighted by the Minister of Coordination, Colonel Macarezos, as "the largest productive project ever undertaken in Peloponnese and one of the largest (projects) on Greece".[63] The construction of the project was made possible by a loan of DM 200 M from the German Government

[60] M. Pelt, *Tying Greece to the West* (Museum Tusculanum Press, 2006): 71–84.

[61] Embros, 1957, 7; Eleftheria, "The inauguration of the interconnection of the Greek and Yugoslavian electricity grids and the second Ptolemaida thermal power plant" (in Greek), *Εμπρός*, September 6, 1960, 8.

[62] Ta Nea, "The execution of ambitious projects begins" (in Greek), *Τα Νέα (Ta Nea)*, January 2, 1968, 10.

[63] Ibid, 10.

from 1961. The project concerned the exploitation and development of the lignite field through the installation of storage and transport equipment, and additionally, the construction of a 250 MV steam fired/power station. This station alone was projected to supply PPC's energy network equal to 1500 M kWh, an amount equal to Greece's electricity capacity back in the early 1960s.

A year later, in Lavrion, Junta officials celebrated the construction of another 300 MW (300.000 kw) liquid fired power plant. This grandiose project was funded by the French government with a loan of $20,8 M, and had a specific meaning behind it. As Colonel Macarezos mentioned in his inaugural speech (the unit was to become available/ integrated to PPC's network in 1973), this unit was a milestone for the Greek electricity sector, since PPC modernized its units by moving from the 70–150 kw to 300 kw. This would result in the company being modernized by establishing energy factories reflecting international standards and capacity.[64] In his mind, all these energy projects aimed to strengthen the country and provided a high standard of living: "We justifiably celebrate every foundation of an energy station ... Because every power plant enlarges and strengthens the national infrastructure, gives the country extra strength and achieves a higher standard of living".[65]

By 1978, lignite production was increasing by 17.7% annually.[66] The conservative government planned the expansion of lignite explorations in North and Northwest Greece, specifically in the region of Ptolemaida. The 1973 and 1978 energy crises contributed to the intensification of investment in the extraction and use of native lignite resources. The ten year development programme of PPC that was publicized in 1978 made provision for the dominance of lignite in Greece's electricity production while, at the same time, included nuclear power as one of the options, too. It was in this period that the American consulting company EBASCO surveyed Greece in order to advise the Greek government on the construction of a nuclear power plant. The project was initially postponed and later cancelled in the 1980s.[67] As we have shown in this section the promotion and policies of lignite extraction and use in the energy mix for electricity production informed the emergence of discourses of autarky and state sovereignty. Yet due to geopolitical power relations and agendas for the empowerment of the linkages of Greece with Western capitalist countries, as well as the lack of machinery production capabilities, the electrical system became technologically dependent in a way that was not mentioned in the public discourses. On the contrary, the self-reliance and sufficiency of Greece were prevailing to indicate the energy- related sovereignty of the state of Greece in the Balkans.

---

[64] To Vima, "The second Steam Electric Unit of Lavrio will operate in the March of 1973" (in Greek), *Το Βήμα*, May 14, 1971, 9.

[65] Ibid, 9.

[66] Macedonia, "Lignite will cover the Greek energy problem" (in Greek), *Μακεδονία,* June 1, 1978, 6.

[67] S. Arapostathis, A. Kandaraki, Y. Garyfallos, A. Tympas, "Tobacco for atoms: nuclear politics, ambivalences and resistances about a reactor that was never built," *Hist. Technol.* Vol. 33 (2017), 205–227; S. Arapostathis and Y. Fotopoulos, "Transnational energy flows, capacity building and Greece's quest for energy autarky, 1914–2010", *Energy Policy* Vol. 127 (April 2019), 39–50. 2.

6.5 The Independence Paradox: Battling for and Against Lignite from 1980 to 2010

During the period between 1968 and 1981, the influence of hydropower stabilized within the energy sector as it came to satisfy one third of electricity production, and oil's contributions drastically decreased from 23.7% to 6.8% with a corresponding increase in the contributions of lignite (Fig. 6.4).[68]

In the very early days of 1980, the conservative government of Constantine Karamanlis (1974–1980) developed a plan for the establishment of a series of lignite-based power production units. The plans were based on new extraction program in the lignite fields of West Macedonia that increased national lignite deposits by 45%. At the same time, there were concerns about the exploitation of water resources and the actual energy capacity they can contribute. Gradually hydropower conceived of as a backup for peak loads.[69]

The substitution of oil was a long-term priority in Greek politics that was influential during the 1980s, too. The Prime Minister of Greece from 1980 to 1981 was George Rallis, who used to be the Minister of Foreign Affairs during the accession process of Greece to the European Economic Community (EEC). Rallis' speech as Prime Minister at the opening Inauguration of the third PPC thermal power

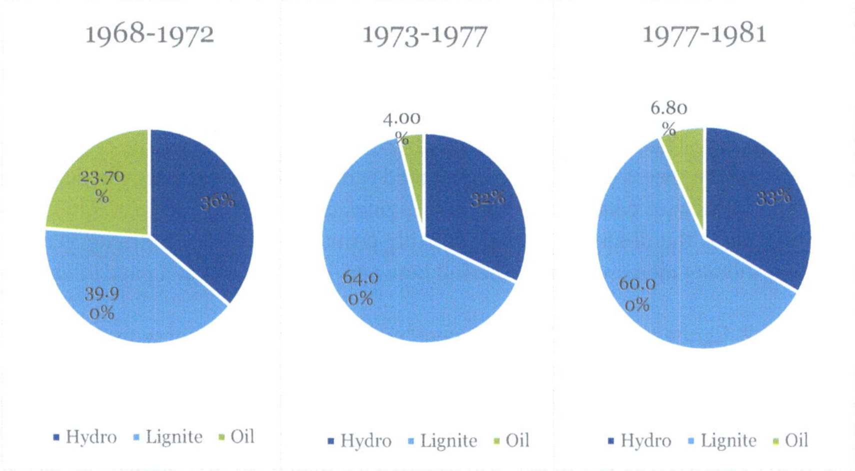

Fig. 6.4 Composition of sources in power production for the interconnected System for the years 1968–1981. (Source: Data adapted from Tsotsoros, 1995, 430)

[68] Tsotsoros, 1995, 168.

[69] Macedonia, "Another step towards the energy independence of the country" (in Greek), *Μακεδονία*, March 27, 1981, 9.

plant (Kardia) in Ptolemaida set the energy plan for his presidency.[70] The replacement of imported oil with cheaper energies, as well as the securitization of the supply of resources and the conservation of stocks were high priorities. The energy policy after the restoration of democracy was patchy, rushed, and haphazard according to Rallis. Previous governments had taken immediate measures to meet the country's short term energy needs, covering the energy demand with only liquid fuels and electricity. Rallis argued that one of the state's top priorities should be research for and investments in extractive industry and infrastructures for the exploitation of lignite resources.

In the aftermath of two oil crises in 1973 and 1979, the EEC directives highlighted the use of coal not only in electricity production, but also for industrial uses as a substitute of crude oil. The use of coal along with that of lignite in power production started due to pressure from the European Economic Community (EEC) over safety and diversification concerns in power production. Yet it complicated an energy policy that was politically and socially legitimized by an ideology grounded in the optimal use of indigenous sources, since it introduced an unacknowledged kind of dependency (that of coal) to a sector that had been structured by the ideals of energy autarky and independence. The use of coal would increase the substitution of petroleum products, as crude oil accounted for the 50% of the oil consumed in the 1980s. PPC had planned the construction of two coal-fired power plants for 1985 and the general plan was the replacement of oil-fired plants with coal-fired power stations. Scientific bodies such as the National Chambers of Greece, the PPC Engineers Union, and some political left-wing parties criticized this approach.[71] Due to these reactions in the early 1980s, PPC started to envision the use of coal in those stations that had already exhausted their lignite reserves, like Aliveri and Ptolemaida, stations whose actual operation could no longer achieve their initial power capacity. Those proposals remained controversial and under the scrutiny of the Union of Engineers of PPC who demanded further facts concerning the potential impacts of the use of coal on the stations' engines and boilers. Lignite continued to be supported as the dominant, unquestionable primary source of electricity generation. Investments increased for the period between 1970 to 2005 at a pace of around 8%.[72]

By 1992, lignite fueled 79.8% of Greece's energy use. This intense use of lignite continued since PPC had planned to increase lignite production by 13–15 million tons in order to fuel the new lignite fired power plants between 1991 and 2000.[73] In

[70] Macedonia, "In Kardia, Ralis analyzed the energy policy" (in Greek), *Μακεδονία,* November 23, 1980, 13.

[71] Rizospastis, "Safety issues of mixed coal-lignite combustion in PPC units" (in Greek), *Ριζοσπάστης (Rizospastis)*, December 28, 1983, 2.

[72] I. Drougas, K. Kavouridis, and N. Akyllas, "The Exploitation and Expansion of PPC Owned Lignite Deposits" (in Greek), Conference Paper in "Electricity Generation under Crisis?, Technical Chamber of Greece" (Athens: 1992), 2; K. Kavouridis, "Lignite industry in Greece within a world context: mining, energy supply and environment," *Energy Poli*cy 36, no.4 (2008): 1265.

[73] Drougas et al., 1992, 13.

the early years of the new millennium, PPC continued to invest in new mining and exploration infrastructures and in excavation technologies, aiming to increase the already very high productivity of the mines. The increased emphasis on lignite did not only reflect energy policy priorities but also a continuously increasing workforce in the mines. By 2000, 6452 of people were working in the mines; five years later, this number was reduced to 5694 in order to reduce cost of employment and improve the utilization of human capital. It was estimated that by 2008 around 15,000 people were employed in jobs directly or indirectly related to the mines.[74] This was an important factor in influencing decisions over the future of the lignite in the energy sector, due to the powerful lobbying of the miners' union and local communities. Since the early years of the millennium, lignite started to be considered a pollutant, mainly due to new European Commission legislation[75] and new Climate Change treaties like the Kyoto Protocol. Despite the fact that the energy sector planned to cease using lignite by 2050, the response from incumbent actors was ambivalent to concerns about pollution and still, in some cases, it remains dubbed the "national fuel" by policy makers and managers of the PPC.[76] In 1997, the Energy Committee of the Technical Chamber of Greece argued that

> The position of lignite for electricity generation in Greece is considered to be very favorable in the new setting. Provided that additional measures are taken by the PPC, such as better planning, cost reduction, etc. In the future, Lignite will be an economical, safe and, of course, domestic option, as there are new and sufficient deposits for exploitation. This choice also helps to tackle unemployment.[77]

It is indicative that the continuous increase of PPC tariffs between 2008 and 2014 resulted also from the cost of emission rights that the company had to pay due to its lignite-fired plants.[78]

Despite the dominance of domestic lignite in the 1980s and 1990s, there were key energy dependencies in the electricity system. The electrical installations in the Greek islands were highly dependent on foreign oil for power production. The imports from the interconnectors with Albania, North Macedonia (then part of Yugoslavia), and Bulgaria provided oil that was critical for the stability and security

---

[74] Kavouridis, 2008, 1265.

[75] European Commission Decision 2002/358/EC.

[76] Arapostathis and Fopopoulos, 2019; D.G. Pagoulatos, 2009, "Development of new thermoelectric units by PPC in western Macedonia" (in Greek), In "Optimal exploitation of lignite in power generation," *Technical Chamber of Greece* - Department of Western Macedonia, 11 May (Ptolemayda, 2009); IEA, *Energy Outlook*, International Energy Agency 2009, 14.

[77] M. Mpeskou, "Concluding remarks from the workshop: Lignite and fossil fuels in Greece" (in Greek), in "Workshop Proceedings: Lignite and fossil fuels in Greece," May 13–14,1997, Athens, in *Technical Chronicles*, 6 (Nov-Dec) 1997, 229.

[78] Ch. Kolonas, 2018. "Pollutants seriously damage the ... PPC fund" (in Greek), *euro2day* June 26, www.euro2day.gr/news/enterprises/article/1622086/oi-rypoi-vlaptoyn-sovara-to-tameio-ths-deh.html; A.G. Christodoulakis, "We will pay dearly ... for PPC's pollutants" (in Greek), *To Βήμα*, December 16, 2008, www.tovima.gr/2008/12/16/finance/tha-plirwsoyme-akriba-toys-rypoys-tis-dei/

of the electricity system. Interconnectors continued to integrate the national grid into transnational networks and increase the technological and energy dependencies Greece had with neighboring countries.[79]

Since the mid-1990s, voices from civil society, like the Non-Governmental Organization (NGO) Greenpeace Hellas, started to question the wisdom of investing in lignite to shape the electricity system. They questioned the efficiency of the lignite-fired plants while stressing the substantially high CO_2 emissions that made the electricity industry as one of the most polluting industries in Greece, contributing through the production of heat almost 50% of overall CO_2 emissions by sector. Natural gas and an expansion of renewable energy resources appeared to be the solution to achieve emission reductions. With the introduction of natural gas and the Renewable Energy Sources (RES) in the power production during the middle 1990s, the sources of dependence that structured Greece's electricity system changed.[80]

Lignite and its 19 lignite/coal-fired plants increasingly lost their dominance in Greece's energy sector, while NG (natural gas) established an elevated position in power production with 12 new power plants (Natural Gas Combined Cycle, Gas Turbine Combined Cycle).[81] The introduction of NG in power production was a priority forged by the European Union (EU) directives that insisted the share of oil in the energy sector must be reduced to less than 15% by 1995.[82] The majority of the (chemical) engineers of that time, who were the technocrats of the transition to the introduction of natural gas in Greece,[83] considered natural gas to not belong to the solid fuel category and therefore to be problematic and polluting for the atmosphere, despite being in line with RES as a "clean"[84] energy for the future. In this context, natural gas was introduced as a 'transit' resource fuel and as mediator for the zero or low carbon transition of electricity generation. This rhetoric and policy were purposive for the introduction of natural gas to electricity generation. Initially, the intention was to convert only the oil-fired PPC units, particularly those near the city center of Athens (St. George back up in Keratsini) that polluted the atmosphere and had been met with indignation and resistance from local civilians. It is not a coincidence that until 2001 all the imported natural gas, that was integrated through the Bulgarian pipeline interconnection and the LNG terminal, was used for electricity generation. In 2010, PPC was consuming more than 57.9% of the natural gas

[79] Tympas et al., 2013.

[80] Arapostathis and Fotopoulos, 2019.

[81] Th. Mpelidis, "Installed power, produced energy" (in Greek), in *Lignite and energy balance: Current situation - prospects, TEE* - Department of Western Macedonia, Kozani, 2–3 (October, 1992).

[82] E. Lekatsas, "The current 10-year PPC program and its revision study," in *Crisis in the electricity production of Greece* (in Greek), *TEE*, (Athens, June 18–19, 1992).

[83] Y. Fotopoulos, S. Arapostathis and P.J.G. Pearson, "Branching points and transition pathways in the Greek Natural Gas Regime, 1966–2016," *Environmental Innovation and Societal Transitions*, 32 (2019), 69–89.

[84] Z. Nivolianitou, "Natural gas in Greece and environment" (in Greek), *Τεχνικά Χρονικά* (July–August, 1991), 53.

imported in Greece.[85] While power generation has essentially established the natural gas industry in Greece over the last 25 years based on an agreement between the PPC and the public gas company (DEPA), its further incorporation into electricity production has been met with resistance from circles within the PPC, namely the lignite-lobby. That human capital, a workforce of 15.000 miners and lignite-based activities, had been stockpiled in the heart of PPC for over 50 years, created a protective net for their expertise and relevant technologies, and retarded and postponed drastic transformation of the electricity sector through at least 2018. The labor force ratio in the lignite units (the workers per kilowatt/hour) was roughly 1 to 4 relative to the gas power generation units. Although the gas fuel is more costly than the lignite alternative, PPC would have benefited from the overall cost of production and the environmentally friendly quality of the fuel.[86] In order to change this, it was deemed necessary for the electricity system undergo an internal organizational, political, and cultural transformation. The Regulatory Agency for Energy (RAE), an independent regulatory authority established[87] in 1999 concerned with electricity and gas, took firm steps towards the liberalization of the electricity and gas markets and the demolition of lignite-based electricity production. Its early Power Adequacy Study from 2003, anticipated a mediocre integration of natural gas and a slow decline of the lignite fire units on the assumption that by 2010, the installed capacity must exceed by 50% the capabilities of the year 2000 – roughly 11,000 MW—and by 2020, it must be expanded by another 26%. For this cause, at ceteris paribus, increased demand for electricity needed substantial investment in new power units. The vision and policy for 2030 in relation to a multiplicity of energy resources and technologies in electricity production that included natural gas, coal, lignite, RES, and small hydropower schemes was publicly formulated.[88] ADMIE,[89] the Independent Power Transmission Operator, in his studies, enacted the transition of the electricity production by the transformation of the lignite and oil power units, either to natural gas-based fueled units or the retirement of the old carbon-based units.[90]

The greening of the grid was implemented through the integration of RES into the electricity system. The idea to make use of Greece's wind potential had existed since at least 1982, when PPC established the first pilot and demonstration wind park on the island of Kythnos. Until the mid 1990s, wind energy had remained marginal since the installations were small scale. The pressures by the EU to liberalize the

---

[85] DEPA, *Annual report* (Athens, 2010), 23.

[86] Ch. Verelis, interview to Stathis Arapostathis and Yannis Fotopoulos (in Greek), (Athens, 24 May, 2017).

[87] Law 2775–007, adopting the Directives 2003/54/EC, 2003/55/EC.

[88] RAE, *General Information on the Greek Electricity Sector for the period 2000–2003: Installed capacity, production & consumption levels, Renewable Energy Sources and Long-Term Energy Planning*, Regulatory Agency for Energy (Athens, 2003): 12, www.hellasres.gr/Greek/THEMATA/REPORTS/rae_elec.greece.00-03.pdf

[89] Law 4001/2011, adopting the Directive 2009/72/EC.

[90] ADMIE, *Efficiency Study for the period 2013–2020 (in Greek)*, Independent Power Transmission Operator, (Athens: October 2013), 15–16.

electricity market and increase the percentage of renewable energy sources to 20,1% by 2010 provided new legitimacy to RES as a valid and economically feasible technical solution.[91] The imaginary for a greener grid was linked to industrial scale wind farms that promoted entrepreneurship, boosted a market-led governance of the energy transition, and configured discourses of economies of scale based on conceptions of the exploitation of natural commons. Industrial scale wind farms have been promoted and envisioned as integrating the wind potential of on-shore and off-shore regions into an interconnected network that also included the islands of the Aegean Sea, yet included no provision for domestic technology production. The emphasis on commodification of natural resources and engagement of the private sector in energy production increased the number of applications by private companies for the establishment of large-scale wind farms in those locations where the wind potential had been identified as appropriate and for wind turbines to function interruptingly and efficiently. The engagement of private companies and the commodification of natural commons with no real deliberation triggered ambivalences and tensions in local communities, particularly on the islands. Through the synergy of expert advice, NGO intervention, and centralized decarbonization policies emerged the hubris of large-scale wind farms whose infrastructures were difficult to integrate into the local environment.[92] The plans for the interconnection of the wind farms shaped a resource space that covered both the Ionian and the Aegean Sea as well as the island of Crete and the Cretan Sea. Discourses emerging in local communities configured a sociotechnical imaginary where technology was designed on a smaller scale.[93]

Since the late 1980s there is an ongoing transition with priorities in the environmental mitigation. Natural gas was enacted as a transit fuel, until the RES industry matures enough in the Greek setting and undertakes the full responsibility (if applicable) of the electricity production. In this ongoing last transition, the

[91] Directive 2001/77/EC of the European Parliament and of the Council of 27 September 2001,' Official Journal of the European Communities, L283/33–40.

[92] J.K. Kaldellis, "Social attitude towards wind energy applications in Greece," *Energy Policy*, 33:5 (March 2005), 595–602; C. Chadjilambrinos, "Development of RES in Greece," *Energy Policy*, 24: 6, 1996, 563–573; N. Danias, J.K. Swales, P. McGregor, "The Greek Electricity Market Reforms: Political and Regulatory Considerations," *Energy Policy* Vol 62 (November 2013), 1040–1047; G. Betzios, "Potential and Prospects of the Use of Renewable Energy Sources in the Autonomous Grids of the Aegean Islands" (in Greek), 5th National Conference on Renewable Energy Sources, *Institute of Solar Technology*, 6–8 (Athens: November 1996), 36–47; J.K. Kaldellis, M. Kapsali, Ev. Katsanou, "Renewable energy applications in Greece—What is the public attitude?," *Energy Policy*, 42 (2012), 37–48; H.D. Kambezidis, B. Kasselouri, P. Konidari, "Evaluating policy options for increasing the RES-E penetration in Greece," *Energy Policy*, 39 no.9 (2011), 5388–5398; V. Aggelopoulou, "*Aspects of the development of Wind energy in Greece: from history of technology to technological policy,*" (Ph. D. Thesis, National and Kapodistrian University of Athens, 2014); S. Arapostathis and V. Aggelopoulou, "Energy networks, Experts and the Management of Natural Commons in Battles over Industrial wind farms in contemporary Greece: Democracy and Technology. Europe under Tensions, 19th-twenty-first C." *Tension of Europe Workshop*, (1–3 October 2014).

[93] Arapostathis and Aggelopoulou, 2014.

transformation of nature to yet another national common is dominant. The natural resources to be exploited are the air, through the large-scale wind-turbines, the sun through the photovoltaic-farms and to a lesser extent the water. In this context, the public deliberation of this ongoing transition is on the sustainable path-creation that would allow Greece to regain its energy independence, since natural gas is an imported fuel, but also to retain a low/zero carbon emissions imprint, through the large establishment of RES. Once again, a hidden technological dependence is enacted through governance that have prioritized the measures for the climate change mitigation and the responses to EU pressures and directives.

6.6 Conclusion: A State in Between

In the present chapter, we study two identified transitions of the electricity system during the 70 years from 1940 to 2010, a period bookended by two major events, beginning with WWII and ending just before the 2010 financial crisis. The first transition that covers the years from 1940 to 1990 is the creation of a centralized network, the National Grid, that was built with the intent to use native natural resources of waterfalls, lakes, and lignite. The second transition has been ongoing since 1990s and is characterized by its emphasis on the decarbonization of the electricity system. We argue that in both transitions, the electrical state was shaped through the imaginaries and the politics of natural resources that emphasized exploitation of commons with large-scale infrastructures, yet without publicly addressing the issues of technological dependences that those infrastructures involve. Statehood as it is enacted through the electrical infrastructures, networks and policies, was built on an energy ideology that prioritized the use of domestic natural resources and the exploitation of natural commons. Yet it was not built as a political entity that was shaped from the top down. We argue that the co-production was materialised whereby experts participated in the technopolitics of system making and the shaping of the social order. Decisions made about energy resources linked energy to technopolitical agendas and to geopolitics in each period. We show how electrical networks and technologies functioned as technopolitical entities and participated in the state formation. This historiographic approach has assisted us to locate our actors in the political setting of each transition, identify their role and their political agendas and to show how their technocratic visions enabled the transformation of the electrical system while empowering the system to participate in the material politics of state-building. We understand better the activity and the agency of actors by embedding them organically in political processes. Our approach is inspired by Hughes's *Large Technical Systems*[94] approach, and shows the making of the national grid and how technology mattered in configuring meanings of natural resources and national natural commons. In this context, our emphasis is on the

[94] T. P. Hughes, *Networks of Power* (Baltimore: Johns Hopkins University Press, 1984).

views of domestic and foreign entanglers and their role in steering public discourses and policies towards resource independence and thus towards the transformation of a national common to a national natural resource. Through technological change natural resources acquired value and potential for exploitation to fuel and reinforce developmental patterns that boosted the increasing energy demand. Concurrently we show how this process was part and parcel of the political transformations that happened in the periods under consideration and we identify the roles, the visions and the technopolitical aspirations and agencies of local actors.

We show that the dominance of lignite originated as a priority in designing the national grid due to the influence of foreign experts and was implemented due to their authoritative function within a context of coproduction. While forms of economic rationalism prevailed in the 1950s and 1960s, the emphasis on domestic natural resources resulted from political agendas that were linked to governmental understanding of the role and potential of the state to direct energy policy for Greece. The ideology prioritizing the exploitation of domestic resources permeated both democratic governments and the seven years of the military junta (1967–1974). The dominance of lignite characterized the policies and informed the network design and architecture of the electrical grid. The dominance of the lignite in the 1980s was forged by political populism that invested in rhetoric about national resources and was reproduced through the establishment of networks of patronage with workers unions. At the same time, the eventual destabilization of lignite grew out of geopolitical dynamics, national diplomatic priorities, and European energy politics that emphasized the plurality of energy sources. This is the context for the legitimization of and investment in natural gas and wind energy and their relevant infrastructures.

The reconfiguration of the architecture of the grid to serve the needs of not only the mainland, but also the islands, and the integration of wind energy resources, triggered the politics of scale in projects and schemes for the establishment of wind parks in the Aegean and the Ionian Sea. Tensions emerged not because of the technology of the wind turbines but due to the sociotechnical plans for grid expansion towards the islands that would provide material infrastructure to set large scale wind parks and to exercise politics of resource nationalism rather than those of resource regionalism with the natural common of wind.

Our study shows that mediators like engineering experts, research centers, and technocratic committees played key roles in promoting and implementing energy policies and configuring energy politics. Following Erik van der Vleuten,[95] who has stressed the importance of entanglers in the making of transitions, we argue that Greek and foreign experts in different periods during the 70 years of our study, acted as entanglers. They triggered the transitions by setting the visions, promoting specific energy ideologies, and configuring the meanings of natural commons. They

[95] van der Vleuten, 2019, 22–32.

entangled electricity within the broader energy sector in Greece. While during the first transition they boosted the exploitation of water and lignite resources, during the second transition and more specifically in the early years of the new millennium, they boosted wind parks as a technology policy priority critical within the electricity sector in order to achieve both the energy autarky and the decarbonization of the National Grid.

The electricity system was coproduced by the privatization and commodification of power production, the commodification of natural commons, and the ideology of energy autarky. Yet the lack of any vision for domestic technology production provided an ironic dimension to the imaginary of energy autarky since the country would be heavily dependent on foreign wind and electrical transmission infrastructures. The expansion of the grid into the Aegean constructed the material articulation of the emergence of the neoliberal state as a new stage of statehood. While the quest for energy autarky influenced the construction of the energy policies in post-WWII Greece, it shaped material continuities of the state that political history[96] has not addressed. The investments in the exploitation of native natural resources that characterized the period from 1960s to 1980s combined with the privatization and commodification of natural commons in the new millennium. Because of the emergence of the neoliberal approach in state affairs, a new paradox has emerged. The financial crisis created the conditions for the empowerment of visions and imaginaries of energy autarky forged with hydrocarbon resources, contributing to the extensive explorations in the Ionian Sea and the diplomatic struggles for the Aegean and the area south east of Crete.[97]

[96] A. Liakos, *The Greek twentieth Century* (in Greek) (Polis: Athens, 2019); Y. Voulgaris, *Greece: A country paradoxically modern* (Polis: Athens, 2019); K. Kostis, *The spoiled children of history* (in Greek) (Pataki, 2018); N. Alevizatos, *Two steps forwards, one backwards* (in Greek) (Metxmio, 2020); S. Kalyvas, *Disasters and triumphs. The 7 cycles of modern Greek history* (in Greek) (Papadopoulos, 2016); S. Kalyvas, *Modern Greece: What Everyone Needs to Know* (Oxford University Press, 2015); G.B. Dertilis, *The history of modern and contemporary Greece* (in Greek), 1750–2015 (ΠΕΚ, 2019); W.H. McNeill, *The Metamorphosis of Greece Since World War II* (University of Chicago Press, 1st edition, October 1, 1978).

[97] A. Stergiou and M. Karagianni, "Does Energy Cause Ethnic War? East Mediterranean and Caspian Sea Natural Gas and Regional Conflicts" (Newcastle upon Tyne, United Kingdom: Cambridge Scholars Publishing, 2019); E. Konofagos, "Geopolitics as remedy to Attila-3: Large-scale lincensing for hydrocarbon in the Greek Exclusive Economic Zone," 2019, https://www.academia.edu/39052277/ΓΕΩΛΟΓΙΚΟ_ΦΙΑΣΚΟ_ΟΙ_ΠΑΡΑΝΟΜΕΣ_ΤΟΥΡΚΙΚΕΣ_ΓΕΩΤΡΗΣΕΙΣ_ΣΤΗΝ_ΚΥΠΡΙΑΚΗ_ΑΟΖ, accessed 5/1/2021; E. Konofagos, "Energy Surprises and Geopolitical Upheavals in the Eastern Mediterranean" (in Greek), Άμυνα και Διπλωματία (November 2015): 18–25, https://www.academia.edu/17949850/Ενεργειακές_Εκπλήξεις_και_Γεωπολιτικές_Ανατροπές_στην_Ανατολική_Μεσόγειο

Bibliography

ADMIE. 2013. *Efficiency Study for the period 2013–2020 (in Greek)*, Independent Power Transmission Operator. Athens: October.

Aggelopoulou, V. 2014. *Aspects of the Development of Wind Energy in Greece: From History of Technology to Technological Policy*. Ph. D. Thesis, National and Kapodistrian University of Athens.

Alevizatos, N. 2020. *Two Steps Forwards, One Backwards* (in Greek). Metxmio.

S. Kalyvas. 2016. *Disasters and Triumphs: The 7 Cycles of Modern Greek History* (in Greek), Papadopoulos.

American Committee of Marshall Plan in Greece – ACMPG. 1952. *The Marshall Plan in Greece: the full account of the Marshall Plan aid to Greece: July 1948 – January 1952* (in Greek), Athens.

Arapostathis, S., Aggelopoulou, V. 2014. *Energy Networks, Experts and the Management of Natural Commons in Battles over Industrial Wind Farms in Contemporary Greece: Democracy and Technology. Europe Under Tensions, 19th–21st C.* Tension of Europe Workshop, 1–3 October.

Arapostathis, S., and Y. Fotopoulos. 2019. Transnational Energy Flows, Capacity Building and Greece's Quest for Energy Autarky, 1914–2010. *Energy Policy* 127: 39–50.

Arapostathis, S., A. Kandaraki, Y. Garyfallos, and A. Tympas. 2017. Tobacco for Atoms: Nuclear Politics, Ambivalences and Resistances About a Reactor That Was Never Built. *Historical Technology* 33: 205–227.

Arapostathis, S., and S. Karas. 2017. Water Management, Expertise and Techno-politics in Energy and Agriculture in Greece, 1940–2014: the Case of Acheloos River. *Historical Technology* 33: 179–204.

Betzios, G. 1996. *Potential and Prospects of the Use of Renewable Energy Sources in the Autonomous Grids of the Aegean Islands* (in Greek). 5th National Conference on Renewable Energy Sources. Institute of Solar Technology, 6–8. Athens: November, 36–47.

Broumis, N. 1937. *The White Coal: The National Economy* (in Greek). Eleftheron Vima. June 26.

Central Bank of Greece. 2008. *Report of the Commander of Central Bank of Greece for 2008* (in Greek), National Bank of Greece-Eurosytem. Athens: April. http://www.bankofgreece.gr/BogEkdoseis/ekthdkth2008.pdf.

Chadjilambrinos, C. 1996. Development of RES in Greece. *Energy Policy* 24 (6): 563–573.

Christodoulakis, AG. 2008. *We will pay dearly . . . for PPC's pollutants* (in Greek). To Vima. https://www.tovima.gr/2008/12/16/finance/tha-plirwsoyme-akriba-toys-rypoys-tis-dei/.

Danias, N., J.K. Swales, and P. McGregor. 2013. The Greek Electricity Market Reforms: Political and Regulatory Considerations. *Energy Policy* 62 (November): 1040–1047.

DEPA. 2010. *Annual report*. Athens.

Dertilis, G.B. 2019. *The history of modern and contemporary Greece (in Greek), 1750–2015*. ΠΕΚ.

Dimitriadou-Loumaki, M. 2010. *The German economic and political invasion in Greece the decade 1920–1930* (in Greek). Ph.D. Diss., University of Panteio.

Drakopoulos, S. 1949. The treasures of the Greek subsoil - Lignite as a factor in raising the standard of living (in Greek). *Embros* 20: 5.

Drougas, I., K. Kavouridis, N. Akyllas. 1992. The Exploitation and Expansion of PPC Owned Lignite Deposits (in Greek). Conference Paper in "Electricity Generation under Crisis"? *Technical Chamber of Greece*. Athens.

EBASCO. 1950. Electric Power Program of the Kingdom of Greece. *Electric Bond and Share Company (January)*: 2–1.

Economikos Taxydromos. 1938. *The Greek White Coal: Contract for the Exploitation of the Power of Acheloos*, February, 19: 3.

Embros. 1949a. The Program of Hydroelectric Projects is Compiled on the Basis of Which the Use of Lignite is Required (in Greek). *Embros,* August 30: 6.

———. 1949b. From Kozani. The Exploitation of Lignite (in Greek). *Embros,* 18 October: 4.

———. 1950a. The report for the electrification was submitted for publication to the government by the Americans (in Greek). *Embros*, February 28: 6.
———. 1950b. 100 million tons of lignite depocits, exist in the Kozani area (in Greek). *Embros,* 13 May: 5.
———. 1950c. The mineral wealth of Greece. Scientific research is being conducted in all areas (in Greek). *Embros,* November 9: 4.
———. 1951. The Implementation of the reconstruction work is accelerated) (in Greek). *Embros,* April 27: 4.
———. 1957. Greece on the road to technological advancement (in Greek). *Embros,* December 14: 7.
———. 1960a. Two symbols of national strength and wealth (in Greek), *Embros.* January 2: 6.
———. 1960b. 1950-1960: A decade of cosmogony. The contribution of the PPC to the struggle for prosperity (in Greek). *Embros,* December 31: 18.
———. 1960c. The inauguration of the interconnection of the Greek and Yugoslavian electricity grids and the second Ptolemaida thermal power plant (in Greek). *Embros,* September 6: 8.
Economikos Taxydromos. 1940. *Achelooss hydroelectric capabilities* (In Greek). February 26: 1.
Floudopoulou, Ch. 2020. PPC: The reasons that led to + 315% from the low of 2019 (in Greek). *Capital.gr.* https://www.capital.gr/epixeiriseis/3481923/dei-oi-logoi-pou-odigisan-sto-315-apo-ta-xamila-tou-2019.
Fotopoulos, Y., S. Arapostathis, and P.J.G. Pearson. 2019. Branching points and transition pathways in the Greek Natural Gas Regime, 1966–2016. *Environmental Innovation and Societal Transitions* 32: 69–89.
Genidounia, D. 1937. The utopias about Aliakmonas: Our technical problems (in Greek). *Eleftheron Vima* 27: 1.
Hughes, T.P. 1984. *Networks of Power.* Baltimore: Johns Hopkins University Press.
IEA. 2009. *Energy Outlook.* International Energy Agency.
———. 2021. *Data and statistics: Greece.* https://www.iea.org/data-and-statistics?country=GREECE&fuel=Electricity%20and%20heat&indicator=ElecGenByFuel (Reproduced courtesy of IEA).
Jasanoff, S., ed. 2004. *States of Knowledge: The Co-production of Science and the Social Order.* London: Routledge.
Jasanoff, S., and S.H. Kim. 2009. Containing the Atom: Sociotechnical Imaginaries and Nuclear Power in the United States and South Korea. *Minerva* 47 (2): 119–146.
Kadda, D. 2020. Which Investments "Unlock" the Package of 32 Billion Euros (in Greek). *Capital. gr.* https://www.capital.gr/oikonomia/3482014/poies-ependuseis-xekleidonoun-to-paketo-ton-32-dis-euro.
Kaldellis, J.K. 2005. Social Attitude Towards Wind Energy Applications in Greece. *Energy Policy* 33 (5): 595–602.
Kaldellis, J.K., M. Kapsali, and Ev. Katsanou. 2012. Renewable Energy Applications in Greece—What is the Public Attitude? *Energy Policy* 42: 37–48.
Kalyvas, S. 2015. *Modern Greece: What Everyone Needs to Know.* Oxford University Press.
Kambezidis, H.D., B. Kasselouri, and P. Konidari. 2011. Evaluating Policy Options for Increasing the RES-E Penetration in Greece. *Energy Policy* 39 (9): 5388–5398.
Kanellopoulos LA. 1937. White Coal: The Waterfalls of Tripotamos, a Letter (in Greek). *Eleftheron Vima* June 17:3.
Kavvadia, F. 1966. All Rivers are Exploited! A Huge Project to Boost the National Economy (in Greek). *Eleftheria. September* 18: 7.
Kavouridis, K. 2008. Lignite Industry in Greece Within a World Context: Mining, Energy Supply and Environment. *Energy Policy* 36 (4): 1265.
Kegel, K. 1939. The Investigation of the Lignite Problem of Greece. Summary of the Report of prof. Kegel of the Freiberg / Sa Braunkohlen-Forschungs-Institut. *Technical Chronicles* (in Greek). April 1939.

Kolonas, Ch. 2018. Pollutants Seriously Damage the ... PPC Fund (in Greek). *euro2day*. https://www.euro2day.gr/news/enterprises/article/1622086/oi-rypoi-vlaptoyn-sovara-to-tameio-ths-deh.html.

Konofagos, E. 2015. Energy Surprises and Geopolitical Upheavals in the Eastern Mediterranean. (in Greek). *Defense and Diplomacy*. November: 18-25. https://www.academia.edu/17949850/Ενεργειακές_Εκπλήξεις_και_Γεωπολιτικές_Ανατροπές_στην_Ανατολική_Μεσόγειο.

Konofagos, E. 2019. *Geopolitics as remedy to Attila-3: Large-scale lincensing for hydrocarbon in the Greek Exclusive Economic Zone*. https://www.academia.edu/39052277/ΓΕΩΛΟΓΙΚΟ_ΦΙΑΣΚΟ_ΟΙ_ΠΑΡΑΝΟΜΕΣ_ΤΟΥΡΚΙΚΕΣ_ΓΕΩΤΡΗΣΕΙΣ_ΣΤΗΝ_ΚΥΠΡΙΑΚΗ_ΑΟΖ. Accessed 5/1/2021.

Stergiou, A., and M. Karagianni. *Does Energy Cause Ethnic War? East Mediterranean and Caspian Sea Natural Gas and Regional Conflicts*. Newcastle upon Tyne, UK: Cambridge Scholars Publishing.

Kostis, K. 2018. *The Spoiled Children of History (in Greek)*. Pataki.

Laganas, K. 1938. The Energy Economy of Greece. *Technical Chronicles,* 1–15 May.

Lekatsas, E. 1992. The Current 10-year PPC Program and its Revision Study. *Crisis in the Electricity Production of Greece* (in Greek). *TEE*. Athens, June 18–19.

Liakos, A. 2019. *The Greek 20th Century (in Greek)*. Polis: Athens.

Macedonia. 1957. The Electrification Program is now Entering a New Phase (in Greek). *Macedonia* August 15: 4.

———. 1978. Lignite Will Cover the Greek Energy Problem (in Greek). *Macedonia* June 1: 6.

———. 1980. In Kardia, Ralis analyzed the energy policy (in Greek). *Macedonia* November 23: 13.

———. 1981. Another step towards the energy independence of the country (in Greek). *Macedonia* March 27: 9.

Maurogiorgi, F. 2019. The end of lignite-fired plants until 2028" (in Greek). *Energyin.gr*. https://www.energyin.gr/2019/09/24/τέλος-οι-λιγνιτικές-μονάδες-μέχρι-το-2028/.

McNeill, W.H. 1978. *The Metamorphosis of Greece Since World War II*. University of Chicago Press.

Mpelidis, Th. 1992. Installed Power, Produced Energy (in Greek). *Lignite and Energy Balance: Current Situation – Prospects, TEE*. Department of Western Macedonia, Kozani, 2–3 October.

Mpeskou, M. 1997. Concluding Remarks from the Workshop: Lignite and Fossil Fuels in Greece. In Workshop Proceedings: Lignite and fossil Fuels in Greece, May 13–14. Athens, *Technical Chronicles*, 6 (Nov–Dec).

Nivolianitou, Z. 1991. Natural Gas in Greece and Environment (in Greek). *Technical Chronicles* July–August.

Papadimitriou I. 2002. The Energy Prospect of Greece (in Greek). *ΕΝΕΡΓΕΙΑ 2002* Athens: 10–12.

Pagoulatos, D.G. 2009. Development of new thermoelectric units by PPC in western Macedonia (in Greek). In *Optimal Exploitation of Lignite in Power Generation*. Ptolemayda: Technical Chamber of Greece – Department of Western Macedonia, 11 May.

Pantelakis, N. 1991. *The electrification of Greece from the private initiative to the state monopoly, 1889-1956 (in Greek)*. Athens: MIET.

Pelt, M. 2006. *Tying Greece to the West*. Museum Tusculanum Press.

RAE. 2003. *General Information on the Greek Electricity Sector for the Period 2000–2003: Installed Capacity, Production & Consumption Levels, Renewable Energy Sources and Long-Term Energy Planning*. Regulatory Agency for Energy. Athens. www.hellasres.gr/Greek/THEMATA/REPORTS/rae_elec.greece.00-03.pdf.

Raftopoulos Th.I. 1942a. The energy economy and the sources of energy in Greece (in Greek). *Economicos Taxydromos*. April 13: 1.

Raftopoulos, Th.I. 1942b. Energy economy and sources of energy in Greece / lecture by Th. Raftopoulos (in Greek). *Technical Chronicles* 1–15 Jan. – 1–15 Feb; iss. 241–244, σ. 28–34.

———. 1946. *The National Electricity Network of Greece (in Greek)*. Athens: Argyris Papazisis.

Rizospastis. 1983. Safety Issues of Mixed Coal-Lignite Combustion in PPC Units (in Greek). *Rizospastis*. December 28:2.
Romaidis, I., K. Laganas, and O. Michalopoulos. 1947. *Report on Energy Economy and Irrigation Projects (in Greek). Reports for the Exploitation of Resources in Greece, V.I Water and Energy Economy*, 183–232. Athens: United Nations Relief and Rehabilitation Administration (UNRRA).
Ta Nea. 1968. The Execution of Ambitious Projects Begins (in Greek). *Ta Nea* January 2: 10.
Technika Xronika. 1954. The Inauguration of the Hydroelectric Project of Louros. *Technical Chronicles* 61: 28–31.
To Vima. 1949. The Problem of Hydroelectric Power. Mr. Cisler Recommends the Use of Lignite (in Greek). *To Vima*. August 30: 6.
———. 1950. The Embasco Report on the Electrification of Our Country" (in Greek). *To Vima*, May 8: 3.
———. 1971. The Second Steam Electric Unit of Lavrio will Operate in the March of 1973 (in Greek). *To Vima*. May 14: 9.
Tsotsoros, S.N. 1995. *Energy and Development in the Postwar Period. The Power Public Corporation 1950–1992 (in Greek)*. Athens: National Institution of Research.
Tympas, A., S. Arapostathis, K. Vlantoni, Y. Garyfallos. 2013. Border-Crossing Electrons: Critical Energy Flows to and from Greece. In *The Making of Europe's Critical Infrastructure*, eds. Per Hogselius, Anique Hommels, Arne Kaijser and Erik van der Vleuten. Palgrave: 157–181.
Tzamouranis, E. 1943. Water and Our Culture – The Energy Economy (in Greek). *Eleftheron Vima* August 15: 1.
Verelis, Ch. 2017. Interview with Stathis Arapostathis and Yannis Fotopoulos (in Greek). Athens May 24.
Van der Vleuten, Erik. 2019. Radical Change and Deep Transitions: Lessons from Europe's Infrastructure Transition 1815–2015. *Environmental Innovation and Societal Transitions* 32: 22–32.
Voulgaris, Y. 2019. *Greece: A Country Paradoxically Modern*. Polis: Athens.
WWF. 2019a. *Even the market turns its back on lignite* (in Greek). https://www.wwf.gr/ta_nea_mas/?uNewsID=365925.
———. 2019b. Chronicle of previously announced lignite damage in the Greek economy (in Greek). *World Wide Fund for Nature (Greece)*. https://www.wwf.gr/ta_nea_mas/?uNewsID=824045.
Zafeirouli, Ch. 2020. Public Power Company (PPC) ... whipped off lignite (in Greek). *Business Daily*. https://www.businessdaily.gr/oikonomia/30285_i-dei-exafanise-ligniti-epese-se-77-meridio-toy.

Stathis Arapostathis is Associate Professor in the National and Kapodistrian University of Athens, Greece. He has published extensively on the history of intellectual property in energy and telecommunication technologies, the history of energy transitions, the governance of sustainability transitions and on science and technology policy. With Graeme Gooday he coauthored the book *Patently Contestable* (MIT Press, 2013).

Yannis Fotopoulos is a PhD student at the National and Kapodistrian University of Athens, Greece, where he is researching the Transition and Governance of Socio-technical Infrastructures. His research has been funded by the Sylff Association and the Hellenic Foundation of Research and Innovation (No.16232).

Chapter 7
Large-Scale Renewables and Infrastructure Gatekeepers: How Local Actors Shaped the Texas Competitive Renewable Energy Zones (CREZ) Initiative

Julie A. Cohn

Abstract In the early twenty-first century, Texas leads the country in renewables for electric power. In 2005, under a Republican governor and a closely divided legislature, the state defined priority areas for renewables development called Competitive Renewable Energy Zones; and invested heavily in new high-voltage transmission infrastructure to bring wind power into the market. The entire process, from the embrace of renewable energy resources to the activation of the final transmission line, involved citizen engagement. An examination of this initiative indicates that the move toward renewables will not be a straightforward transition, that multiple perspectives undermine the notion of a single technical imaginary, and that it is time to introduce new categories of stakeholders into the history of electrification. Noting the role of local governments, landowners, and regional organizations, two themes emerge. First, stakeholders may have broadly shared explicit costs and implicit benefits, but their intervention in energy system development is built around decidedly local concerns. Second, this broadened group of stakeholders became gatekeepers for one of Texas' most far-reaching state-sponsored projects. The evidence from contemporary pipeline and powerline projects suggests that this will be increasingly true.

Keywords Texas · Renewable Energy resources · Infrastructure · Power grid · Transmission · Energy imaginary · Stakeholders

7.1 Introduction

In the early twenty-first century, Texas, a state notorious for its oil and gas industries, its high per-capita energy use, and its conservative government, leads the country in renewables for electric power. In 2005, under a Republican governor and a closely divided legislature, the state defined priority areas for renewables development

J. A. Cohn (✉)
Center for Public History, University of Houston, Houston, TX, USA

W. B. Carlson, E. M. Conway (eds.), *Electrical Conquest*, Archimedes 67,
https://doi.org/10.1007/978-3-031-44591-0_7

called Competitive Renewable Energy Zones (CREZ); and invested heavily in new high-voltage transmission infrastructure to bring wind power into the market.[1] The entire process, from the embrace of renewable energy resources to the activation of the final CREZ line, involved citizen engagement. Of those who aggressively endorsed this initiative, some envisioned a transformed climate-friendly future while others pursued a more locally imagined economic boost. Others resisted wind turbines and transmission towers, citing generalized anxiety about ecological change as well as specific objections to lost property values, disrupted landscapes, and displaced natural gas revenues. An examination of the CREZ initiative, within a longer history of power line development in the United States, indicates that the move toward renewables will not be a straightforward transition, that multiple perspectives undermine the notion of a single technical imaginary, and that it is time to introduce new categories of stakeholders into the history of electrification.

From experimenters to advocates to users, many actors shaped energy transitions across human history. In the past two centuries, inventors, investors, marketers, regulators, and customers together realized shifts from one energy resource to another for lighting, heat, and motive power. As David Nye argues elsewhere in this volume, these shifts were not wholesale transitions but rather complex aggregations of resource opportunities, technologies, and social choices determining when, where, how much, and for what any form of energy was used. The Texas CREZ case illustrates quite strikingly that the advent of utility-scale wind represented an expansion of the array of resources used to generate electricity, in a context of increased overall energy use, and that a relatively unseen category of stakeholders – those located along the routes of power lines – deeply informed the process. In a similar vein, Stathis Arapostathis and Yannis Fotopoulos in their chapter describe the growing variety of energy resources used for electrification in Greece between 1940 and 2010, in that nation's quest to become both modern and energy independent. They document the roles of multiple individuals and entities, particularly technocrats and island residents, who influence the direction and pace of change.

Taken together, the Texas and Greece cases suggest that the concept of an energy transition within a few decades, as Nye explains, is misguided. A more complex process is underway in which there may be displacement of an undesirable energy source used to generate electricity, for example coal in Texas and lignite in Greece, by a different energy source. But the legacy energy sources remain, and multiple factors influence the reorganized energy regime. Growing demand overall, for example, may entail continued use of particular energy resources because they are cheaper, more fuel efficient, or better suited to a task. In Texas, a boom in hydrofracturing pushed natural gas prices down starting in 2008. This cheaper and

[1]Research for this project was funded, in part, through a contract with Rice University's Baker Institute for Public Policy. A related publication, with some parallel information, may be found at Julie Cohn and Olivera Jankovska, "Texas CREZ Lines: How Stakeholders Shape Major Energy Infrastructure Projects," Center for Energy Studies, Rice University's Baker Institute for Public Policy (2020). https://doi.org/10.25613/261m-4215

newly abundant resource started to displace coal in the state's energy mix at the very same time that new renewables came online. During these years, total power generation grew. More significantly, CREZ-supported utility-scale wind helped electrify the burgeoning hydrofracturing operations that produced the cheap natural gas. As Cyrus Mody offers in his chapter, deep entanglements across energy sectors have been largely unexplored, yet they offer insight into how transitions might unfold. In this study of CREZ, the multiple economic, environmental, and aesthetic concerns of local communities further complicate the transition story.

Within a broad technical imaginary there are many visions of what the future holds. In 2005, Texas legislators, wind and environmental lobbyists, and investors proposed a wind-enriched future in which there was cleaner electricity to be generated, money to be made, and continued ready power to be enjoyed. Texans living, working, and recreating along the routes of the future CREZ lines already accessed relatively inexpensive, abundant, and very reliable electricity from the state's grid. They might have conceptually favored renewables, or resented the arrival of power lines that seemingly bypassed them to deliver electricity to urban and industrial centers, but these were not the most compelling concerns they discussed before state agencies. Instead, rural Texans in the Panhandle actively sought CREZ zones and lines for very local economic reasons, while rural Texans in the Hill Country fought the new infrastructure for equally local economic and aesthetic reasons. In a sense, the Texans responded to the slow and complex energy transitions underway out of local preference over grander visions for the future.

Laws, policies, and practices have made an increasingly broad array of stakeholders important in determining the size, shape, cost, location, and meaning of transmission technologies. Just as transmission lines tend to fade from view over time, so those who live face-to-face with this physical infrastructure tend to be absent from power histories. Their interventions over the past have been less well documented than, say, opponents of dams or fans of solar power. Yet their actions can be found. In Texas, the abundant public documentation of lawmaking, regulatory action, system operator planning, and transmission service provider reporting offers a wealth of perspectives on the state's CREZ project. Under a 1987 statute, the state regulator maintains a growing online database of thousands of transmission line cases. Within the seventy-two cases filed for the CREZ initiative, intervenors from traditional power generators to government officials to local landowners and others offer their ideas about where, when, and how power lines should run across the state.

Historians of electricity, electrification, and power systems have drawn larger and larger circles around the constituents who make and maintain those systems. One strand of scholarship focuses on inventors, the electricity companies, and utility politics.[2] Another strand offers a growing analysis and understanding of large-scale

---

[2] A few examples include W. Bernard Carlson, *Tesla: Inventor of the Electrical Age* (Princeton: Princeton University Press, 2013); Sarah Elkind, *How Local Politics Shape Federal Policy: Business, Power, and the Environment in Twentieth Century Los Angeles*, The Luther H. Hodges Jr. And Luther H. Hodges Sr. Series on Business, Society, and the State (Chapel Hill: The University of North Carolina Press, 2011); Philip J. Funigiello, *Toward a National Power Policy;*

technological systems.[3] In parallel, historians have examined urbanization, suburbanization, industrialization, and regional change in the context of electrification and energy transitions.[4] Others have revealed the pace and effects of rural electrification.[5] Environmental historians have illustrated how power projects spawned political activism that resulted in expansion of environmental laws and citizen standing in court.[6] More recently, others highlight social injustices perpetuated through electrification.[7] Scholars of literary criticism, American Studies, and media studies have

---

the New Deal and the Electric Utility Industry, 1933–1941 (Pittsburgh: University of Pittsburgh Press, 1973); Richard F. Hirsh, *Power Loss: The Origins of Deregulation and Restructuring in the American Electric Utility System* (Cambridge, MA: MIT Press, 1999); Paul Israel, *Edison: A Life of Invention* (New York: John Wiley, 1998); Jill Jonnes, *Empires of Light: Edison, Tesla, Westinghouse, and the Race to Electrify the World* (New York: Random House, 2003); Harold C. Passer, *The Electrical Manufacturers, 1875–1900; a Study in Competition, Entrepreneurship, Technical Change, and Economic Growth*, Technology and Society (New York: Arno Press, 1972); Joseph A. Pratt, *A Managerial History of Consolidated Edison, 1936–1981* (New York: Consolidated Edison Company of New York, 1988); Abby Spinak, "'Not Quite So Freely as Air': Electrical Statecraft in North America," *Technology and Culture* 61, no. 1 (2020): 71–108; Leah Cardamore Stokes, *Short Circuiting Policy: Interest Groups and the Battle Over Clean Energy and Climate Policy in the American States*, Studies in Postwar American Political Development (Oxford: Oxford University Press, 2020).

[3]The seminal work is Thomas Parke Hughes, *Networks of Power: Electrification in Western Society, 1880–1930* (Baltimore: Johns Hopkins University Press, 1983).

[4]Examples include Christopher F. Jones, *Routes of Power: Energy and Modern America* (Cambridge, MA: Harvard University Press, 2014); Andrew Needham, *Power Lines: Pheonix and the Making of the Modern Southwest*, Politics and Society in Twentieth-Century America (Princeton: Princeton University Press, 2014); Harold L. Platt, *The Electric City: Energy and the Growth of the Chicago Area, 1880–1930* (Chicago: University of Chicago Press, 1991); Mark H. Rose, *Cities of Light and Heat: Domesticating Gas and Electricity in Urban America* (University Park, PA: Pennsylvania State University Press, 1995); Ruth W. Sandwell, ed., *Powering up Canada: A History of Power, Fuel, and Energy from 1600* (Montreal: McGill-Queens University Press, 2016).

[5]Examples include Leah S. Glaser, *Electrifying the Rural American West: Stories of Power, People, & Place* (Lincoln, NE: University of Nebraska Press, 2009); Ronald R. Kline, *Consumers in the Country: Technology and Social Change in Rural America*, Revisiting Rural America (Baltimore: Johns Hopkins Press, 2000); Sarah T Phillips, *This Land, This Nation: Conservation, Rural America, and the New Deal* (Cambridge: Cambridge University Press, 2007); Ronald C. Tobey, *Technology as Freedom: The New Deal and the Electrical Modernization of the American Home* (Berkeley: University of California Press, 1996).

[6]Examples include Robert Lifset, *Power on the Hudson: Storm King Mountain and the Emergence of Modern American Environmentalism* (Pittsburgh: University of Pittsburgh Press, 2014); Daniel Pope, *Nuclear Implosions: The Rise and Fall of the Washington Public Power Supply System* (New York: Cambridge University Press, 2008); Thomas Raymond Wellock, *Critical Masses: Opposition to Nuclear Power in California, 1958–1978* (Madison: The University of Wisconsin Press, 1998).

[7]Examples include Conor Harrison, "Race, Space, and Electric Power: Jim Crow and the 1934 North Carolina Rural Electrification Survey," *Annals of the American Association of Geographers* 106, no. 4 (2016): 909–31; Sarah Mittlefehldt, "Wood Waste and Race: The Industrialization of Biomass Energy Technologies and Environmental Justice," *Technology and Culture* 59, no. 4 (2018): 875–98.

illustrated the ways communities interpret the technologies of electrification, and how those understandings have reframed the technologies in turn.[8] Historians have focused on users to comprehend technology adoption and maintainers to understand long-term use.[9] Through these studies, the definition of stakeholders in electrification has expanded, an understanding of how we got where we are today has deepened, and our interest in the meanings of our electricity-dependent lives has grown.

Where do we look next to better explain the development of power infrastructure in our world and perhaps peek into our electrical future? The CREZ initiative offers a useful perspective on who has a stake in electricity infrastructure, how those stakeholders might be understood, and where to find them. By examining the role of local governments, landowners, and regional organizations, two themes emerge. First, stakeholders may have broadly shared explicit costs and implicit benefits, but their intervention in energy system development is built around decidedly local concerns. Enlightened self-interest is not a motivating factor. Second, this broadened group of stakeholders became gatekeepers for one of Texas' most far-reaching state-sponsored projects.[10] The evidence from contemporary pipeline and powerline projects suggests that this will be increasingly true.

7.2 Background Cases

For more than a century, long-distance transmission lines engaged support and opposition from both local residents and distant interest groups. In this sense, the Texas CREZ case is not new. Some issues repeat themselves in these stories: private property rights, local boosterism, economic protectionism, modernization, NIMBY,

[8]Examples include Jennifer L. Lieberman, *Power Lines: Electricity in American Life and Letters, 1882–1952* (Cambridge, MA: MIT Press, 2017); David E. Nye, *Electrifying America: Social Meanings of a New Technology, 1880–1940* (Cambridge, MA: MIT Press, 1990); Daniel Wuebben, *Power-Lined* (Lincoln, NE: University of Nebraska Press, 2019).

[9]Examples of works focused on user adoption include Ruth Schwartz Cowan, *More Work for Mother: The Ironies of Household Technology from the Open Hearth to the Microwave* (New York: Basic Books, 1983); Graeme Gooday, *Domesticating Electricity: Technology, Uncertainty and Gender, 1880–1914* (London: Pickering & Chatto, 2008); Abigail Harrison Moore and Ruth W. Sandwell, eds., *In a New Light, Histories of Women and Energy* (Montreal: McGill-Queens University Press, 2021).

[10]"Texas' Wind Transmission Project Keeps Rolling," *The Texas Tribune*, updated September 8, 2010, accessed November 20, 2013, http://www.texastribune.org/2010/09/08/texas-wind-transmission-project-keeps-rolling-/. Journalist Kate Gilbraith described Hill Country landowners as *gatekeepers* while both former PUCT chairman Barry Smitherman and former ERCOT senior transmission planner Warren Lasher argue they were influencers rather than gatekeepers. Barry Smitherman, communication with the author, July 27, 2020; Warren Lasher, communication with the author, April 23, 2020. The dictionary's second definition of *gatekeeper*, "a person who controls access," makes it an apt term for these stakeholders. Merriam-Webster Online Dictionary (https://www.merriam-webster.com/dictionary/gatekeeper). Accessed August 18, 2020.

IMBY, fear of new technologies, environmental transformation, and preservation of natural beauty. Through the twentieth century, the technical literature, the newspapers, the government records, all reflect nearly unanimous enthusiasm for electrification, extension of long-distance lines, and interconnection between systems.[11] But individuals and groups contested power lines at many levels.

An early example of controversial long-distance power lines encompassed concerns about governance, personal property rights, and fear of new technologies. In 1910, the Massachusetts legislature asked the state's Gas & Electric Light Commission to examine the conditions and laws affecting power transmission. A rising number of companies had formed for the express purpose of generating power at one location and transmitting it across a distance, in some cases from hydroelectric plants – the renewables of their day. The new transmission lines traversed both open countryside and small towns. The companies complained to legislators that they encountered opposition from towns, which resulted in longer and more expensive routes. Elsewhere, they were forced to pay private landowners exorbitant fees to cross their property. Both situations undermined the intent to lower the cost of electricity while increasing profits. The transmission companies sought the power of eminent domain.[12]

A number of matters caught the commission's attention. The existing laws required the companies to build transmission lines in public rights-of-way, avoid entering towns that already had a power provider, and protect the public from any unsafe characteristics of the lines. But the laws did not address the question of eminent domain for utilities. The power companies argued the "statutes are absolutely silent in regard to much that relates to the construction and maintenance of lines," while other states granted utilities the power of eminent domain. Some of the public opposition reflected fear that the higher voltage of these new lines posed an even greater safety hazard than existing lower-voltage distribution lines. The journal *Electrical World* noted there was the "utmost confusion and uncertainty" about using the right of way for transmission lines. The editor also commented that the

[11] For a history of the development of North America's interconnected transmission networks, see Julie Cohn, *The Grid: Biography of an American Technology* (Cambridge, MA: MIT Press, 2017).

[12] "Investigation of High-Tension Transmission in Massachusetts," *Electrical World* 56, no. 3 (July 21, 1910): 139–40; "Annual Meeting of Massachusetts Electric Lighting Association," *Electrical World* 56, no. 4 (July 28, 1910): 194–95; "Eminent Domain for Power Transmission," *Electrical World* 56, no. 16 (October 20, 1910): 919–20; "Investigation of Energy Transmission in Massachusetts," *Electrical World* 56, no. 16 (October 20, 1910): 939–41; "Investigation of Energy Transmission Continued in Massachusetts," *Electrical World* 56, no. 19 (November 10, 1910): 1114–5; "Final Hearings in Massachusetts Transmission Investigation," *Electrical World* 56, no. 21 (November 24, 1910): 1219; "Massachusetts Commission News," *Electrical World* 56, no. 25 (December 22, 1910): 1463; "Public Control of Electricity: Interesting Hearing by State Board. W.R. Peabody Urges Bill for Eminent Domain Right. Predicts Doom of All the Small Companies," *The Boston Globe* (Boston), September 16 1910, 2; "Massachusetts Commission News," *Electrical World* 57, no. 4 (January 26, 1911): 223–24.

public fear of danger was "ill-grounded."[13] Opponents of eminent domain cited constitutional issues as well as the likelihood that utilities would abuse this new-found authority.

In the end, the commission determined that some of the company complaints were spurious, at best. At that time, there were two principal high-voltage transmission lines crossing Massachusetts and both had been constructed without exercise of eminent domain, no major re-routing took place in response to local opposition, and no excessive fees were charged. The commission also determined that current laws required sufficient public notices and hearings to ensure that local concerns would be adequately heard. The commission resisted the proposal to grant eminent domain to the transmission companies. In this instance, local communities won in the sense that they retained control over the placement of power lines in their towns and villages and on their private land. However, it was a temporary victory, as the commission noted it would probably be necessary to grant eminent domain to the utilities in the future.

Sometimes farmers fought farmers. Through the Rural Electrification Act of 1935, Congress created a revolving loan fund that allowed rural cooperatives to establish local electric service and link into larger power networks through long-distance power lines. Other federal programs aided the development of rural electrification.[14] For landowners and regional groups that formed electric cooperatives, rural electrification represented an "In My Backyard" phenomenon. Yet at times rural landowners engaged in fights over long-distance transmission lines that would directly cross their properties but only indirectly served their power needs.[15]

A notorious case occurred in Minnesota in the 1970s. The late Senator Paul Wellstone and physics professor Barry M. Casper documented their work with rural landowners to stop an interstate power line.[16] Oddly, this case pitted two rural generation and transmission cooperatives in eastern Minnesota against farmers in the western part of the state. The cooperatives proposed to build a lignite coal-fired plant in North Dakota, and then carry the electricity on an extra-high voltage line straight across Minnesota to bring power to the Twin cities and the eastern part of the state. This triggered a classic NIMBY response among western farmers, but it also linked the local residents to national discussions of energy policy, arguments over

---

[13] "Eminent Domain for Power Transmission"; "Annual Meeting of Massachusetts Electric Lighting Association."

[14] Jay L. Brigham, *Empowering the West: Electrical Politics before FDR*, Development of Western Resources (Lawrence: University Press of Kansas, 1998); Martin V. Melosi, *Coping with Abundance: Energy and Environment in Industrial America*, 1st ed. (Philadelphia: Temple University Press, 1985); Ronald C. Tobey, *Technology as Freedom: The New Deal and the Electrical Modernization of the American Home;* Public Works in Prosperity and Depression (New York: National Bureau of Economic Research, 1935).

[15] *Brief History of the Rural Electric and Telephone Programs* (Washington, DC: Rural Electrification Administration, U.S. Department of Agriculture, 1986).

[16] Paul David Wellstone and Barry M. Casper, *Powerline: The First Battle of America's Energy War* (Minneapolis: University of Minnesota Press, 2003).

what type of landscape deserves environmental protection, and fears about health issues associated with high voltage power lines. The project even stirred a group of opponents to attempt a creation of Nikola Tesla's imagined wireless electrical transmission technology.[17] Significantly, the fight over this project highlighted inadequacies in federal, state, and local regulations when it came to consideration of community concerns.

At the outset, the two cooperatives – Cooperative Power Association (CPA) and United Power Association (UPA) – seemed to be complying with national goals for energy conservation and environmental protection. The project would produce electricity at the mine mouth – thus avoiding the high transportation and energy costs associated with moving coal to a power plant. The plant would use abundantly available lignite, relieving pressure on other energy resources. And, to determine the path of the power line, the cooperatives relied on a computer-generated map that established environmental values for each square mile of the state. Areas with high scores were deemed most sensitive to an impact from the power lines. The computer model gave high scores to airports, highways, and wildlife areas. Farms received the lowest scores, and farmland became the targeted route for the powerline.

When the farmers in the western part of the state realized that a giant transmission line would soon be crossing their land, they organized to protest. They pursued multiple angles. Each rural cooperative in that area had one member that sat on the boards of CPA and UPA, but their combined voices were insufficient to change the power line plans. The farmers protested at town hall meetings, before state regulators, and in court. Outside individuals and organizations came to their aid. Their stated objections included questions about the immediate need for more electricity, absence of conservation initiatives, little representation in the decision-making process, frustration that farms were deemed environmentally expendable, concern that high-voltage power lines might cause cancer, and lack of direct access to the electricity.

Of each of these concerns, the lack of representation in decision-making was probably the most cogent. None of the laws at local, state, or federal levels offered a reasonable avenue for local groups to address the location of power lines. Scientific research at that time undermined the theory that power lines caused cancer. The concern about lack of direct access to electricity was a bit spurious because the networks linking power stations across the region ultimately brought the benefits of the power back to the farmers. In the end, the protest movement really was about power lines in the back yard. As one farmer put it, "every time I see the towers, every time I walk in the fields, it kind of brings back all the memories of fighting the thing, and in a way every time I see it, I feel more bitter."[18] The farmers lost at every stage of the regulatory and permitting process. But they didn't give up. As the cooperatives built the line, the farmers held midnight picnics around the poles and used any

[17] Pete Schmidt, "Grassroots Technological Resistance: The People's Power Project and the Impossible Dream of Wireless Transmission of Energy," *Endeavor* 41, no. 3 (2017): 146–9.

[18] Wellstone and Casper, *Powerline: The First Battle of America's Energy War.*

means at their disposal to sabotage the construction. Civil disobedience failed as well, but at least provided a briefly satisfying outlet for the protestors' frustration.

From the towns of 1910s Massachusetts to the farms of 1970s Minnesota, locals sought a role in decision-making about large transmission infrastructure. They exhausted every legal avenue to combat projects, and a few illegal ones as well. These stakeholders, who are somewhat invisible in the historical literature, sought and fought power lines before regulators with varying degrees of influence. Across this same period, local, state, and federal agencies sorted out regulatory authority, and created a patchwork of entities that together determine the routing and costs of interstate transmission lines. The Texas CREZ case illustrates that local voices have become loud, legitimate, and highly influential in shaping electricity infrastructure.

7.3 The Texas CREZ Case

While Texas exceptionalism may be a myth in most respects, it is a reality with respect to the majority of the state's electric power system. Ninety percent of Texas electricity customers acquire power from a grid called the Texas Interconnection that is internal to the state and does not regularly exchange power with its neighbors. The Electric Reliability Council of Texas (ERCOT) operates the Texas Interconnection, one of the three major grids that provides power in the continental United States. The other two are the Eastern Interconnection (east of the Rocky Mountains) which also serves customers in Canada; and the Western Interconnection (west of the Rocky Mountains) which also serves customers in Canada and parts of Mexico. There are direct current links between the Texas grid and surrounding states and Mexico, but these are not considered interconnections.[19] This means that fewer agencies and lawmakers set policy and regulate the Texas grid, compared to all other bulk power networks in the continental United States. The map in Fig. 7.1 delineates the region served by the Texas grid.

Texas is unique in other ways. In 1975 Texas was the last state to establish a commission to regulate electric utilities, while most states completed this process by 1920. Texas differs as well in the matter of cost allocation for transmission services. Through most of the twentieth century, regulated utility monopolies owned most of the country's power networks, from the generating plants to the transmission lines to

[19] *Grid* is the term used informally to refer to the Bulk Power System, defined under the Federal Power Act as "facilities and control systems necessary for operating an interconnected electric energy transmission network (or any portion thereof); and electric energy from generation facilities needed to maintain transmission system reliability." (16 U.S.C. §824) *Interconnected* refers to the synchronous operating of alternating current elements. In the United States, all three grids are linked with direct current interties, which allow scheduled and controlled exchanges of power in one direction, without requiring synchronous operation between the grids. Julie Cohn, "When the Grid Was the Grid: The History of North America's Brief Coast-to-Coast Interconnected machine," *Proceedings of the IEEE* 107, no. 1 (January 2019): 232–233.

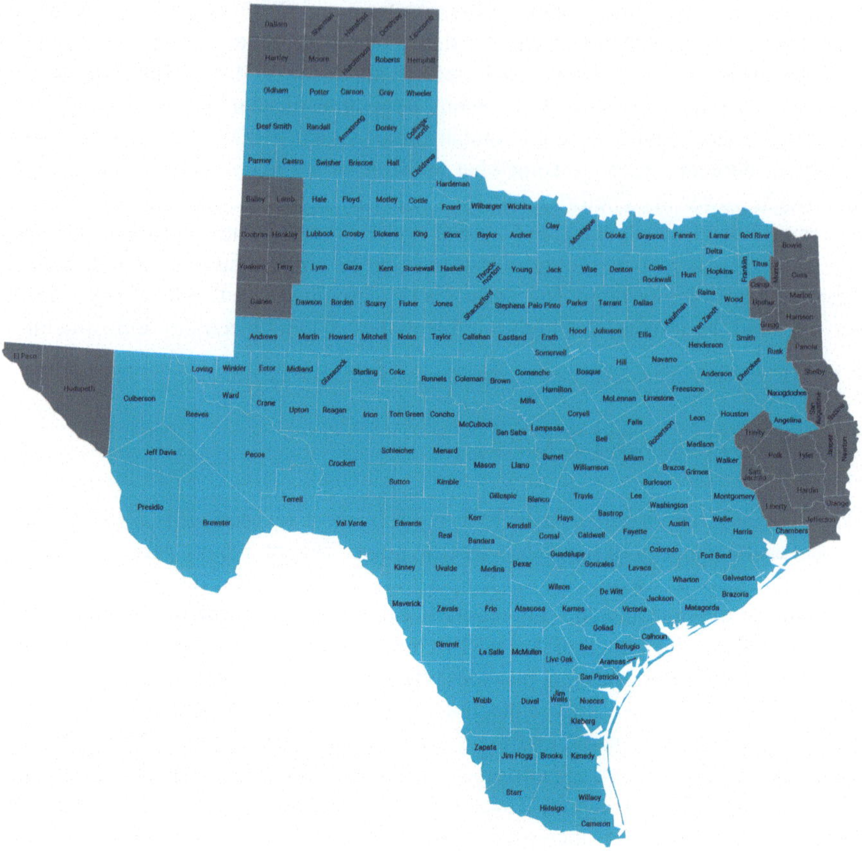

Fig. 7.1 Region served by the Texas grid. Areas shown in gray are served by power companies that are not connected to the grid. (Source: Electric Reliability Council of Texas website, https://www.ercot.com/files/assets/2022/12/13/ERCOT-Maps_Area-by-county.jpg?)

distribution systems. Utilities incorporated the cost of building, maintaining, and operating intrastate transmission lines into their rate requests before state utility commissions.[20] This was true in Texas as well, until the 1990s, when the state instituted a postage stamp rate to recover transmission costs.[21] ERCOT calculates the combined cost of wholesale power transmission across its entire network,

[20] States and regions have implemented a variety of transmission cost recovery mechanisms as each state has chosen to either restructure segments of its power market or continue to regulate vertically integrated monopoly utility companies. Some independent system operators use postage stamp rates to recover costs on specific types of transmission projects, but none other than ERCOT apply this method across the board.

[21] Public Utility Commission of Texas, Substantive Rules, Chapter 25 – Electric, § 25.192 - Transmission Service Rates (July 5, 2016). https://www.puc.texas.gov/agency/rulesnlaws/subrules/electric/25.192/25.192.pdf

including the investment costs for new or retrofitted power lines. This cost is divided by total demand, and then paid on the basis of how much power a transmission customer uses, regardless of the distance between generating sources and customers.

The postage stamp rates are significant for several reasons. First, calculation of power loss along transmission lines is no longer critical to dispatch decisions. As a result, ERCOT faces a somewhat simplified task in determining which increment of power should come from which generating facility next. Second, the postage stamp rate lowers the entry barrier for wind farms located at a distance from end users. For a Houston customer, for example, an electron from a Panhandle wind farm does not cost more by virtue of the distance it travels than an electron from, say, the much closer South Texas Nuclear Project. The postage stamp rate has been advantageous to wind developers entering the market, though periodically protested by traditional fossil fuel generators.[22] Third, every customer within the ERCOT region shares the cost of new power lines, regardless of who provides distribution services locally or which generating source a customer might prefer. For the CREZ projects, everyone connected to the Texas grid shared explicitly in the cost of new power lines.

Just as is the case on every other power network, each Texas grid customer potentially accesses electricity from any and all connected generating sources. Alternating current electricity literally follows the path of least resistance and impedance. Power system operators do not direct electricity from a particular generating source to a particular customer; instead, they strive to match overall generation to overall demand. Thus, every customer on the Texas grid might light a lamp or start a motor using electricity generated at a wind farm, even if that customer didn't actively seek renewables. To the extent that power from renewables was desirable to power customers, then, the CREZ project benefited everyone on the Texas grid.

The Texas legislature began to reshape the power industry in 1995 by establishing a competitive wholesale market and decoupling transmission operations from generation.[23] The new law also required public input into the transmission planning process.[24] Eight large utilities used a process called deliberative polling to learn about public power concerns. They conducted town hall meetings at which a panel of experts presented energy issues and participants had the opportunity to discuss their electricity concerns. The findings were combined with results from traditional telephone surveys to access input from a broad cross-section of hundreds of Texans. Participants surprised the utility managers, as well as legislators and regulators, by expressing a desire for renewables and a willingness to pay more for them. Along with lobbying from wind developers and environmental groups, and a vision of

---

[22] L.M. Sixel, "Transmission is Latest Front in Fossil Fuels v. Renewables Battle," *Houston Chronicle,* November 30, 2018, Updated December 6, 2018, https://www.houstonchronicle.com/business/energy/article/Transmission-is-latest-front-in-fossil-fuels-v-13433013.php

[23] Texas S.B. 373, 74th Legislature, R.S. (1995).

[24] R.L. Lehr et al., *Listening to Customers: How Deliberative Polling Helped Build 1000 MW of New Renewable Energy Projects in Texas*, National Renewable Energy Lab. (June 2003).

Texas as an energy producing state, the deliberative polling results persuaded legislators to commit to renewables.[25]

In 1999, the legislature set hard goals for increasing power generation from renewables through a Renewable Portfolio Standard.[26] With 184 MW of renewable capacity already in place, the standard called for an increase to 400 MW by 2003 and 2000 MW total by 2009. With significant state and federal tax incentives and credits already in place, the standard created a favorable environment for rapid wind development. By 2001, Texas had exceeded the 2003 goal, and by early 2005, it was clear that the state would meet the 2009 goal within a year.[27] In addition, the state's grid operator, ERCOT had to force curtailment of wind generation from time to time as a result of bottlenecks on the transmission infrastructure.

With these issues in mind, Texas legislators voted in 2005 to increase the Renewable Portfolio Standard goal to 5880 MW by 2015 - with an additional target of 10,000 MW total installed capacity by 2025. Governor Rick Perry signed Senate Bill 20 on August 2, 2005. The legislators also ordered the state's Public Utility Commission (PUCT) to establish the Competitive Renewable Energy Zones and to plan for related improvements in transmission infrastructure. Legislators intended to direct investment in wind power to areas of the state with both good wind potential and plenty of land, and at the same time to assure that new power generation could be added to the state's grid without causing bottlenecks. Nathan Kapoor, in his essay in this volume, illustrates a similar, though much earlier, effort by New Zealand's central government to exploit resources unique to the region. New Zealand's provincial government envisioned electrification as an expression of modernity, independence, and economic forward thinking. Likewise, the CREZ initiative seemed to expand upon Texas' reputation as an independent and entrepreneurial state with extraordinary natural resources and a growing energy economy.

---

[25] Becky H. Diffen, "Competitive Renewable Energy Zones: How the Texas Wind Industry is Cracking the Chicken & Egg Problem," *Rocky Mountain Mineral Law Foundation Journal* 46, no. 1 (2009): 47–98; Kate Galbraith and Asher Price, *The Great Texas Wind Rush: How George Bush, Ann Richards, and a Bunch of Tinkerers Helped the Oil and Gas State Win the Race to Wind Power*, The Peter T. Flawn Series in Natural Resource Management and Conservation (Austin, TX: University of Texas Press, 2013); L. Lynne Kiesling and Andrew N. Kleit, eds., *Electricity Restructuring: The Texas Story* (Washington, DC: AEI Press, 2009); R. Ryan Staine, "CREZ II, Coming Soon to a Windy Texas Plain Near You: Encouraging the Texas Renewable Energy Industry through Transmission Investment," *Texas Law Review* 93, no. 2 (December 2014): 521–56; Leah Cardamore Stokes, *Short Circuiting Policy: Interest Groups and the Battle over Clean Energy and Climate Policy in the American States,* Studies in Postwar American Political Development (Oxford: Oxford University Press, 2020).

[26] Texas S.B. 7, 76th Legislature, R.S. (1999); Madeline Claire Gould, "Everything's Bigger in Texas: Evaluating the Success and Outlook of the Competitive Renewable Energy Zone (CREZ) Legislation in Texas" (masters thesis, University of Texas, 2018); States with Renewable Portfolio Standards by 1999: Nevada, Iowa, Wisconsin, Maine, Massachusetts, Connecticut, New Jersey, Texas. https://www.ncsl.org/research/energy/renewable-portfolio-standards.aspx

[27] David Hurlbut, "A Look Behind the Texas Renewable Portfolio Standard: A Case Study," *Natural Resources Journal* 48, no. 1 (2008): 129–61.

Over the next several years, the PUCT took steps to first establish the CREZ, and to then delineate the corridors for the CREZ transmission lines. The commission opened a series of contested hearings that allowed interested parties to have input. During this process, prospective wind developers and transmission service providers, existing power generators and utilities, ranchers, local governments, state elected officials, private citizens, environmental groups, alliances of interested parties, and state agency representatives all offered opinions, data, and comments for and against different CREZ areas and different transmission corridors. By 2014, 3600 miles of new 345 kV transmission lines, built at a cost of $6.9 billion, carried wind power – as well as electricity generated by other sources – from five CREZ areas into the ERCOT grid.[28] In addition, wind developers had installed more than 12,000 MW of wind power capacity by the end of that year, exceeding the state's 2025 target with 11 years to go.

In 2020, the ERCOT region boasts nearly 24,000 MW installed wind capacity and averages 20 percent of its actual power generation from wind.[29] This puts Texas at the forefront of renewables development in the country – with the largest installed wind capacity and the largest percentage of electricity generated by wind of any state. In addition, Texas leads all other states in generation of power from renewables in general, including conventional hydroelectric power generation.[30]

Texas has made environmental progress as well. Notably during a period of both population and economic growth, "Texas air quality has made major strides in the past few decades."[31] Measurements in virtually every category of criteria pollutants has improved over the past 15 years. While air quality metrics do not necessarily reflect changes in a single industry, emissions data are informative. Between the enactment of CREZ (2005) and the most recent year for which data is available (2018), power plant emissions of three pollutants – carbon dioxide, sulfur dioxide, and nitrous oxides – fell by 13 percent, 63 percent, and 31 percent respectively.[32] These changes may be attributable in part to the increased portion of power

[28] Warren Lasher, *The Competitive Renewable Energy Zones Process*, Electric Reliability Council of Texas (Austin, TX, August 11 2014), https://www.energy.gov/sites/prod/files/2014/08/f18/c_lasher_qer_santafe_presentation.pdf

[29] *Fact Sheet*, Electric Reliability Council of Texas (Austin, TX, March 2020), http://www.ercot.com/content/wcm/lists/197391/ERCOT_Fact_Sheet_3.25.20.pdf

[30] "Electric Power Monthly data for January 2020," EIA.gov, U.S. Department of Energy, https://www.eia.gov/electricity/monthly/, accessed April, 6, 2020.

[31] "Air Quality Successes – Criteria Pollutants," Texas Commission in Environmental Quality website, last modified July 20, 2020, https://www.tceq.texas.gov/airquality/airsuccess/airsuccesscriteria, accessed July 22, 2020; "2018 Estimated population of Texas, its counties, and places," Texas Demographic Center, https://demographics.texas.gov/Resources/publications/2019/20191205_PopEstimatesBrief.pdf; "Total Gross Domestic Product for Texas," Economic Research, Federal Reserve Bank of St. Louis, https://fred.stlouisfed.org/series/TXNGSP

[32] EIA.gov Electricity Data Browser (Beta), Number of plants for multiple sectors, Texas, multiple fuel types: Plant Level Data, Annual Emissions, 2005 and 2018, https://www.eia.gov/beta/electricity/data/browser/#/topic/1?agg=2,0,1&fuel=vtvv&geo=g&sec=g&freq=M&start=200101&end=201710&ctype=linechart&ltype=pin&maptype=0&rse=0&pin=.

generated by wind (up 1787 percent since 2005); but must also be attributed to the significant shift from coal (down 25 percent) to natural gas (up 21 percent).[33] Environmental Protection Agency data also indicate that power plant contributions have fallen as a share of total greenhouse gas emissions in Texas.[34] Changes in air quality in Texas are undoubtedly due to multiple factors: overall economic activity; improvements in plant operations of all types; changes in primary energy resource; and the addition of wind power. But CREZ made at least a small contribution to environmental protection in the state.

Taken together, the increase in wind power and the improvement in air quality suggest that the Texas legislature made a smart move by authorizing CREZ. At the same time, however, the CREZ process took 9 years, and resulted in an extra 1200 circuit miles of transmission lines at a cost four times the legislature's original conjectures, and 30 percent more than solid estimates made in 2008.[35] Through seventy-two separate cases, the Public Utility Commission of Texas considered more than 24,000 filings, and tens of thousands of individual letters, postcards, and emails within those filings. To quote the movie *Casablanca*, the "usual suspects" – such as owners of fossil-fuel fired power plants, incumbent transmission line owners, national and statewide environmental advocacy groups, wind investors, and technical consultants – provided extensive input into the process.[36] Within the dockets, subsidiaries of more widely known Texas energy companies appear. Shell Windenergy, a subsidiary of Royal Dutch Shell, BP Solar, a subsidiary of BP, and Mesa Water, owned by T. Boone Pickens, all sought to influence the location of the priority zones for wind development. In his chapter, Cyrus Mody details earlier links between the oil and gas sectors and electrification. Similar links appear in the CREZ story, as fossil fuel investors sought to capitalize on a wind boom. CREZ offers other compelling entanglements between these energy sectors. In 2013 and 2014, as the CREZ initiative reached fruition, hydrofracturing activity in Texas boomed and resulted in increased demand for electricity. Thus, wind powered fossil-fuel extraction through the CREZ initiative. The investment and energy entanglements extend well beyond the state's borders as well, as investors, producers of turbines, and

[33] EIA.gov Electricity Data Browser, Net generation for all sectors, selected fuel sources, https://www.eia.gov/electricity/data/browser/.

[34] Environmental Protection Agency website, EPA Facility Level GreenHouse gases Tool (FLIGHT), http://www.ghgdata.epa.gov/ghgp/main.do#, accessed July 22, 2020. Greenhouse gas emissions measured by EPA include CO2, CH4, N2O, and several Fluorinated GHGs.

[35] Lasher, *The Competitive Renewable Energy Zones Process*; Docket 33672, *Open Meeting; July 20, 2007 - Agenda Item No. 22 - Memo from Commissioner Parley* [*sic*], Public Utility Commission of Texas, July 20, 2007. Hereinafter, all documents retrieved from the PUCT case dockets will be cited with the docket number, the document title, and the file stamp date. All case docket filings are available at https://interchange.puc.texas.gov/.

[36] Michael Curtiz, dir., *Casablanca* (Hollywood: Warner Brothers, 1942).

installation operators from around the globe intervened in the CREZ deliberations.[37]

Notably, however, local individuals and organizations directly influenced the duration of the hearing process; the determination of the CREZ zones; the size, location, and cost of the final transmission infrastructure; and even the elimination of certain power lines and related cost-savings. In this sense, landowners and local entities served as gatekeepers for the development of power system infrastructure. Two examples illustrate stakeholder gatekeeping. In one, addressing pro-CREZ interests in the Texas Panhandle, the PUCT added two geographic areas to the CREZ and selected more capacious transmission corridors. In the second, addressing antipathy from landowners and local organizations in the scenic Texas Hill Country, the PUCT eliminated two transmission lines while navigating highly contentious route options for a third.

The PUCT went to work immediately after the law took effect. In October 2005 the commission initiated rulemaking and ordered ERCOT to undertake a wind study. The very first docket quite simply asked "interested persons" for input on how to get the whole project done.[38] Hundreds responded within 217 filings recorded over the next year.[39] Meanwhile, ERCOT contracted with a consulting firm to assess and map the state's wind potential, while wind developers filed expressions of interest in specific regions.[40]

The vast majority of comments indicated support for wind energy, and many begged for establishment of CREZ in specific locations. State senators and legislators forwarded messages from the areas they represented calling for wind farms and transmission lines.[41] Towns, county commissioners courts, and landowners in the Texas Panhandle were especially insistent.[42] A typical resolution stated: "citizens of the Texas Panhandle generally welcome large energy and infrastructure projects and the increased property tax base and economic development benefits such projects produce, and have limited alternative opportunities for economic development."[43]

For areas that had been hard hit economically in recent years, the potential for jobs and tax revenues outweighed all other considerations. Indeed, by 2005, reports

---

[37] Sarah Stanford McIntyre, "The Saudi Arabia of Wind: Deregulation and the Rise of Wind Power in Texas," *The Journal of Energy History/Revue d'Histoire d'Énergie (Online)* vol. 7, 15, December 2021. energyhistory.eu/en/node/281.

[38] Docket 31852, *PUC ADM: Request for Comments*, January 10, 2005.

[39] "Interchange Filing Search: Filings for 31852," Public Utility Commission of Texas, 2020, accessed various, https://interchange.puc.texas.gov/Search/Filings?ControlNumber=31852. In Texas county administrative bodies are called County Courts and the chief administrator is called County Judge.

[40] AWS Truewind provided wind consulting to ERCOT for CREZ.

[41] Docket 31852, *State Representative David Swinford: Comments*, October 3, 2006.

[42] The 25,000+ square mile region known as the Texas Panhandle is largely rural, with very low population. Historically, the area's economic base included ranching, farming, and oil and mineral development. Handbook of Texas Online, Frederick W. Rathjen, "PANHANDLE," http://www.tshaonline.org/handbook/online/articles/ryp01, accessed August 12, 2020.

[43] Docket 31852, *Swisher County: Resolution in Support*, April 27, 2006.

indicated that wind power installations had already added nearly $15 million in annual tax revenues for local school districts in other regions.[44] The state's Office of Rural and Community Affairs emphasized that criteria for selecting each zone should include economic impact on local communities and generation of new school tax revenue.[45] An enthusiastic high school student offered that CREZ in Silverton, Texas "means we, students could get a job back here in our hometown after college."[46] Briscoe County Judge Wayne Nance hand-wrote "I am just a country boy with country ways, but I know how vital the wind chargers are to Briscoe County. It is a ray of hope for every aspect of our lives. Hope for our school (which is a 'Recognized' school), hope for new jobs, and hope for a better life. This all depends upon the transmission lines. We have the best wind available – with thousands of acres already leased to Shell, Horizon, and R.E.S. Our hope lies with you and your decision."[47]

ERCOT filed its analysis of wind potential and transmission alternatives with the PUCT in December 2006. The analysis identified 25 potential zones and narrowed further study to ten that also had received expressions of interest from wind developers. Notably, these ten included only half of the aforementioned Briscoe County and did not include the zone with the very best wind in the state – an area of the Panhandle that contained nearly all of Deaf Smith County. Figure 7.2 illustrates the 25 potential zones, with red indicating the ten study zones (2, 4, 5, 6, 7, 9, 10, 12, 14, 24). This brought the first docket to a close and the commission moved on to selecting the actual CREZ.

7.4 "Drawing Circles, Drawing Lines": Selecting Zones and Transmission Corridors

Even more exciting than the question of *how* PUCT should proceed, the petition to designate the CREZ focused on *where* the CREZ would be. Action in the case began in January of 2007 and continued through 2008 with nearly 1500 filings, several of which represented dozens, if not hundreds, of pieces of correspondence from citizens in various communities.

---

[44] Environment Texas, "Texas Legislature Boosts State's Renewable Energy Goal," news release, July 14, 2005, https://environmenttexas.org/news/txe/texas-legislature-boosts-states-renewable-energy-goal.

[45] Docket 31852, *Office of Rural Community Affairs: Comments*, February 21, 2006.

[46] Docket 31852, *Various Parties: Comments*, November 30, 2006; *Various Parties: Comments*, December 5, 2006.

[47] Docket 31852, *Various Parties: Comments*, December 5, 2006. In Texas, the state education agency ranks school districts based on several criteria including the results of state mandated achievement tests. In 2005, those categories were: Exemplary, Recognized, Academically Acceptable, Academically Unacceptable, and Not Rated.

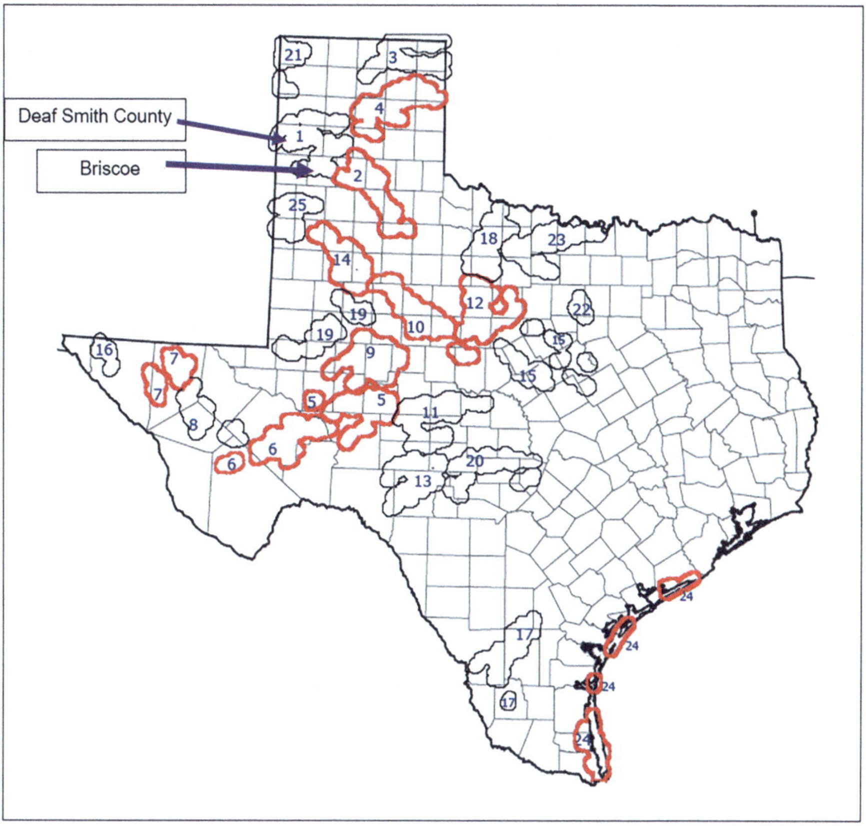

Fig. 7.2 Map showing 25 potential CREZ and 10 study areas. (Source: Analysis of Transmission Alternatives for Competitive Renewable Energy Zones in Texas. (Electric Reliability Council of Texas (Austin: December 2006). P. ES-2))

Citizens made their case to the commissioners. Sherry Phillips, Mayor of the City of McCamey, for example, strongly endorsed designation of the surrounding area as a CREZ.[48] McCamey was already a center of wind power development, which had been curtailed by congestion on the primary power lines leading from there to the east. Mayor Phillips noted "Wind power plays a role in creating energy independence and improving air quality . . . McCamey desires to facilitate additional economic development by private and public entities to strengthen our West Texas communities." Cities, counties, and legislators in the Panhandle sent nearly identical letters, stating, "The flat, open terrain of the Texas Panhandle allows for relatively low cost construction of transmission lines, and local community support for these efforts should also help facilitate their successful completion which will be the most

[48] Docket 33672, *Horizon Wind Energy, LLC's List of Issues*, February 26, 2007. McCamey, Texas is about 220 miles northwest of Kerrville, and about 195 miles almost due west of Mason.

expedient area to build transmission of all the potential CREZs."[49] Indeed, State Representative David Swinford assured the PUCT that every single one of the more than 100 County Commissions and City Councils in his Panhandle district had approved a resolution stating support for designation of a CREZ.[50] He further encouraged the commission to be bold, "This really came to me while I was walking down a row of broken electric poles during a huge ice storm last year. I saw a date nail on a pole and it was dated 1950 model. Those poles had served us for 57 years. My point is this: our forefathers were bold. They built an infrastructure for their grandchildren and great-grandchildren."

Commenters and intervenors introduced other issues. Rancher Gene Scivally, in the Panhandle, expressed concern about international competition.[51] A coalition of environmental groups called for substantial new transmission infrastructure.[52] John Rappolo, representing the 825,000-acre King Ranch, filed a list of threatened, endangered, or special concern animal species in a migration corridor along the Texas Gulf Coast (Zone 24).[53] This followed earlier pleas from King Ranch for the PUCT to eliminate Zone 24 from further consideration.[54]

At a July 20, 2007 open meeting, the PUCT made several key decisions, including preliminary designation of the CREZ. Commissioner Julie Parsley, in advance of the meeting, proposed that the PUCT limit its focus, and all further research, to just three CREZ: Zone 2 (but adding all of Briscoe County), Zones 5 & 6 (the McCamey area), and Zone 9 (just north of McCamey in central West Texas and expanded to include Zone 19).[55] Commissioner Parsley expressed concern that the legislature expected the cost of new transmission lines to be no more than $1.3 billion, though she was willing to see the projected cost go up to $1.75 billion. PUCT Chair Paul Hudson, describing the commission's work as "drawing circles, drawing lines," offered a list with two additional zones: 4 and 10.[56] He was ready to eliminate all the others. During the open meeting, Commissioner Barry Smitherman made a strong case for including Zone 1 – covering Deaf Smith County - in the final CREZ

[49] Docket 33672, *Comments of County Judges of Dallam, Sherman, Oldham, Swisher, Lipscomb, and Parmer Counties*, April 25, 2007; Docket 33672, *Comments of County Judges of Lamar* [*sic*] *and Hall Counties*, April 26, 2007. (The comments were actually from the County Judges of Parmer and Hall Counties.)

[50] Docket 33672, *Representative Swinforf* [*sic*] *Comments on Designation*, June 12, 2007.

[51] Docket 33672, *Emails Received by Commissioner Smitherman's Office*, June 20, 2007.

[52] Docket 33672, *Public Citizen's Comments*, July18, 2007.

[53] Docket 33672, *Direct Testimony John Rappole OBO King Ranch, Inc.*, April 24, 2007; Docket 33672, *CD Containing List of Rare, Threatened and Endangered Species and Maps of the Location*, May 16, 2007.

[54] Docket 33672, *King Ranch Inc.'s List of Issues*, February 26, 2007; Docket 33672, *King Ranch Inc. Motion to Sever Zone 24*, March 8, 2007.

[55] Docket 33672, *Open Meeting; July 20, 2007 - Agenda Item No. 22 - Memo from Commissioner Parley* [*sic*], July 20, 2007.

[56] Docket 33672, *Open Meeting; July 20, 2007 - Agenda Item No. 22 - Chairman Hudson Memo*, July 20, 2007.

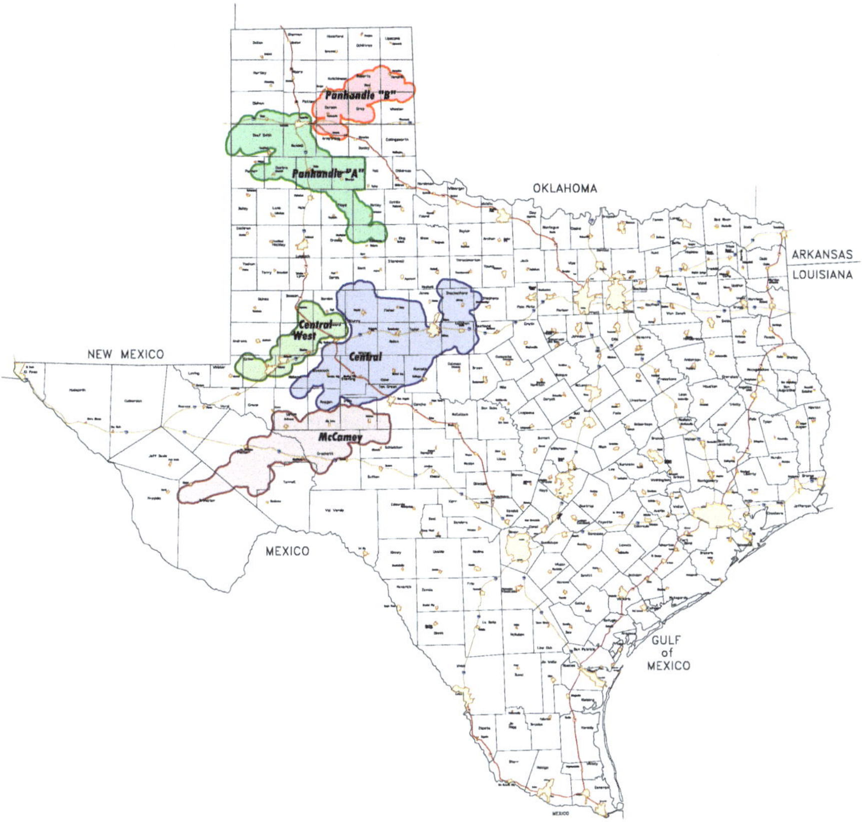

Fig. 7.3 ERCOT Map of CREZ Zones November 2007. (Available from ERCOT website: http://www.ercot.com/news/presentations/2007)

designation.[57] Despite the tepid expressions of interest from wind developers and the lack of additional transmission studies by ERCOT, Smitherman argued the commission would be hard-pressed to defend leaving out the best wind in the state. In an unusual on-the-spot decision, the commission voted two to one to expand Zone 2 to include Zone 1, and also approved Hudson's preliminary list.[58] The five CREZ appear in Fig. 7.3 above:

The commission renamed the zones as follows:

Panhandle "A" = Zone 1 and expanded Zone 2.
Panhandle "B" = Zone 4
Central West = Zone 19

[57] *Public Utility Commission of Texas Open Meeting*, July 20, 2007; 100.

[58] Thank you to Warren Lasher who brought this important vote to my attention. Lasher, Communication with the author.

	Scenario 1 (MW)	Scenario 2 (MW)	Scenario 3 (MW)	Scenario 4 (MW)
Panhandle A	1,422	3,191	4,960	6,660
Panhandle B	1,067	2,393	3,270	0
McCamey	829	1,859	2,890	3,190
Central	1,358	3,047	4,735	5,615
Central West	474	1,063	1,651	2,051
CREZ Wind Capacity	5,150	11,553	17,956	17,516

Fig. 7.4 ERCOT's CREZ Transmission Optimization Study, Public Utility Commission of Texas. (2008). P.2

Central = Zones 9 & 10
McCamey = Zones 5 & 6

While commissioners may have considered the comments, letters, and other filings from private citizens, local governments, and alliances of interested environmental groups, these were not mentioned in the commission's order. The PUCT noted that "stakeholder input" would be considered in ERCOT's CREZ Transmission Optimization Study, but the nature of the stakeholders was not identified. The overall sense of the order suggests that the input of local landowners and similar interested parties was not, ultimately, of as much consequence as the legislature's mandates for CREZ selection. Commissioner Smitherman's comments at the July 20th open meeting, however, indicate some concern for backlash from citizens and legislators if the commission failed to bring all the best wind into CREZ.

The PUCT next considered where to locate the CREZ transmission corridors, the essential arteries for bringing wind power into the ERCOT market. In early April 2008, ERCOT presented its evaluation of transmission options.[59] ERCOT credited stakeholders such as existing and potential transmission owners, market participants, and vendors with providing key input. The report noted that 276 individuals joined the project mailing list and at least fifty stakeholders, including many private citizens, attended each of twelve project meetings.[60] The optimization study outlined four scenarios for transmission development, each anticipating a different total installed wind capacity, as shown in Fig. 7.4. Note that Scenarios 3 and 4 would support the most aggressive growth of wind power, while Scenario 2 would support

[59] Docket 33672, *ERCOT's CREZ Transmission Optimization Study*, April 2, 2008.

[60] The records for these meetings are no longer available from ERCOT, which follows a ten-year records retention policy. Lasher, Communication with the author.

the wishful 2025 target for renewables, and Scenario 1 would ensure only the state's hard goal of 5880 MW installed wind capacity by 2015.[61]

With the interim designation of CREZ, and the ERCOT transmission study, other parties appeared before the PUCT. For example, commenters from Odessa in west Texas expressed support for including a planned Integrated Gasification Combined Cycle project in the Permian Basin as a potential contributor to a new CREZ transmission line.[62] Odessa lies at the heart of Texas oil country. Panhandle ranchers described plans for wind farms on their land.[63] Individuals encouraged higher levels of wind capacity to halt global warming, while environmental groups and consumer advocates made the case for Scenario 3, which promised maximum development of new wind power.[64] A group of twelve state legislators representing various parts of the state urged the PUCT to move forward quickly.[65] Briscoe County Judge Nance wrote again to the commissioners urging their consideration of the economic benefit new wind farms and transmission lines might bring to his region, one of the poorest in the state.[66] A class of third graders from Silverton, Texas (in Briscoe County) submitted letters and drawings in support of Scenario 3 (an example appears in Fig. 7.5 below).[67] While the PUCT staff recommended adoption of the conservative Scenario 1, legislators from across the state, including a group of mostly Republicans, called the Texas Conservative Coalition, urged Scenario 3 as a "bold option."[68]

Stakeholders from different regions, and with differing economic and environmental considerations, offered multiple imaginaries of the state's energy future. For Panhandle residents, the future looked windy, with new infrastructure driving economic renewal and diversification. For Odessans, Texas's next energy system included fossil fuels modified to be less polluting through technical innovation. For citizens across the state, maximizing wind power through investment in transmission

---

[61] Interestingly, as of the end of 2008, Texas already reported 8005 MW installed wind capacity. "Impact of Increased Wind resources in the ERCOT Region," ERCOT, August 2018, http://www.ercot.com/content/wcm/lists/144927/Wind_One_Pager_FINAL.pdf.

[62] Docket 33672, *Odessa Chamber of Commerce Comments*, April 18, 2008; Docket 33672, *Summit Power Comments*, April 11, 2008; Docket 33672, *Response to Monahans Chamber of Commerce Executive Director Teresa Burnet's Letter*, June 17, 2008; Docket 33672, *Response to Comments from Odessa College President*, June 5, 2008; Docket 33672, *Response to Comments from University of Texas of the Permian Basin*, June 5, 2008.

[63] Docket 33672, *Response to Letters Regarding Staff's Petition*, June 18, 2008. (Irion County)

[64] Docket 33672, *Tim Mock: Comments*, June 24, 2008; Docket 33672, *Public Citizen of Texas et al: Statement of Position*, June 6, 2008.

[65] Docket 33672, *State Representative David Swinforf* [*sic*], June 25, 2008.

[66] Docket 33672, *Response to Comments from Briscoe County Judge Wayne Nance*, June 12, 2008.

[67] Docket 33672, *Copy of Emails Sent to Commissioner Hudson's Office*, June 26, 2008.

[68] Docket 33672, *Commission Staff's Post Hearing Brief*, June 26, 2008; Docket 33672, *Letter to the Commissioners Supporting Competitive Renewable Energy Zone (CREZ) Scenario #3*, June 26, 2008; Docket 33672, *Texas Conservative Coalition: Comments*, July 15, 2008; Docket 33672, *Representative Mike Villarreal: Additional Supporters of Scenario 3*, July 15, 2008. Representative Villarreal's submission included the signatures of 5471 Texans, and a list of 49 supportive members of the legislature from both sides of the aisle.

Fig. 7.5 Drawing of wind turbines by Briscoe County Third Grader. (Source: PUCT Interchange Filings Search, Case 33672–1384, Tina Nance: Comments)

infrastructure equaled a more sustainable tomorrow with a reduced threat to the environment today.

At a July 17th open meeting, Barry Smitherman, by now commission chair, urged his colleagues to thread the needle. He stated, "If you will join me in selecting CREZ Scenario 2, I believe that we can be both bold and cautious, both visionary and practical."[69] He acknowledged that Panhandle communities, wind developers, environmentalists, and West Texas citizens supported Scenario 3, though some Panhandle residents opposed Scenario 4. He also noted that a number of cities, as well as certain other generators and transmission companies, and notably, the PUCT staff, opposed the more aggressive transmission build-out. Concerns included increased cost, lack of sufficient analysis, and, in the case of Scenario 4, the possibility of selling CREZ wind power into the Eastern Interconnection. He argued that Scenario 2 offered the best possibility of fulfilling the CREZ mission cost-effectively, while also providing capacity for future energy development. Years later, Smitherman reflected that the letters from the third graders had struck him as a demonstration of the importance of wind power to the region.[70] On August 15, 2008, the PUCT issued an order designating Scenario 2 as the preferred plan for transmission corridors, thus assuring that wind power developed in the Panhandle would find its way into the

[69] Docket 33672, *Memo to Commissioners Re: The Benefits of Scenario 2*, July 17, 2008.

[70] Smitherman, Communication with the author.

Texas grid.[71] During dozens of ensuing cases, the PUCT then selected transmission service providers and designated specific routes.

7.4.1 The Hill Country Lines: A "Huge Slugfest"

While residents of the Texas Panhandle had expressed unbridled enthusiasm for wind power and its related infrastructure, the citizens of the Texas Hill Country felt differently. This region of Texas is heralded for its scenic beauty; and the local economies, although primarily agricultural, depend upon the revenues from tourists, weekend farmers, and retirees. The area lies within a half-day drive from Houston, Dallas, Fort Worth, San Antonio, and Austin – all the major population centers of the state. While many in this area would no doubt style themselves as environmentalists in favor of protection of ecosystems, other matters were of graver concern during the CREZ process. Notably, the most cost-effective way to connect the Central and Central West CREZ zones to the Texas grid required new transmission lines across the Hill Country region. In other words, residents of the Hill Country would host large power lines on their land with no off ramps to their local electricity providers. While, in reality, new wind generation would serve all customers on the grid, conceptually it seemed to bypass the Hill Country.

In January 2009, the PUCT selected the Lower Colorado River Authority Transmission Service Corporation (LCRA), an incumbent provider, to build several lines in the Hill Country. Figure 7.6 illustrates the location of the LCRA lines in red, and arrows point to the ones that became highly contentious. LCRA initiated communications with affected landowners almost immediately.[72] The entity contracted with a consultant to develop route alternatives, invited the public to open house meetings for informal review of the alternatives, solicited landowner input, returned the information to the consultant, and repeated the process – all before submitting proposed routes to the PUCT.[73] The process, however, did not always result in favorable reactions from the public participants, as evidenced in numerous letters to Governor Perry.[74]

At the outset of the case to consider Hill Country lines, Chairman Smitherman delineated the issues the commission would consider, with community values at the top of the list. In correspondence with area citizens, he wrote, "The siting of high capacity transmission lines, while of critical importance to our state, can be a difficult and emotional process for landowners; therefore it is very important that we adhere

[71] Docket 33672, *PUC CADM: Order*, August 15, 2008.

[72] Jonathan Greene, Senior Vice President, Transmission Operations, Lower Colorado River Authority, Communication with the author, August 14, 2020.

[73] Greene, Communication with the author.

[74] Docket 35665, *Letter to Gov. Rick Perry Regarding Signatures Petition Opposing Transmission Line in Gillespie County*, January 14, 2009.

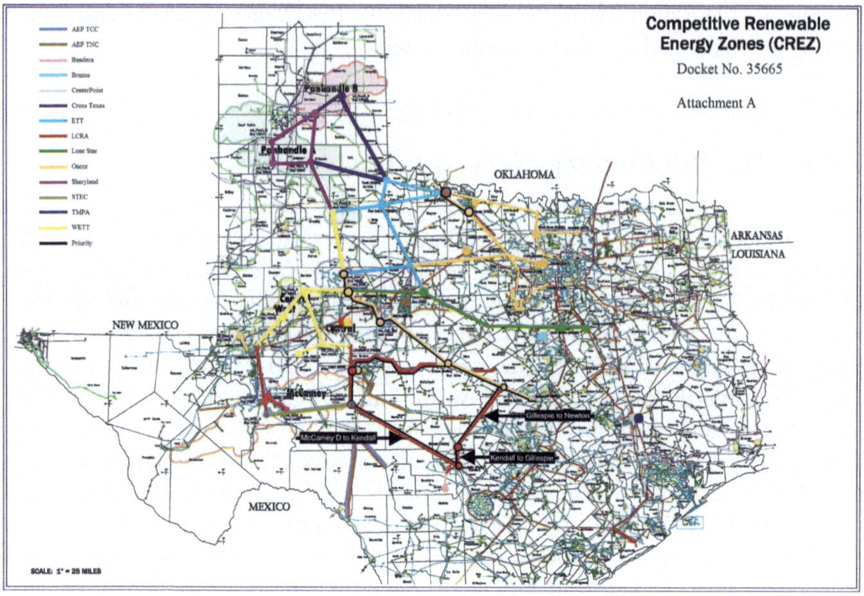

Fig. 7.6 Map showing Assignment of CREZ Lines to Transmission Service Providers. (Source: PUCT Final Order in Case 356655)

to the criteria outlined in both statute and Commission rule."[75] Two Hill Country lines – McCamey D to Kendall and Gillsepie to Newton, attracted a great deal of attention.

From counties central to LCRA's study area for the McCamey D to Kendall route, commenters raised questions about health impacts, environmental impacts, and disruption of scenic areas. They encouraged the use of monopoles or buried lines if the transmission route came through their areas. And they begged the commission to expand the study area to include existing rights of way to the north, through the city of Mason, and to the south, through Kerrville. Notably, they passionately defended the "pristine environment of the Hill Country of Texas."[76] In late July, the PUCT acquiesced to the pressure of so many commenters and ordered LCRA to expand its study area.[77] Shortly thereafter, LCRA requested permission to divide its Hill Country lines into two dockets, one with a deadline of October 2009 for the Gillespie to Newton line, which was unaffected by the expanded study area; and

[75] Docket 37409, *Chairman Barry Smitherman: Various Comments and Responses*, May 28, 2009.

[76] Docket 37409, *Commissioner Smitherman: Response to Letter from Mr. And Mrs. Merritt*, July 20, 2009.

[77] Docket 37409, *Chairman Smitherman: Response to Comments from the City of Junction*, August 4, 2009.

another for the McCamey D to Kendall and Kendall to Gillespie lines, with a deadline for submission of June 2010.[78] The PUCT agreed to this revised schedule.

This marked the beginning of a north-south divide in the battle over the Hill Country CREZ lines. The city of Junction, as an example, specifically encouraged the commission to select a route for the McCamey D to Kendall line along the northern border of the new study area, directly through Mason (hereafter the Mason-Menard Route).[79] In contrast, the mayor of Mason wrote to express the dismay of his entire area at the possibility that a high voltage line might transect their community.[80] He noted that the original study area included the most direct and least expensive routes for new power lines while the new study area included important scenic and historic sites.

Mason residents began a full court press for a southern CREZ route (hereafter the I-10 Route).[81] With the help of land rights non-profit American Stewards of Liberty, the city and county formed a subregional planning commission under the Texas Regional Planning Act of 1965.[82] This purportedly gave them standing before both state and federal agencies, potentially allowing them to demand a full Environmental Impact Statement for a transmission line through their area.[83] Others along the Mason-Menard Route established the Texas Hill Country Heritage Association to oppose the power line.[84] Many from outside the region, including the Sierra Club, supported Mason's position.[85]

While the comment phase for Hill Country CREZ lines continued, the PUCT opened the case to address the proposed routing of the Gillespie to Newton line.[86] This would ultimately link with the McCamey D to Kendall and Kendall to Gillespie lines, and would add a long-sought north-south high-voltage line between San Antonio and west Austin. The PUCT turned the case over to an Administrative

---

[78] Docket 37409, *LCRA TSC: Joint Motion of LCRA TSC and Commission Staff to Delay Filing Date of Competitive Renewable Energy Zones (CREZ) Priority Projects and Request for Emergency Relief*, September 16, 2009.

[79] *Chairman Smitherman: Response to Comments from the City of Junction*, August 4, 2009.

[80] Docket 37409, *Response to Comments of Mayor of the City of Mason Brent Hinckley*, September 25, 2009; City Commission Minutes, September 14, 2009 and attached letter from Brent Hinckley to Barry Smitherman, dated September 16, 2009 (made available via email, courtesy of Mason City Secretary).

[81] Minutes Special Meeting/Town Hall Meeting, September 28, 2009 (made available via email, courtesy of Mason City Secretary).

[82] City Commission Minutes, October 19, 2009 (made available via email, courtesy of Mason City Secretary); "Texas Local Government Code, Title 12. Subtitle C. Chapter 391. Regional Planning Commissions"; "American Stewards of Liberty," 2020, accessed various, https://americanstewards.us/.

[83] NEPA, 42 U.S.C. sections 4321, 4331, 4332.

[84] Docket 37049, *Texas Hill Country Heritage Association*, November 3, 2009; "Texas Hill Country Heritage Association," 2020, accessed various, 2020, http://thcha.org/.

[85] Docket 37049, *Sierra Club, Lonestar Chapter: Comments*, September 21, 2009.

[86] "Interchange Filings Search: Filings for 37448," Public Utility Commission of Texas, 2020, accessed various, http://interchange.puc.texas.gov/Search/Filings?ControlNumber=37448.

Law Judge before whom LCRA presented several options. This new CREZ line would cross a scenic region with important recreation and wilderness amenities. Through 1287 filings, hundreds of area landowners, realtors, chambers of commerce, and others expressed opposition to every section of each route. In early spring 2010, LCRA indicated a preferred route, the commission's Legal department promoted a second route, and the judge chose a third.[87] In a surprise move, the commission then rejected all the proposed routes and asked ERCOT to reconsider the need for both the Kendall to Gillespie and the Gillespie to Newton Lines.[88]

As might be expected, those commenting on the overall Hill Country CREZ project took note. If the PUCT might eliminate two Hill Country lines, why not eliminate a third? In June, Kerr County Commissioners passed a resolution specifically endorsing the Mason-Menard Route while opposing the I-10 Route.[89] One month later, the court took up a resolution asking the PUCT to reconsider the necessity of any McCamey D to Kendall line.[90] Mason, meanwhile, continued strong opposition to the northern route. Through resolutions and other communications, the city, the county, and the subregional planning commission indicated their position to the PUCT.[91] The American Stewards of Liberty touted progress in opposing power lines.[92] Like the Kerr County advocates, the Mason advocates also called for reconsideration of the McCamey D to Kendall line.[93] Over the summer, local governments across the Hill Country attempted, unsuccessfully, to coalesce around a regional approach to submit to the PUCT.[94]

In August 2010, ERCOT reported that the Kendall to Gillespie and Gillespie to Newton lines were not needed and the PUCT removed them from the infrastructure plan. Years later, PUCT chairman Smitherman recalled that both legislators and local citizens objected to the lines and it appeared that existing infrastructure might

[87]Docket 37448, *SOAH: PFD*, March 18, 2010; Docket 37448, *PUC Legal: Commission Staff's Reply Brief*, March 2, 2010; Docket 37448, *LCRA-TSC's Initial Post-Hearing Brief*, February 26, 2010.

[88]Greene, Communication with the author.

[89]Kerr County Commissioners Court Special Session, June 21, 2010, Minutes, 83, http://www.co.kerr.tx.us/commcrt/minutes/2010/052410cc.txt.

[90]Kerr County Commissioners Court Regular Session, July 12, 2010, Minutes, http://www.co.kerr.tx.us/commcrt/minutes/2010/071210cc.txt.

[91]City Commission Minutes, April 12, 2010; City Commission Minutes, May 17, 2010 (both made available via email, courtesy of Mason City Secretary); Public Utility Commission of Texas, *Mason County Appraisal District: Comments*, April 7, 2010.

[92]"The Politics of Transmission Lines," Scripps Interactive Newspapers Group, updated June 29, 2010, accessed November 20, 2013, http://www.greentechmedia.com/articles/read/in-texas-big-wind-jumps-on-new-transmission; Dan Byfield, "Protecting Private Property Locally," *Ranch and Rural Living* 91, no. 87 (May, 2010): 8–10.

[93]City Commission Minutes, July 12, 2010 (made available via email, courtesy of Mason City Secretary); Docket 38354, *Save Our Scenic Hill Country Environment, Inc.: Comments*, August 12, 2010.

[94]City Council Minutes, Regular Meeting, August 24, 2010. https://www.kerrvilletx.gov/ArchiveCenter/ViewFile/Item/203.

support the increased wind power transmission.[95] Most importantly, by eliminating the lines, the PUCT saved well over $100 million, and "took opponents off the table."[96] ERCOT did, however, state that there was no viable alternative to the highly contested McCamey D to Kendall Line, disappointing landowners and local groups from Kerrville to Mason, and beyond.

Hill Country residents switched their attention to the reviled McCamey D to Kendall route, which became the lengthiest and most hotly disputed case in the CREZ process.[97] New landowner groups added their voices. The Clear View Alliance, Inc., for example, representing at least 240 landowners, specifically endorsed the I-10 Route through Kerrville.[98] Clear View was aggressive in its efforts to protect the interests of its membership, authoring more than ninety filings in the case. Likewise, residents of Mason County organized the P-Line Intervention Association and provided direct testimony from thirteen paid and volunteer experts to the PUCT.[99] These individuals opined upon the effects of a transmission line on the area's cultural, archeological, environmental, financial, and business resources. They addressed property values, aesthetics, and ecological concerns. One specifically called out Kerr County for expressing "in the strongest possible terms, their opposition to even one transmission line in Kerr County because of the claimed devastating impact that a single transmission line would allegedly have upon the area" when three to four lines could be expected elsewhere.[100] The back and forth between the northern and southern route landowners continued through the fall.

In the end, Mason won, and Kerrville lost, when the PUCT selected a modified I-10 Route (as shown in Fig. 7.7).[101] Mason residents touted their victory and credited the role of the subregional planning commission.[102] In its final order, the PUCT called for collaboration with landowners and municipalities on materials used, and height and spacing of structures, and further specified that landowner views prevailed when disputes arose. This did not palliate Kerr residents, who first

---

[95] Smitherman, Communication with the author.

[96] Ibid.

[97] "Interchange Filing Search: Filings for 38354," Public Utility Commission of Texas, 2020, accessed various, http://interchange.puc.texas.gov/Search/Filings?ControlNumber=38354.

[98] Docket 38354, *Clear View Alliance, Inc.: Direct Testimony of Bill Neiman*, September 28, 2010.

[99] Docket 38354, *P-Line Intervention Association: Direct Testimony of Scott Zesch, Ron Crocker, Michael J. Heaney, III, Ph.D., Mayor Brent Hinckley, Marvin G. Pipkin, Kevin Ramberg, Judge Jerry Bearden, G.T. Bohmfalk, Ph.D., Don Daniel, Dean Lagle Kothmann, Bobby Gierisch, Ann M. Scott*, September 28, 2010.

[100] *P-Line Intervention Association: Direct Testimony of Marvin G. Pipkin*, 2, September 28, 2010.

[101] Docket 38354, *PUC ADM Final Order*, 24, January 24, 2011; *LCRA Transmission Services Corporation Competitive Renewable Energy Zone Transmission Lines: Environmental Assessment*, SWCA Environmental Associates (Austin, TX, April 2012).

[102] Docket 38354, *PUC ADM Final Order*, 24, January 24, 2011; "Coordination Stops Transmission Line Deep in the Heart of Texas," American Stewards of Liberty, updated February 2, 2011, accessed November 20, 2013, http://www.americanstewards.us/previous-articles/43-coordination-works-previous-articles/637-coordination-stops-transmission-line-deep-in-the-heart-of-texas.

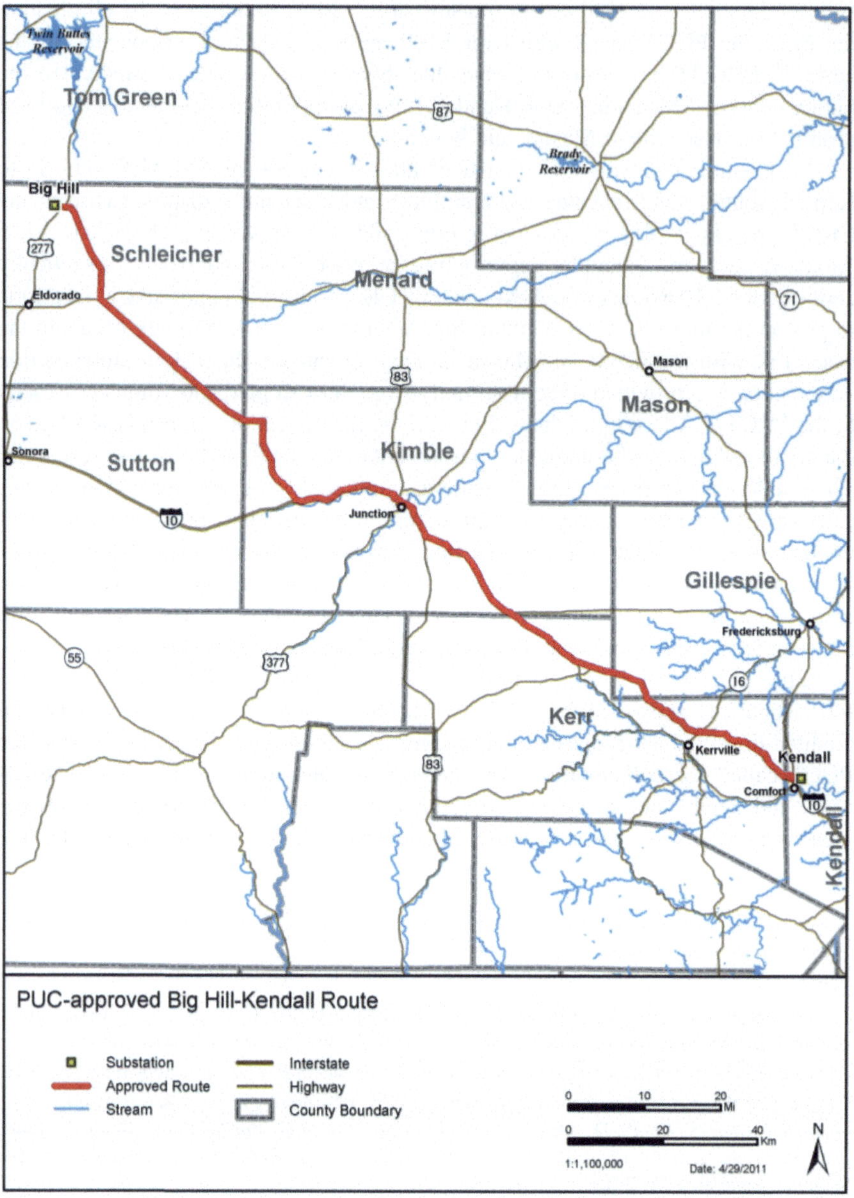

Fig. 7.7 Map illustrating final route of McCamey D to Kendall CREZ transmission line through Kerrville, Texas. (Source: LCRA TSC Environmental Assessment, p. 22)

appealed the PUCT's decision and later sued the PUCT, then LCRA, over details of the final order.[103] Ultimately the city and county settled the case.[104] LCRA placed the CREZ line into service in 2014, three years later than originally planned.[105]

When retiring from his position in 2014, the Kerr County Attorney offered sanguine reflections on the fight to oppose the I-10 Route, "We went through the huge slugfest over the CREZ transmission lines, and even though ... we weren't successful in that endeavor, I absolutely believe that we did the right thing in

[103] Kerr County Commissioners Court Special Session, February 3, 2011. http://www.co.kerr.tx.us/commcrt/minutes/2011/020311cc.txt; City Council Minutes, Regular Meeting, February 2, 2011. https://www.kerrvilletx.gov/ArchiveCenter/ViewFile/Item/387; LCRA CREZ Correspondence Files, *Email from Robert Henneke to Cojudge@Co.Kerr.Tx.Us; 'Buster Baldwin'; 'Guy Overby'; 'Jonathan Letz'; Boehler@Co.Kerr.Tx.Us Re: LCRA TSC's Response Brief*, Kerr County Commissioners Court, May 17, 2011; Docket 38354, *Motion for Rehearing of City of Kerrville, Kerr County, Kerrville Public Utility Board, and City of Junction*, February 16, 2011; "Fresh Tactic Tried in Power-Line Tiff," San Antonio Express-News, updated June 22, 2011, 2011, accessed November 20, 2013, http://www.mysanantonio.com/news/local_news/article/Fresh-tactic-tried-in-power-line-tiff-1434055.php.

[104] Kerr County Commissioners Court Special Session, September 7, 2011. http://www.co.kerr.tx.us/commcrt/minutes/2011/090711cc.txt; City Council Minutes, Regular Meeting, November 8, 2011. https://www.kerrvilletx.gov/ArchiveCenter/ViewFile/Item/657.

[105] "Power Fight Amps up Again," San Antonio Express-News, updated September 19, 2012, 2012, accessed November 20, 2013, http://www.mysanantonio.com/news/local_news/article/Power-fight-amps-up-again-3872420.php; Kerr County Commissioners Court Regular Session, September 10, 2012. http://www.co.kerr.tx.us/commcrt/minutes/2012/091012cc.txt; City Council Minutes, Regular Meeting, August 28, 2012. https://www.kerrvilletx.gov/ArchiveCenter/ViewFile/Item/2336; City Council Minutes, Regular Meeting, September 11, 2012. https://www.kerrvilletx.gov/ArchiveCenter/ViewFile/Item/977; City Council Minutes, Regular Meeting, July 24, 2012. https://www.kerrvilletx.gov/ArchiveCenter/ViewFile/Item/939; "Fighting Power with Power," Braun & Gresham, PLLC, 2013, accessed November 21, 2013, http://www.braungresham.com/wp-content/uploads/2013/10/BG_Success_WhiteDeer-UPDATED-10.1.13.pdf.

standing up for this community."[106] Figure 7.8 illustrates the view approaching Kerrville from the west, both before and after the CREZ lines.[107]

7.5 Who Are the Gatekeepers?

Throughout the CREZ process, tens of thousands of putative stakeholders contributed opinions and comments. The legislators who had enacted CREZ pushed to see the process completed swiftly. State and national environmental groups promoted wind power, endorsed CREZ, called for a fast and aggressive transmission plan, and failed to acknowledge local concerns. Meanwhile, state agencies, ranchers, newly formed landowner associations, and others argued against transmission infrastructure that would endanger various species. Residents of the Panhandle and their elected officials made a dramatic case for CREZ and claimed to welcome transmission lines. In contrast, many residents of the Texas Hill Country defended their "pristine" landscapes. None saw themselves as direct beneficiaries of wind development and few seemed concerned with the cost of new transmission. Instead, the interests of Hill Country intervenors were decidedly local, focused on scenic beauty, protection of land rights, promotion of tourism, and a willingness to burden their neighbors with the power line. Mason County residents gave lip service to environmental issues, but more importantly used the tools of an unrelated state law and the National Environmental Protection Act to strengthen their position vis-à-vis other counties.

Agencies and investors participated as well. Incumbent power generators generally lobbied for decisions that would minimize new wind development, while wind producers lobbied for even more than the law required. The transmission service providers expressed neutrality with respect to particular routes for their power lines, knowing they would ultimately build something, somewhere, and reclaim the costs (and future profits) under the state's postage stamp policy. They were sensitive, however, to accusations of back room deals and failure to listen to the people they had served for decades. ERCOT provided technical input, theoretically addressing reliability and stability issues above all else. But ERCOT planning included nontechnical stakeholders as well.

The ultimate authority rested with the Public Utility Commission. With clear directives from the legislature, the PUCT had to answer only to their extensive public record. While some accused the commission of bending to politics, the many

[106] Kerr County Commissioners Court Regular Session, March 10, 2014, Short.

[107] *Google Maps Street View*, https://www.google.com/maps/@30.0720139,-99.1245463,3a,75y,301.64h,90t/data=!3m8!1e1!3m6!1sCCLMUD7BagFGk5CZJuEp4w!2e0!5s20191001T000000!6s%2F%2Fgeo1.ggpht.com%2Fcbk%3Fpanoid%3DCCLMUD7BagFGk5CZJuEp4w%26output%3Dthumbnail%26cb_client%3Dmaps_sv.tactile.gps%26thumb%3D2%26w%3D203%26h%3D100%26yaw%3D359.3298%26pitch%3D0%26thumbfov%3D100!7i16384!8i8192, accessed August 18, 2020.

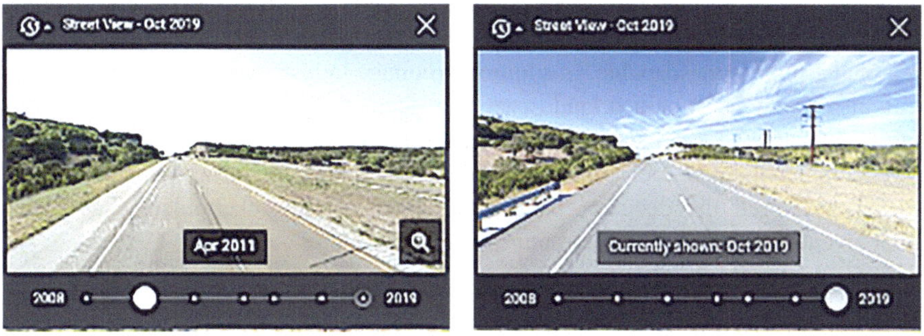

Fig. 7.8 Images showing Interstate 10 approaching Kerrville from the West before and after CREZ lines. (Source: Google Maps Street View)

dockets for each CREZ line suggest that the technical requirements of the law, the commission's policies, and the input of directly affected citizens framed PUCT decisions. The PUCT added to ERCOT's recommended CREZ in recognition of the outstanding wind resources in both Deaf Smith and Briscoe Counties, but also to protect their decision-making in front of legislators and area landowners. The commission selected Scenario 2 for transmission corridors, acknowledging both the technical and cost concerns of state agencies and the economic and environmental interests of many others. The PUCT deleted two Hill Country power lines following extensive citizen input that brought attention to a "no good choices" scenario.[108] When faced with Hill Country landowners promoting opposing plans, the PUCT chose a route that appeased some, appalled others, and cost more to statewide ratepayers.

As former Chairman Smitherman recalled, both the statute and the commissions' rules gave direction for how to think about routing.[109] The legislation called for new power lines, ERCOT's technical review established the need for each line, and through the postage stamp rate everyone would pay for it. The statute required the commission to consider community values, environmental integrity, and aesthetics. The commission's rules stated that the priority for routes went first to the unused side of existing transmission towers, then to existing rights of way, then along property boundaries. The last choice was to "go straight across grandma's farm ... from one side to the other." In particular, Smitherman noted, when LCRA conducted open houses, the number one concern was preserving the integrity of the Hill Country. In the case of the McCamey D to Kendall line, the southern route along I-10 was the one that matched those requirements most closely.

The PUCT split the baby on several occasions – when delineating the CREZ, when selecting the transmission scenario, and when determining the routing of power lines. In each case, the written record reflects extensive input from individuals

[108] Smitherman, Communication with the author.

[109] Smitherman, Communication with the author.

and entities wrestling with the physicality of giant transmission towers on their landscapes. The role of these thousands of landowners, local governments, and regional groups is inescapable. In addition, a cottage industry of lawyers specializing in transmission cases helped landowners and local governments undertake increasingly sophisticated approaches to project negotiation.[110] As the transmission service providers formulated their options, ERCOT reviewed technical issues, and the PUCT navigated the decision-making for routes, local stakeholders shaped the final extent of the CREZ infrastructure.

7.6 Looking Backward, Looking Ahead

From the early days of electrification, landowners, local governments, and regional groups influenced the development of electricity infrastructure. In the 1910s, local communities in Massachusetts blocked the development of power lines through their towns, even though their own utilities hoped to bring down electricity rates through interconnections. But later Massachusetts granted utilities the power of eminent domain. In the 1970s, Minnesota farmers tore down transmission towers because they could find no legal means to stop a giant transmission line. But transmission line owners, in this case rural cooperatives, replaced the towers; and the transmission lines later contributed power to the pool serving those farms. In the early 2000s, Texas landowners and local governments helped establish the outlines of new zones for wind power development and the routes and distance traveled for new transmission lines to bring the wind into the state's grid.

What is the same and what has changed? Two things stand out as being the same. In the first place, the opponents to electricity infrastructure in the farther past, just like the CREZ intervenors, based their activism on the immediate physicality of the planned infrastructure: would it mar their view? damage their property? cause them health problems? interfere with the ecosystem? Or, as in the case of Texas Panhandle constituents, would new power lines bring jobs and tax revenues to improve their daily lives? Second, the implicit benefit of the infrastructure – a sustainable future built upon renewables – was of no interest. The access to power itself, or cheaper power, or power generated with more desirable resources played little role in the decisions of these advocates.

Many things have changed. The federal government explicitly regulates interstate power transmission, and now Texas is the only state in the continental United States with its own autonomous grid. Further, ERCOT is the only U.S. system on which transmission costs are shared on a postage stamp basis. In the CREZ case, the legislature, the PUCT, ERCOT, and even the transmission developer all deliberately

[110] Thank you to Warren Lasher and Michael Greene, former ERCOT board chair and senior executive for several Texas utilities, for emphasizing this point. Michael Greene, Communication with the author, August 12, 2020.

sought public input. This was baked into changes in Texas power markets from 1995 onward. Beyond Texas, federal and state laws and energy-regulator policies explicitly call for stakeholder involvement at many stages of power system development. Yet the definition of stakeholder is fluid, and typically refers to power generators, transmission providers, rate-payer groups, environmental organizations, and regulators. The term stakeholder less often explicitly includes the thousands and thousands of landowners and local government groups directly affected by the physical presence of transmission lines. Instead, these participants are described as "others." At the same time, laws and court cases, have created a space for individuals and new entities to have standing in infrastructure development. In the Texas case, Mason leaders availed themselves of state and federal laws to create a new entity that some – though not all – credited with adding clout to their advocacy.

By searching out these historically invisible stakeholders, scholars have the opportunity to better understand how and why power lines occupy certain spaces and stretch certain distances. We can disentangle the explicit and implicit shared costs and benefits, and we can then underscore how local concerns matter. We can look for the patterns that might frame future electricity development. We can more thoroughly understand who will determine where and when and at what costs and for what benefits technology infrastructure is built. The importance of local stakeholders has grown over the past century. In the twenty-first century, those facing the physicality of transmission infrastructure in their daily lives are already keeping the gate for utility-scale renewables.

Bibliography

American Stewards of Liberty website. https://americanstewards.us/.

Electrical World, various volumes, available online at HathiTrust Digital Library. https://www.hathitrust.org/.

Brief History of the Rural Electric and Telephone Programs. 1986. *Washington*. DC: Rural Electrification Administration, US Department of Agriculture.

Braun & Gresham, PLLC website. https://braungresham.com/.

Brigham, Jay L. 1998. *Empowering the West: Electrical Politics before FDR. Development of Western Resources*. Lawrence: University Press of Kansas.

Byfield, Dan. 2010a. *The Politics of Transmission Lines. San Angelo Standard-Times* (San Angelo, TX), June 29 2010.

———. 2010b. Protecting Private Property Locally. *Ranch and Rural Living* 91 (87): 8–10.

Carlson, W. Bernard. 2013. *Tesla: Inventor of the Electrical Age*. Princeton: Princeton University Press.

City Commission Minutes, City of Mason, Texas, 2009–2010.

City Council Minutes, City of Kerrville, Texas, 2010–2012.

Cohn, Julie. 2017. *The Grid: Biography of an American Technology*. Cambridge, MA: MIT Press.

Cowan, Ruth Schwartz. 1983. *More Work for Mother: The Ironies of Household Technology from the Open Hearth to the Microwave*. New York: Basic Books.

Diffen, Becky H. 2009. Competitive Renewable Energy Zones: How the Texas Wind Industry Is Cracking the Chicken & Egg Problem. *Rocky Mountain Mineral Law Foundation Journal* 46 (1): 47–98.

Electric Reliability Council of Texas website. https://www.ercot.com/.

Funigiello, Philip J. 1973. *Toward a National Power Policy; the New Deal and the Electric Utility Industry, 1933–1941*. Pittsburgh: University of Pittsburgh Press.

Glaser, Leah S. 2009. *Electrifying the Rural American West: Stories of Power, People, & Place*. Lincoln, NE: University of Nebraska Press.

Gooday, Graeme. 2008. *Domesticating Electricity: Technology, Uncertainty and Gender, 1880–1914*. London: Pickering & Chatto.

Gould, Madeline Claire. 2018. *Everything's Bigger in Texas: Evaluating the Success and Outlook of the Competitive Renewable Energy Zone (Crez) Legislation in Texas*. Master of Arts, University of Texas.

Hirsh, Richard F. 1999. *Power loss: The origins of deregulation and restructuring in the American electric utility system*. Cambridge, MA: MIT Press.

Hughes, Thomas Parke. 1983. *Networks of Power: Electrification in Western Society, 1880–1930*. Baltimore: Johns Hopkins University Press.

Hurlbut, David. 2008. A Look Behind the Texas Renewable Portfolio Standard: A Case Study. *Natural Resources Journal* 48 (1): 129–161.

Israel, Paul. 1998. *Edison: A Life of Invention*. New York: John Wiley.

Jones, Christopher F. 2014. *Routes of Power: Energy and Modern America*. Cambridge, MA: Harvard University Press.

Kerr County Commissioners Court Minutes, Kerr County, Texas, 2010–2014.

Kiesling, L. Lynne, and Andre N. Kleit, eds. 2009. *Electricity Restructuring: The Texas Story*. Washington, DC: AEI Press.

Kline, Ronald R. 2000. In *Consumers in the Country: Technology and Social Change in Rural America. Revisiting Rural America*, ed. Pete Daniel and Mary C. Neth. Baltimore: Johns Hopkins Press.

Lambert, Jeremiah D. 2015. *The Power Brokers: The Struggle to Shape and Control the Electric Power Industry*. Cambridge: Massachusetts Institute of Technology Press.

LCRA Transmission Services Corporation Competitive Renewable Energy Zone Transmission Lines: Environmental Assessment. 2012. SWCA Environmental Associates (Austin, TX: April 2012). (Available from the Austin Ecological Field Services Office, U.S. Fish and Wildlife Service, https://www.fws.gov/office/austin-ecological-services).

Lehr, R.L., W. Guild, D.L. Thomas, and B.G. Swezey. 2003. *Listening to Customers: How Deliberative Polling Helped Build 1,000 MW of New Renewable Energy Projects in Texas*. National Renewable Energy Lab.

Lieberman, Jennifer L. 2017. *Power Lines: Electricity in American Life and Letters, 1882–1952*. Cambridge, MA: MIT Press.

Lifset, Robert. 2014. *Power on the Hudson: Storm King Mountain and the Emergence of Modern American Environmentalism*. Pittsburgh: University of Pittsburgh Press.

Melosi, Martin V. 1985. *Coping with Abundance: Energy and Environment in Industrial America*. 1st ed. Philadelphia: Temple University Press.

Needham, Andrew. 2014. In *Power Lines: Pheonix and the Making of the Modern Southwest. Politics and Society in Twentieth-Century America*, ed. William Chafe, Gary Gerstle, Linda Gordon, and Julian Zelizer. Princeton, Princeton University Press.

Nye, David E. 1990. *Electrifying America: Social Meanings of a New Technology, 1880–1940*. Cambridge, MA: MIT Press.

Passer, Harold C. 1972. *The Electrical Manufacturers, 1875–1900; a Study in Competition, Entrepreneurship, Technical Change, and Economic Growth*. Technology and Society. New York: Arno Press.

Phillips, Sarah T. 2007. *This Land, This Nation: Conservation, Rural America, and the New Deal*. Cambridge: Cambridge University Press.

Platt, Harold L. 1991. *The Electric City: Energy and the Growth of the Chicago Area, 1880–1930*. Chicago: University of Chicago Press.

Pope, Daniel. 2008. *Nuclear Implosions: The Rise and Fall of the Washington Public Power Supply System*. New York: Cambridge University Press.
Pratt, Joseph A. 1988. *A Managerial History of Consolidated Edison, 1936–1981*. New York: Consolidated Edison Company of New York.
Public Utility Commission of Texas Interchange Filing Search, Control Numbers: 31852, 33672 35665, 37409, 37448, 38290, 38354, 45624. https://interchange.puc.texas.gov/.
Rose, Mark H. 1995. *Cities of Light and Heat: Domesticating Gas and Electricity in Urban America*. University Park, Pa: Pennsylvania State University Press.
Sandwell, Ruth W., ed. 2016. *Powering up Canada: A History of Power, Fuel, and Energy from 1600*. McGill-Queen's Press-MQUP.
Staine, R. Ryan. 2014. Crez II, Coming Soon to a Windy Texas Plain Near You: Encouraging the Texas Renewable Energy Industry through Transmission Investment. *Texas Law Review* 93 (2): 521–556.
Stokes, Leah Cardamore. 2020. *Short Circuiting Policy: Interest Groups and the Battle over Clean Energy and Climate Policy in the American States. Studies in Postwar American Political Development*. Oxford: Oxford University Press.
Texas Hill Country Heritage Association. http://thcha.org/.
Texas Local Government Code, Title 12. Subtitle C. Chapter 391. Regional Planning Commissions.
Tobey, Ronald C. 1996. *Technology as Freedom: The New Deal and the Electrical Modernization of the American Home*. Berkeley: University of California Press.
Wellock, Thomas Raymond. 1998. *Critical Masses: Opposition to Nuclear Power in California, 1958–1978*. Madison: The University of Wisconsin Press. http://www.h-net.org/review/hrev-a0b3a3-aa.
Wellstone, Paul David, and Barry M. Casper. 2003. *Powerline: The First Battle of America's Energy War*. 1st ed. Minneapolis: University of Minnesota Press. http://www.loc.gov/catdir/toc/ecip044/2003011745.html.
Wuebben, Daniel. 2019. *Power-Lined*. Lincoln, NE: University of Nebraska Press.

Julie A. Cohn is a research historian in the Center for Public History at the University of Houston and a non-resident scholar at the Center for Energy Studies in Rice University's Baker Institute for Public Policy. Her work focuses on energy infrastructures, environmental history, technological change, and the relationships between government, business, and the public. Cohn's book, *The Grid, Biography of an American Technology*, examines the history of electrification in North America, and especially the story of how and why power companies chose to interconnect.

Chapter 8
Co-ops Against Castroism: USAID and the Electrification of the Global Countryside

Abby Spinak

Abstract This chapter explores electricity programs financed by the United States Agency for International Development (USAID) as an instrument of Cold War diplomacy. In the 1960s–70s, USAID promoted and funded the development of numerous community-owned electric cooperatives across the global South. As an explicitly anti-communist intervention, however, USAID electric co-ops were not promoted as collective ownership as much as a form of individualistic "self-help." I show here that USAID built on lessons from a federally-funded cooperative program implemented in the United States during the New Deal to promote a cooperative model that – much as federal agents had done in the U.S. after the Great Depression – attempted to reframe collective ownership as a capitalist institution in the global South during the Cold War. In exploring USAID's early promotion of electric cooperatives in rural places outside of American borders, the chapter revisits the role of cooperative thought in Cold War era conceptions of liberal democracy and developmental state capitalism. Beyond the question of electricity ownership, the chapter further traces the symbolic role electrification came to hold in USAID ideology as a panacea for a wide range of development goals, despite contradictory evidence from programs on the ground. The contradictions raised by the history of USAID cooperative electricity ultimately demonstrate the importance of separating ideology – the stories people tell about electricity – from practice – the social and spatial relations electrification has helped manifest – in electricity history.

Keywords United States Agency for International Development (USAID) · Cold war diplomacy · Community-owned electric cooperatives · Global south · Developmental state capitalism · Electrification and ideology

A. Spinak (✉)
Graduate School of Design, Harvard University, Cambridge, MA, USA
e-mail: spinak@gsd.harvard.edu

W. B. Carlson, E. M. Conway (eds.), *Electrical Conquest*, Archimedes 67,
https://doi.org/10.1007/978-3-031-44591-0_8

> The word 'mine' is a powerful stimulus and works magic. It causes houses to be painted, streets to be paved, fields to become more productive, children to be educated, and battles to be won. We minimize it here where some property owned is taken for granted. This is not so with the peasant, land starved for centuries, or with the shifting part-time laborer. To both, a bicycle may be an out-of-reach luxury. To these people, the feeling of ownership and belonging which evolves from their cooperative can be a strong stabilizing ballast to citizenship. . . . A country covered with local cooperatives and their members – understanding owners – does not fear revolutions and military coups.
>
> –Senator Hubert Humphrey, 1962[1]

> Development without electricity is difficult to imagine.
>
> –Gary Wasserman and Alice Davenport, USAID, 1983[2]

8.1 Introduction

In August of 1980, an American lawyer and two economists spent the afternoon in the home of a small rice farmer in the rural town of Daule in western Ecuador. It was a Sunday, a day of rest for the farmer. Not so for the inquiring visitors, however, who were there on the job. They had come to evaluate a rural electrification program in Daule and its environs that had been organized and financed by the United States Agency for International Development (USAID) – initially as a community-owned cooperative (co-op), though by 1980 the utility had been subsumed by a public power authority. The Ecuadorian farmer relaxed in a hammock in his living room while he hosted his industrious guests. Nearby, his wife, two sons and a daughter, and his elderly father watched television. "As we talked," the USAID team recorded, "a look of satisfaction crossed the farmer's face when his aged toothless father, at one point, jumped up grinning and gestured at the screen to call attention to a funny scene." Very moved by the farmer's intimate experience of residential electricity, the Americans concluded, "We don't know how to value such events."[3]

This moment of familial contentment was one of many that eluded the capacity of the USAID team's evaluation metrics, as they strove to assess an electrification program implemented in the 1960s under a different diplomatic directive. In its formative stage as an electric co-op, the Daule utility was a macroeconomic intervention forged in Cold War politics and owing its institutional legacy to American

[1] Senate Document No. 112, 87th Congress, 2d Session, USAID, "Implementation of the Humphrey Amendment to the Foreign Assistance Act of 1961," July 1962, Washington, D.C.: U.S. Government Printing Office, III, Hathi Trust (henceforth HT).

[2] Gary Wasserman and Alice Davenport, "Power to the People: Rural Electrification Sector," AID Program Evaluation Report No. 11, December 1983, 1, HT.

[3] Judd L. Kessler, Janet Ballantyne, Robert Maushammer, and Nelson Romero Simancas, "Ecuador: Rural Electrification," A.I.D. Project Impact Evaluation Report No. 21, June 1981, 13, HT.

New Deal ideologies about energy and directed economies. By the 1980s, however, the USAID research team was in Ecuador on a "New Directions" mandate focused on "basic human needs," which meant that aid funds were supposed to prioritize disbursements that would "directly improve the lives of the poorest," targeting areas such as nutrition, health, and education.[4] From that perspective, the team assessed, "it's not likely that [the Daule co-op] would have been funded by A.I.D. using today's standards." Electrification, particularly by cooperative, had been touted in early USAID policy discourse as a tried-and-true method of raising the quality of life in impoverished areas, but, by the 1980s, USAID assessments of their electrification programs increasingly found contradictory and inconclusive evidence to support that assumption. In a survey of nearby Santo Domingo, for example, the Ecuadorian team found "surprising numbers" of appliances purchased on credit, which, they noted, posed major financial burdens on many of these households and indicated "significant sacrifices by the purchasers in terms of foregoing other immediate needs." This was the case even where residential surveys showed that other priorities – such as clean drinking water – had been higher on the list of locals' hopes for infrastructural development aid.[5] The team was further dismayed to find rural Ecuadorians engaging with electricity primarily for recreation instead of business investment. "[E]lectricity does not seem to be causing little 'Abe Lincolns'

[4]Kessler et al., "Ecuador," 3; Davenport, "Power to the People," Appendix A-1, 5; "S. 1443 — 93rd Congress: Foreign Assistance Act of 1973." www.GovTrack.us. 1973. December 29, 2020: https://www.govtrack.us/congress/bills/93/s1443. Though critiques of the USAID "New Directions" mandate argue that the "basic human needs" focus was largely rhetorical, shielding funding allocations that in reality "emphasized U.S. national security interests and assistance to political allies." See Mark F. McGuire and Vernon W. Ruttan, "Lost Directions: U.S. Foreign Assistance Policy Since New Directions," *The Journal of Developing Areas* 24, no. 2, 1990: 127–180.

[5]Significant historical work in the North American context has shown that demand for electrical appliances was largely constructed by utilities, policymakers, appliance manufacturers, and other boosters to meet supply, rather than vice versa. Importantly, however, these studies also demonstrate that electrical consumers did not fully allow themselves to be "disciplined" by such top-down electricity boosterism, but rather selectively engaged with electricity and electrical appliances as it fit their lifestyles and desires for the future. In other words, consumer studies show that electricity and its trappings were *both* pushed on many consumers before they wanted it *and* that consumers exhibited agency in their engagement with this highly-advertised market of electrical devices and services. No doubt similar frictions arose between the boosters of electricity in Ecuador and the consumers, who were likely *partially* taught to be electrical consumers, but engaged in electricity consumption selectively and discriminatingly through interpreting electricity boosterism through their own interests and desires. See, for example: Christopher F. Jones, *Routes of Power: Energy and Modern America* (Cambridge: Harvard University Press, 2016); Ronald R. Kline, *Consumers in the Country: Technology and Social Change in Rural America* (Baltimore and London: The Johns Hopkins University Press, 2000); and Ruth Sandwell, "Pedagogies of the Unimpressed: Re-Educating Ontario Women for the Modern Energy Regime, 1900–1940," *Ontario History* 107, no. 1 (2015): 36–59.

to sit up and study later at night," the USAID team observed.[6] USAID electricity in Daule, rather than providing comparative data on how electrification was stimulating "development" around the world, instead raised questions for the researchers about the direction of causality in that assumed relationship. Nonetheless, they were enthusiastic about electricity's secondary effects. Cataloging "substantial indirect benefits to the rural poor in terms of income, employment opportunities, and some social services – including evening literacy programs," they concluded: "One might describe the entire effort as a classic attempt at 'trickle down' economics and, horror of horrors, we are tempted to conclude that it worked."[7]

In revisiting USAID experiments with electric co-ops in the global South during the Cold War, this chapter positions electricity history as a critical lens for understanding the rise of neoliberalism as the ideological currency of economic globalization in the late twentieth century. If one of the goals of this edited volume is to provide a framework through which we can better conceptualize energy transitions, a key aim of this chapter is make political economy central to our understanding of these transitional processes. Though there are many ways electricity could have developed as a social technology, the particular model that gained hegemony in the twentieth century – central-station networked utilities heavily dependent on capital investments and continued growth across large territories with competing property rights and regulatory demands, often serving culturally and economically diverse populations – brought with it a particular set of social relations that depended for its success on naturalizing economic growth, industrial efficiency, and consumerism.[8]

[6]This observation, with its racist and classist undertones, was likely written as satire – that is, an internal critique of 1960s-era USAID programs' shortsightedness – more so than prejudicial judgment about the life choices of rural Ecuadorians in the 1980s. As I will discuss later in the paper, one of the assumptions behind USAID-funded electricity programs in Latin America in the 1960s was that electrification would by itself "naturally" spur small business development and raise the level of "culture" outside of major cities. These assumptions were thoroughly rebuked by the 1980s, in the Ecuador report and others, when research teams started assessing the results of electrification programs both qualitatively and statistically, and concluding that rural electrification needed to be part of a broader, more comprehensive set of rural aid programs tailored towards the specific local cultural and political economic contexts – which is not to say that either 1960s or 1980s USAID programs were appropriate or necessary. The point is simply that their missions differed, and 1980s USAID research thus tended to be critical of 1960s USAID electrification projects.

[7]Kessler, Ballantyne, Maushammer, and Romero Simancas, "Ecuador: Rural Electrification," ii, 5–7, 12–13.

[8]On alternative ways electricity could have been implemented as a social technology, and the non-inevitable reasons why it wasn't, see: Mark Granovetter and Patrick McGuire, "The Making of an Industry: Electricity in the United States," *The Sociological Review* 46, no. 1 (May 1, 1998); Valerie Yakubovich, Mark Granovetter, and Patrick McGuire, "Electric Charges: the Social Construction of Rate Systems," *Theory and Society* 34, 2005: 579–612; John L. Neufeld, *Selling Power: Economics, Policy, and Electric Utilities Before 1940* (Chicago: University of Chicago Press, 2016).

Particularly where electricity has been wielded as a tool of development, in the form of aid or otherwise, the history of electrification programs provides a rich venue through which historians might trace the flows of power (in the social sense) that have attempted to shape the modern world. As ideas of society have changed over time, these electrical systems have played a conservative role – once built, in the words of Thomas Hughes, they have "momentum."[9] One of the most fascinating roles electricity can play for historians, I argue, is therefore to serve as a lens for understanding how past political economies were both foreign to and yet were the foundations out of which were forged the political economic conventions and constraints of the present. They show both origins and artifacts – that is, in electricity systems, we can gleam the seeds of new economic regimes, and we can also see the frictions that they cause when the sociotechnical assemblages of electricity networks continue to assert the economic and governance visions of an earlier era. In other words, studying the history of electrification can widen our periodizations of economic eras – for example, of "trickle down" free market ideologies in foreign aid.[10]

In the case of USAID's global electrification programs in the late twentieth century, this *longue durée* political economic approach to electricity brings two key stories to light:

First, this chapter seeks to understand the political alignment in which the community cooperative served USAID as a tool of economic liberalization, setting the stage for the violent restructurings of neoliberal foreign policy in the 1980s. My aim here is unapologetically presentist: given that American discourse around cooperatives in the twenty-first century tends to paint them as an alternative economic model directly antagonistic to the free-market economics of "trickle down," I argue we need more historical attention to the variety of economic ideologies the cooperative business model has been harnessed to in different historical contexts.[11]

---

[9]Thomas P. Hughes, *Networks of Power: Electrification in Western Society, 1880–1930* (Baltimore and London: The Johns Hopkins University Press, 1983), 15. Also see Christopher F. Jones's discussion of "momentum," along with Donald Worster's idea of an "infrastructure trap" as institutional rather than deterministic reasons why the histories of infrastructure systems often make certain development paths seem "inevitable." Christopher F. Jones, "A Landscape of Energy Abundance: Anthracite Coal Canals and the Roots of American Fossil Fuel Dependence, 1820–1860," *Environmental History* 15 (July 2010), 473.

[10]For a longer discussion of the problems of periodization in the history of development, see: Joseph Morgan Hodge, "Writing the History of Development (Part 1: The First Wave)," *Humanity: An International Journal of Human Rights, Humanitarianism, and Development* 6, no. 3, 2015: 429–463.

[11]My interest here stems from recent discourse around cooperatives (and other institutions of "economic democracy" more broadly), which, in the twenty-first century, have come to be associated with radical economics and socialist politics in a way that, I argue, silences their wide usage within American state programs in the twentieth century. In contrast to the more conservative role that cooperatives played for government institutions like USAID in the mid- to late-twentieth century, activists and scholars promoting co-ops in recent years do not shy away from engaging with Marxist thought. In policy, cooperatives find purchase in the inspirational visioning that advocates use to bolster each other's resolves towards resolutions such as the Green New Deal, but not part of the central messaging crafted for wider audiences. Yet, cooperatives have played a

In exploring USAID's early promotion of electric cooperatives in Daule and other rural places outside of American borders, for example, this chapter revisits the major role of cooperative thought in Cold War era conceptions of liberal democracy and developmental state capitalism. Despite the wide application of the cooperative business model in early USAID programs, there has been relatively little historical

major role in the development of essential American industries. From agricultural corporations such as Land O'Lakes and Sunkist, to national retailers like REI (Recreational Equipment, Inc.), to a multi-billion-dollar network of electricity generation and distribution utilities powering such rural experiments in development as the Colorado ski industry, cooperatives command a significant market share of a variety of everyday products and services. However, most member-owners and certainly most non-member consumers of these cooperative giants feel little if any sense of ownership or inclusion in "economic democracy" in their engagements with this alternative interpretation of capitalist ownership. Scholars of co-ops tend to highlight this discrepancy between the large number of active co-ops in the U.S. and how little these community businesses are on the public's radar – though, they note, the idea of cooperative ownership as community empowerment has been frequently rediscovered at times of economic crises. Given the variety of meanings and practices of cooperative business in the United States and its foreign ventures, the shifting place of cooperatives in political economic thought over the past century is a rich intellectual and political history in its own right.

Recent literature connecting co-ops in the United States to broader critiques of capitalism is too vast and diverse for a single footnote, but see, e.g.: Gar Alperovitz, *America Beyond Capitalism: Reclaiming Our Wealth, Our Liberty, & Our Democracy* (Second Edition, Takoma Park, MD, and Boston, MA: Democracy Collaborative Press, 2011); Kali Akuno and Ajamu Nangwaya (eds.), *Jackson Rising: The Struggle for Economic Democracy and Black Self-Determination in Jackson, Mississippi* (Daraja Press, 2017); James DeFilippis, *Unmaking Goliath: Community Control in the Face of Global Capital* (New York and London: Routledge, 2004); Marjorie Kelly, "The Economy: Under New Ownership," *yes!* (Spring 2013), https://www.yesmagazine.org/issue/issues-how-cooperatives-are-driving-the-new-economy/2013/02/20/the-economy-under-new-ownership/. On the use of cooperative messaging in policy: compare, e.g., Molly Crabapple, "A Message from the Future with Alexandria Ocasio-Cortez," Naomi Klein and *The Intercept* (April 17, 2019): https://theintercept.com/2019/04/17/green-new-deal-short-film-alexandria-ocasio-cortez/; Daniel Aldana Cohen, "We Need a Green New Deal for Housing," *Jacobin*, June 4, 2019; and Audrea Lim, "How the Green New Deal Can Deliver Land Justice," *Jacobin*, May 19, 2019, with the original press conference on the Green New Deal: "House and Senate Democrats on Green New Deal," *C-Span* (February 7, 2019), https://www.c-span.org/video/?457701-1/senator-markey-representative-ocasio-cortez-launch-green-deal. On cooperatives in American history, see: John Curl, *For All the People: Uncovering the Hidden History of Cooperation, Cooperative Movements, and Communalism in America* (Oakland, CA: PM Press, 2009); James DeFilippis, *Unmaking Goliath: Community Control in the Face of Global Capital* (New York: Routledge, 2004); Steven J. Keillor, *Cooperative Commonwealth: Co-ops in Rural Minnesota, 1859–1939* (St. Paul, MN: Minnesota Historical Society, 2000); Kimberly Zeuli, "The Role of Cooperatives in Community Development," *University of Wisconsin Center for Cooperatives Bulletin*, 3, 2002: 1–4. The role and evolution of cooperative thought has developed quite differently in other parts of the world, though, as I discuss in this chapter, it has done so alongside the proliferation of American-style cooperatives across the global countryside at formative moments, and in the broader structural context of an increasingly hegemonic global economy. On co-ops outside of the United States, see, for example: *Business History* 54, no. 6, 2012, Special Issue: The business of Co-operation: National and international dimensions since the nineteenth century; Greg MacLeod, *From Mondragon to America: Experiments in Community Economic Development* (University College of Cape Breton Press, 1998).

attention to its role in foreign aid during the Cold War, and especially how this form of collective-private ownership aligned with (or didn't) the communist-capitalist dichotomy at the heart of the conflict.[12] I show here that USAID built on lessons from a government-funded cooperative program during the New Deal to promote a cooperative model that – much as federal agents had done in the U.S. after the Great Depression – attempted to reframe collective ownership as a capitalist institution in the developing world during the Cold War. In addition to the profound ecological and landscape changes American impositions of energy infrastructure wrought on the global South – particularly in oil extraction and hydroelectric dam building – the institutional influence of American development aid in electricity distribution and utility ownership demands attention as a project of American hegemony.[13]

Electric co-ops were not an invention of USAID, but rather a borrowing. They had already served as a business model for economic development through electrification in the United States as a Depression-era federal aid program. Adopted on a national scale by a New Deal federal rural electrification program as a politically convenient middle ground between public and private power, by the 1960s, electric co-ops represented for many American policymakers the promise of rural development through "self help" and, in that individualistic understanding of cooperative ownership, a grassroots bulwark against the rising tide of communism.

Early USAID programs thus unironically turned to cooperatives to defuse communist organizing in the global South during the Cold War, particularly enthusiastically in the electricity sector. Over the course of its first decade as an international development organization, USAID loaned on the order of $30 million to cooperative electrification projects around the world (see Fig. 8.1). Along with financial aid, USAID sent abroad cooperative consultants and technical guides promising a disparate set of social and economic goods, including "increased production and quality of farm products and at a lower cost; . . .new or stimulated existing businesses or industries; . . .better living conditions and increased income; . . .better health and sanitation; and stimulated education and self-respect."[14] USAID-funded electric cooperatives such as the one in Daule were required to form a board of directors

---

[12] On the need for a more nuanced understanding of development on the ground, a critique of drawing too sharp a dichotomy between high modernist development planning and local practices, and the pre-Cold War ideological influences on development aid, see: See Hodge, "Writing the History of Development (Part 1: The First Wave)."

[13] On American investments in energy infrastructure in the global South, see: Timothy Mitchell, *Carbon Democracy: Political Power in the Age of Oil* (Verso 2011) and "Can the Mosquito Speak," in *Rule of Experts: Egypt, Techno-Politics, Modernity* (University of California Press, 2002); Thomas Robertson, "Cold War Landscapes: Towards an Environmental History of US Development Programmes in the 1950s and 1960s," *Cold War History* 16, no. 4 (November 17, 2016): 417–41; Nick Cullather, "Damming Afghanistan: Modernization in a Buffer State," *The Journal of American History* 89, no. 2, History and September 11: A Special Issue (Sep., 2002): 512–537; David Ekbladh, "Mr. TVA: Grass-Roots Development, David Lilienthal, and the Rise and Fall of the Tennessee Valley Authority as a Symbol for U.S. Overseas Development, 1933–1973," *Diplomatic History* 26, no. 3, 2002: 335–374.

[14] USDA, *The REA Pattern*, REA Bulletin 1–8, April 1963, 12.

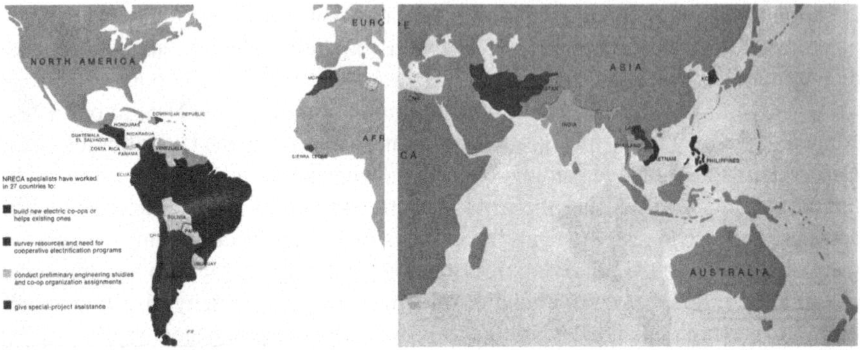

Fig. 8.1 USAID-funded rural electric cooperative projects and consultation services, 1969. (Source: Hearings Before the Committee on Foreign Relations, United States Senate, Ninety-first Congress, First Session, on S. 2347, A Bill to Amend the Foreign Assistance Act of 1961, As Amended, and For Other Purposes, Printed on October 8, 1969, 257–8, HT.)

and hold regular meetings and democratic elections, in the interest of developing social norms of liberal democracy along with the material infrastructures of electrification. As one American consultant unironically praised the institutional influence of USAID cooperative funding to Daule through the 1960s-anti-communist shade of rose-colored glasses: "Here were people participating in the heady experience of economic democracy for the first time."[15]

Yet, in tying the social benefits of cooperative member participation tightly to industrialization and economic growth, and explicitly against communism, 1960s-era USAID electric cooperative aid promoted a vision of economic collectivity that pushed co-op members towards *connecting into* rather than *protecting them from* globalizing markets, through an unabashedly American model of economic participation that reframed cooperative economics as a fundamentally individualist experience. Building on several decades of domestic federal experiments with

[15]Clyde T. Ellis, *A Giant Step* (New York: Random House, 1966), 214. Economic democracy is a historically antimonopolist political philosophy that decisions about industrial production, finance, and trade should be governed by the communities impacted by these activities, directly and deliberately, rather than by shareholders. For a longer discussion of its history and philosophy, see: Gar Alperovitz, "Building a Democratic Economy: Sketch of a Pluralist Commonwealth," *Nonprofit Quarterly* (April 14, 2020), https://nonprofitquarterly.org/building-a-democratic-economy-sketch-of-a-pluralist-commonwealth/; Drew Christie, "Recent Calls for Economic Democracy," *Ethics* 95, no. 1, 1984: 112–128; Andrew Cumbers, "Economic Democracy," in Antipode Editorial Collective et al. (eds), *Keywords in Radical Geography: Antipode at 50* (Wiley Blackwell, 2019), 102–106, https://onlinelibrary.wiley.com/doi/book/10.1002/9781119558071; Robert A. Dahl, *A Preface to Economic Democracy* (University of California Press, 1986); Tom Malleson, *After Occupy: Economic Democracy for the twenty-first Century* (New York: Oxford University Press, 2014).

cooperative capitalism in the rural United States, USAID similarly co-opted collective economic philosophies abroad.[16]

As an instrumental part of the USAID toolkit during the Cold War, American-style electric cooperatives helped to naturalize a global political economy with systemic economic inequality that would subsequently legitimize continuing programs of neoliberal humanitarianism – i.e., non-partisan "poverty alleviation."[17] As opposed to earlier Cold War development projects as an existential defense of the American way of life, global "poverty alleviation"-style aid began to redefine energy development into a humanitarian imperative, refashioning access to high-consumption energy modernity into a human right, as opposed to a political economy. Even as USAID programs continued to intervene in Cold War geopolitics, their reorientation in the 1970s towards basic needs was a step away from the liberal democracy framework of early Cold War aid. As the Cold War came to a close, USAID had already reframed aid into a more prosaic service focused on material conditions of the "poor majority," and thus "deflected concerns over structural inequalities in land, capital, and access to decision-making authority within development processes."[18] Among other things, this reorientation has been critiqued for inappropriately universalizing the particular set of material relations international development projected as the pinnacle of modernization.[19]

The second aim of this chapter, therefore, is to trace the privileged role electrification has held in development ideology as a stimulus for both economic growth and democracy. Few historians now view electricity technologies and infrastructure so narrowly, but the aura of electricity's automatic social uplift is a policy narrative that has shaped foreign aid for decades, even despite its barely-disguised

[16]On the New Deal origins of neoliberal economics, and American foreign aid projects as an intermediary site of economic experimentation, see Amy Offner, *Sorting Out the Mixed Economy: The Rise and Fall of Welfare and Developmental States in the Americas* (Princeton University Press, 2019); David Ekbladh, "'Mr. TVA': Grass-Roots Development, David Lilienthal, and the Rise and Fall of the Tennessee Valley Authority as a Symbol for US Overseas Development, 1933–1973," *Diplomatic History* 26, no. 3, 2002, 335–374.

[17]On rural electrification and economic inequality, see V. Ranganathan, "Rural electrification revisited," *Energy Policy* 21, no. 2, 1993, 6. On the New Directions mandate and its role in Cold War politics and beyond, see: Mark F. McGuire and Vernon W. Ruttan, "Lost Directions: U.S. Foreign Assistance Policy Since New Directions," *The Journal of Developing Areas* 24, no. 2, 1990: 127–180; Jamey Essex, *Development, Security, and Aid: Geopolitics and Geoeconomics at the U.S. Agency for International Development* (Athens, GA: University of Georgia Press, 2013), Chapter 3: "Geoeconomics Ascendant: Development, Interdependence, and Neoliberalization."

[18]Essex, 56.

[19]I am not taking the stance here that electricity is a human right. For critical histories of humanitarianism as culturally-situated engagements as opposed to a framework representing "universal" ideals, see: Thomas L. Haskell, "Capitalism and the Origins of the Humanitarian Sensibility," *The American Historical Review*, 1985 – Part 1, 90(2): 339–361 and Part 2, 90, no. 3: 547–566; Michael Barnett, *Empire of Humanity: A History of Humanitarianism* (Ithaca: Cornell University Press, 2011); David Kennedy, *The Dark Sides of Virtue: Reassessing International Humanitarianism* (Princeton: Princeton University Press, 2011).

weaponization during the Cold War.[20] In the twenty-first century, the idea of access to electricity as a human right deserving of international humanitarian aid continues to garner widespread support.[21] This narrative seems to cling to electrification projects throughout modern American history, both at home and abroad, despite contradictory evidence. Its narrative persistence has moved capital and resources around the world in ways that make sense only in context, in the specific beliefs, hopes, and power struggles of the changing casts of electricity consumers, producers, and policymakers at a variety of scales.

I argue that such "folk understandings" of electrification as a benign and neutral human right are part of an enduring mythology that has supported and continues to support American global hegemony, and I therefore interpret such narratives as a structural act of *depoliticization*. That is, such narratives shift inherently political human endeavors (e.g., the sociotechnical design of distributed electricity infrastructures) from the political sphere in which dissensus is possible to a seemingly neutral framework of "technocratic management and consensual policymaking," conducted by experts on behalf of the "people," which precludes the possibility for recognition or inclusion of "the radical heterogeneity and antagonisms" within and between communities actually impacted by development and governance decisions.[22]

It is my intention here to *denaturalize* and *repoliticize* the global distribution of central-station networked electricity as a project of American hegemony during the Cold War, as well as to expand the political history of cooperatives and other institutions of "economic democracy" as critical institutions in the evolution of

---

[20] On the Cold War discourse of electrification aid as a "weapon," see: James H. Williams and Navroz K. Dubash, "Asian Electricity Reform in Historical Perspective," *Pacific Affairs* 77, no. 3, 2004: 416–17.

[21] Examples of rhetoric that I argue problematically defines energy access as a human right include current USAID policy (see, e.g.: https://www.usaid.gov/infrastructure), as well as humanitarian frameworks such as the United Nations Millennium Development Goals (see, e.g.: https://www.un.org/sustainabledevelopment/energy/). On the complexities of energy justice in practice, see: Kirsten Jenkins, Darren McCauley, Raphael Heffron, Hannes Stephan, and Robert Rehner, "Energy justice: A conceptual review," *Energy Research & Social Science* 11, 2016: 174–182; Bethel Tarekegne, "Just electrification: Imagining the justice dimensions of energy access and addressing energy poverty," *Energy Research & Social Science* 70, 2020: 1–6.

[22] I'm taking the concept of "folk understandings" from Hillary Angelo's use of the term to critique enduring city-nature binary narratives in urban studies, planning, and design – in other words, "folk" for her refers to the unarticulated frameworks that constrain professional fields. See Hillary Angelo, "From the City Lens Toward Urbanisation as a Way of Seeing: Country/City Binaries on an Urbanising Planet," *Urban Studies* 54, no. 1, 2016: 158–78. The language of depoliticization and dissensus comes from: Erik Swyngedouw, "Apocalypse Forever?," *Theory, Culture & Society* 27, no. 2, 2010), 214, 223; Maria Kaika, "'*Don't call me resilient again!*': the New Urban Agenda as immunology…or…what happens when communities refuse to be vaccinated with 'smart cities' and indicators," *Environment & Urbanization* 29, no. 1, 2017: 89–102. On critiques of technocracy and expert-rule as forms of depoliticization, see also: Gabrielle Hecht, *The Radiance of France: Nuclear Power and National Identity after World War II* (Cambridge: MIT Press, 1998), and Timothy Mitchell, *Rule of Experts: Egypt, Techno-Politics, Modernity* (University of California Press, 2002).

twentieth-century global capitalism. Revisiting global rural electrification through the lens of American international development aid, and vice versa, illuminates the ways cooperative electricity has been crafted and contested as an extension of state power, and how its distribution has created new arenas for political struggle, since at least the mid-twentieth century.[23]

8.2 Economic Democracy, American-Style

When the Daule project began in 1964, USAID was a young agency making its first loans to the so-called Third World with an explicitly anti-communist agenda. Building on the lessons of the Marshall Plan and the Point Four program, USAID was President John F. Kennedy's attempt to streamline American international development efforts under one umbrella office and reorient American foreign aid towards funding institutions and technologies of liberal democracy, a focus that has since been critiqued for making the world safer for Western capitalism by entrenching and widening structural inequalities abroad.[24] In addition to reorganizing the institutional structure of foreign aid within the American government, the 1961 Foreign Assistance Act, which created USAID, also formalized a *rescaling* of American foreign aid projects. In place of the large infrastructure projects funded by the Marshall Plan and the later Point Four Program, perhaps best represented in the proliferation of hydroelectric dams across the world from Afghanistan to Nepal to Uruguay, USAID programs targeted institutional reform and community development.[25] USAID programs also explicitly encouraged private

---

[23] This chapter thus aligns with histories of electricity interested in revealing the ideological work that constructed such a depoliticized public engagement with energy as an important part of American continental and global political economic hegemony in the twentieth century. See, for example: Mitchell, *Carbon Democracy;* Jones, *Routes of Power*; and Andrew Needham, *Power Lines: Phoenix and the Making of the Modern Southwest* (Princeton: Princeton University Press, 2014).

[24] On the evolution of foreign aid from the Marshall Plan to the Point Four Program, see Stephen Macekura, "The Point Four Program and U.S. International Development," *Political Science Quarterly*, Vol. 128, No. 1 (Spring 2013), pp. 127–160; Essex, *Development, Security, and Aid.* On the unintended consequences of American foreign aid, see: Thomas Robertson, "Cold War Landscapes: Towards an Environmental History of US Development Programmes in the 1950s and 1960s," *Cold War History* 16, no. 4 (November 17, 2016): 417–41; Megan Black, "Interior's Exterior: The State, Mining Companies, and Resource Ideologies in the Point Four Program," *Diplomatic History* 40, no. 1, 2014: 81–110; David Ekbladh, *The Great American Mission: Modernization and the Construction of an American World Order* (Princeton University Press, 2011).

[25] Though in practice, Stephen Macekura argues, Point Four prioritized relatively small projects. See: Macekura, "Point Four and the Crisis of U.S. Foreign Policy Aid in the 1970s," in Raymond H. Geselbracht, ed., *Foreign Aid and the Legacy of Harry S. Truman* (Truman State University Press, 2015). On hydroelectric dams, see: Christopher Sneddon, *Concrete Revolution: Large Dams, Cold War Geopolitics, and the US Bureau of Reclamation* (Chicago: University of

businesses in the United States to expand operations abroad and invest in burgeoning industries in aid-recipient countries: "Economic and social growth cannot be accomplished by governments alone," JFK emphasized to USAID legislators in 1963.[26] By the 1970s, USAID would drop its explicitly Cold War agenda (rhetorically, anyway[27]) in favor of a new orientation towards "poverty alleviation," a depoliticized framing that continued to normalize Western political economic logics of development through widening access to consumer prosperity and increased inclusion of the poor in cash economies.[28] But initially USAID was very much a macroeconomic project, if not a macro-infrastructural one. Its founding legislation did in fact assume a "trickle down" effect from development projects, both in terms of broader market access in a globalized market society and in the widespread institutionalization of the logics and practices of liberal democracy.

In Ecuador, Latin America, and around the world, American policymakers in the 1960s saw themselves in an existential competition for the "hearts and minds" of the Third World in two ways[29]: First, U.S. officials worried about the ideological "extension of Fidel Castro's influence to other countries in the hemisphere and [the] grave danger of a [M]arxist revolution."[30] The Kennedy administration's fears of "Castroism" underwrote well-funded diplomatic programs like the Alliance for Progress, signed at Punta del Este, Uruguay, on August 17, 1961, to "improve and strengthen democratic institutions through application of the principle of self-determination by the people."[31] In practice, this meant American international aid programs began to target and tinker with the basic social institutions of Latin America, imposing American-incubated visions of modern infrastructure, housing, education, health care, tax reforms, and local governance along with aid funds –

Chicago Press, 2015), Appendix: Geographical Scope of Bureau of Reclamation Activities, 1933–1975; Richard P. Tucker, "Containing Communism by Impounding Rivers: American Strategic Interests and the Global Spread of High Dams in the Early Cold War," in J. R. McNeill and Corinna R. Unger, eds., *Environmental Histories of the Cold War* (Cambridge University Press, 2010), 139–164. On community development as ideology, see: Daniel Immerwahr, *Thinking Small: The United States and the Lure of Community Development* (Harvard University Press, 2015).

[26] Foreign Operations Appropriations for 1964: hearings before a subcommittee of the Committee on Appropriations, House of Representatives, 88th Congress, 1st session, published 1963, 1773, HT.

[27] Though foreign aid was obviously still a proxy battle for the Cold War through the 1980s. See, e.g., McGuire and Ruttan, "Lost Directions."

[28] Essex, "Geoeconomics Ascendant: Development, Interdependence, and Neoliberalization."

[29] I'm influenced here by Simo Laakkonen, Viktor Pál, and Richard Tucker's reflections on "hearts and minds" discourse and the weaponization of nature in the Cold War. See: Laakkonen, Pál, and Tucker, "The Cold War and environmental history: complementary fields," *Cold War History* 16, no. 4, 2016, 378.

[30] Kessler et al., Annex 4–2.

[31] "Survey of the Alliance for Progress: The Political Aspects," Committee on Foreign Relations, United States Senate, September 18, 1967, 7, 210–211, HT.

which, as they did in the United States, allowed for the entrenchment and widening of social inequalities, even as they targeted more aggregate forms of economic growth.[32]

Second, American programs found themselves in competition with Soviet bloc development funding for similar projects – rural electrification, for example.[33] As historians working at the intersection of environmental and diplomatic history have recently argued, similarities between American and Soviet funded infrastructure projects in the third world are one of the clearest demonstrations that mid-twentieth century ideas of liberal democratic capitalism and Soviet-style industrial communism *materially* were not so much opposites as they were different paths to high modernism.[34] As political economies trying to make claims on unaligned

[32]V. Ranganathan, "Rural electrification revisited," *Energy Policy* 21, no. 2, 1993, 6; James Bronwyn, "The impacts of rural electrification: exploring the silences," *EDRC Report Series*, Energy & Development Research Centre, University of Cape Town (July 1995); Wasserman and Davenport, "Power to the People," 11, 17.

[33]For a more nuanced discussion of Soviet development aid during the Cold War, see: Joseph Morgan Hodge, "Writing the History of Development (Part 2: Longer, Deeper, Wider)," *Humanity: An International Journal of Human Rights, Humanitarianism, and Development* 7, no. 1, 2016, 149–51. In an assessment of the literature, Hodge writes: "Soviet development programs and projects faced many of the same complexities and limitations on the ground as those of the United States or the former European colonial states. Like them, the Soviet project was far from monolithic and the outcomes far from what had originally been planned. Despite their sharp ideological differences at the level of rhetoric, in practice both superpowers tended to support strong state sectors with a reliance on centralized planning and to favor industrialization, infrastructure, and other large showcase projects." In the case of USAID electrification aid in particular, I contest this framework slightly: I argue here that cooperative electrification was a departure from this framework, promoting instead a decentralized, private-ownership approach to electrification (though USAID electric co-op projects did continue to promote industrialization and infrastructure development).

[34]On "high modernism," I'm evoking James C. Scott's canonical critique of authoritarian development in *Seeing Like a State: How Certain Schemes to Improve the Human Condition Have Failed* (Yale University Press, 1998). Scott defines high modernism as an ideology (and therefore *not a practice*) based on faith in the linearity and inherent rationality of scientific and technological progress, and, building from that, in the possibility for a top-down "comprehensive planning of human settlement" – large orderly cities reliant on massive dam projects, centralized infrastructures, and standardized production, for example in industrial agriculture – that erased human diversity by assuming universal public goods, and thus made citizens legible and countable individuals in a scalable world. (See Scott, pp. 4–6.) Of course, the *social relations* these material worlds were built to support were quite different, but that was, for Scott, beside the point: "High-modernist faith was no respecter of traditional political boundaries," he wrote; "it could be found across the political spectrum from left to right but particularly among those who wanted to use state power to bring about huge, utopian changes in people's work habits, living patterns, moral conduct, and worldview" (5). *Seeing Like a State* has been critiqued for a too-dichotomous treatment of state "knowledge-power regimes" and local knowledge and practices (see, e.g., Hodge, "Writing the History of Development (Part 1)," who defines "high modernism" as "the dream of administratively ordering nature and society and the unrestricted use of modern state power as an instrument for fulfilling such an ambition," 443–6). Nonetheless, Scott's theoretical critique continues to inform scholarship on the contradictions within twentieth century development planning – see, e.g., Cullather, Robertson, and Ekbladh.

nations (for either humanitarian or military reasons or both[35]), these infrastructure projects sought similar tangible results: prosperity through technology, the reorganization of national landscapes for industrial development, and residential comfort through centrally-managed infrastructural networks.[36] As much as, in Lenin's famous words, "Communism is Soviet government plus the electrification of the whole country,"[37] electrification of the global countryside was evoked by American policymakers as equally necessary for the construction of a capitalist social order at the height of the Cold War.[38] Both political imaginaries relied on high-energy futures. In pursuit of them, the Cold War also became an exercise, in Thomas Robertson's words, "of environmental management on a global scale."[39]

I frame USAID interventions in this way in order to shed new light on how the community cooperative - an institution that, ironically, has historically often had to defend itself from aspersions of communism – came to be used in this Cold War development battle on the side of American capitalism. Inspired in large part by agricultural cooperatives in the United States, Senator Hubert Humphrey successfully proposed an amendment to the 1961 Foreign Assistance Act "to encourage the development and use of cooperatives, credit unions, and savings and loan associations" in USAID recipient countries. The committee report behind the Humphrey Amendment paraphrased FDR's four freedoms for a developing country context in a time of consumer capitalism to justify the cooperative model:

> The goals of the U.S. foreign assistance program are peace, security, progress, and freedom for the people of the newly developing countries – freedom from want, freedom from oppression, freedom from ignorance, and freedom from disease.

---

[35] Obviously, the lines between the two were blurry in practice. For an environmental perspective, see Laakkonen, Pál, and Tucker's special issue on militarized landscapes in *Cold War History*, in which they write that the Cold War was, among other things, "a global power struggle over the control of natural resources," and that "the militarisation of nature became an undeniable part of the Cold War," even as both sides learned from and emulated each other on matters of environmental protection. Simo Laakkonen, Viktor Pál, and Richard Tucker, "The Cold War and environmental history," 378, 387.

[36] See, e.g.: Cullather, "Damming Afghanistan: Modernization in a Buffer State;" Robertson, "Cold War Landscapes: Towards an Environmental History of US Development Programmes in the 1950s and 1960s;" Kate Brown, *Plutopia: Nuclear Families, Atomic Cities, and the Great Soviet and American Plutonium Disasters* (Oxford University Press, 2015); Tucker, "Containing Communism by Impounding Rivers," 139–141, 146–7; Williams and Dubash, "Asian Electricity Reform in Historical Perspective," 417–18; Odd Arne Westad, *The Global Cold War: Third World Interventions and the Making of Our Times* (Cambridge University Press, 2005), 33.

[37] V. I. Lenin, "Our Foreign and Domestic Position and Party Tasks," Speech Delivered to the Moscow Gubernia Conference of The R.C.P.(B.), November 21, 1920, Marxists.org, https://www.marxists.org/archive/lenin/works/1920/nov/21.htm.

[38] See, e.g., United States Senate, Hearings before the Committee on Foreign Relations, 91st Congress, 1st Session, S. 2347, July 14, 15, 18, and August 6, 1969, 249–63.

[39] Robertson, "Cold War Landscapes," 422.

> Essential to these goals is a dynamic increase in the general productivity and wealth of the economies of these nations, as well as a positive increase in the standards of living through a more equitable distribution of income and more widespread ownership and enjoyment of wealth.
>
> Foreign aid programs should encourage people to help themselves and give them a stake in their countries' economic welfare.
>
> ... All of these people – including workers in larger mines, mills, and plantations – as consumers need the *distributive service cooperatives, credit unions, savings banks, and savings and loan associations*, to help them combat exploitation and monopoly practices, to get the most for their hard-earned income in food, clothing, housing, and other services, and to encourage thrift and aid them in their desire and need for homeownership and adequate housing.[40]

In other words, freedom in the developmental context was less about free speech than about economic efficiency, self-help, and material prosperity. In 1962, USAID designated a little over two million dollars for twenty-five USAID projects to promote credit, marketing, savings and loan, housing, agricultural, and other co-ops across Latin America, Africa, Asia, and the Middle East.[41]

USAID's commitment to cooperatives was perhaps nowhere so dogmatically applied as in its rural electrification programs. In that sector, the American experience with its own "development" was recent, much celebrated, and inextricably tied to cooperatives. As late as the 1930s, most rural Americans had minimal if any experience with networked infrastructure.[42] The spatial disparities in access to central-station electricity, in particular, contributed to anxieties about a growing rural-urban divide for decades. Many of the architects of New Deal programs, including Franklin D. Roosevelt himself, had forged their ideas of public service in political fights with private utility companies in the 1920s. Roosevelt's 1932 presidential campaign promised utility reform, and, by the mid-1930s, his administration had committed significant funding and resources to rural electrification as a necessary point of intervention for addressing rural poverty and economic recovery during the Great Depression.[43]

---

[40] Senate Document No. 112, 87th Congress, 2d Session, USAID, "Implementation of the Humphrey Amendment to the Foreign Assistance Act of 1961," July 1962, Washington, D.C.: U.S. Government Printing Office, 3, HT, italics in original.

[41] Ibid., 7–9.

[42] While other forms of networked public works would have been more common to city dwellers, rural residents in that era would have been less familiar with sewers, piped gas, paved roads maintained by a public authority, etc. They were, however, familiar with many of the applications of central station electricity, due to the popularity of small generators and gas-powered versions of home and farm appliances. For a longer discussion of these alternatives to central station electricity, see: Kline, *Consumers in the Country*, and Neufeld, *Selling Power*.

[43] Morris Llewellyn Cooke, "The Early Days of the Rural Electrification Idea: 1914–1936," *The American Political Science Review* 42, no. 3, 1948: 431–447; Abby Spinak, "'Not Quite So Freely as Air': Electrical Statecraft in North America," *Technology and Culture* 61, no. 1, 2020: 71–108.

In 1935, President Roosevelt issued an executive order to create a new agency, the Rural Electrification Administration (REA), which was initially allocated $100 million in reconstruction funding to loan to rural electrification projects.[44] After negotiations with for-profit utilities stalled during the first summer of the REA's operations, however, and navigating aspersions that public electricity was socialism, the REA turned to promoting the electric cooperative model as a "middle way" that was private, but nonprofit, and from which – they promised – taxpayers would see a return on investment.[45] The REA quickly reorganized to provide low-interest loans and technical advice for cooperative electricity and encouraged communities to organize and apply for REA funding. Within 10 years, REA-funded co-ops provided power to over a million rural Americans.

Elsewhere, I have explored the contradictions within the REA's top-down program of supposedly grassroots cooperative. Particularly, in seeking to understand how the REA reimagined the cooperative utility model as an institutional alternative to private, for-profit electric companies and public, municipal or state-run power, I argue that the REA forged a historically-specific conception of cooperatives that challenged earlier radical implementations of cooperatives in Gilded Age labor movements. The novel REA-style cooperative de-emphasized the antimonopolist politics of economic democracy that co-ops had previously represented and instead promoted them as engines of industrial growth that would reshape the rural economy, disciplined largely by their debt responsibilities to the federal government.[46] Under the REA's managerial tutelage, REA co-ops incorporated many of the same practices for-profit utilities had developed around load-building, electricity pricing, employee salaries, service extensions, and the like – many of which, as they did for private industry, exacerbated systemic racial and class inequalities under the logic of perpetual business growth. Nascent co-op communities learned to emphasize their growth potential, speculating in their REA loan applications about future industries such as creameries and sugar factories, and discussing in detail the quality of their

---

[44] Initially through Executive Order 7037. The REA became a permanent federal agency in 1936 through the Rural Electrification Act, and was later subsumed into the Department of Agriculture.

[45] Mary Hilson, "Consumer Co-operation and the Economic Crisis: The 1936 Roosevelt Inquiry on Co-operative Enterprise and the Emergence of the Nordic 'Middle Way,'" *Contemporary European History* 22, no. 2, 2013, 186–8, 194–8.

[46] Spinak, "'Not Quite So Freely as Air'"; On the contradictions inherent in REA economics, see also: David E. Nye, *Electrifying America: Social Meanings of a New Technology* (Cambridge: MIT Press, 1990), 331–4. On REA attitudes towards cooperatives, see: Transcript of Mr. Boyd Fisher's talk at a meeting of representatives from rural electric member cooperatives of Statewide, Wisconsin, January 27, 1938, OF 1570, FDR Library; Judson King, "The REA – A New Deal Venture in Human Welfare," 1938, Gifford Pinchot collection, Box No. 682, Library of Congress; Udo Rall, "Cooperative Rural Electrification in the United States," *Annals of Public and Cooperative Economics* 18(2), 1942; Stephen J. Keillor, *Cooperative Commonwealth: Co-ops in Rural Minnesota, 1859–1939* (St. Paul, MN: Minnesota Historical Society, 2000), 325.

soil for intensified agriculture.[47] Co-op members volunteered their labor to keep costs down, granted easements for lines to cross their land without compensation, and promised to meet personal electricity consumption targets. After distributing loans, the REA made sure co-ops followed up on their promises: field agents visited new co-ops regularly and often chided them on their rates; utilization specialists showed up with free lamps and flashy kitchen demonstrations; and, more importantly, the REA made sure farm families of "moderate income" had access to personal loans for all the electrical appliances they desired.[48] As their cooperatives were borrowing to extend infrastructure, co-op members were encouraged to take out loans for home wiring and plumbing, and to buy refrigerators, electric ranges, water pumps, and washing machines on federally-insured installment plans. In short, I argue that we should see the REA as a context-specific experimentation with growth-oriented "cooperative capitalism."[49]

The electric cooperative in the hands of the American state was not exactly communal. The REA explicitly celebrated the democratic spirit of individuals coming together as an effective tool for improving private property and catalyzing industrial development. They celebrated a kind of bootstrap "togetherness" of otherwise independent individuals in small towns and family farms, which, as Daniel Immerwahr has argued, in the mid-twentieth century replaced the frontier as the

[47] Correspondence between E. W. Beckman and John Carmody, regarding Fremont, Wyoming, 12 March 1938; Application for an REA Loan, regarding Washington 5 Benton, 24 June 1938, Project Case Files Concerning Loans to REA Borrowers, 1938–1939, Box No. 7, REA, NARA.

[48] Utilization Progress Report Memorandum from G. D. Munger, January 1938, Utilization Division Correspondence re: More Effective Uses of Elec. Energy by Members of Co-ops, 1938–1940, Kentucky; Correspondence between C. O. Falkenwald and G. A. Lewis, 9 May 1938, Project Case Files Concerning Loans to REA Borrowers, 1938–1939, Box No. 7, REA, NARA. Electric Home and Farm Authority, Activities Review, 12 December 1934; Aid to Purchasers of Electrical Appliances Offered by Electric Home and Farm Authority, Thomas G. Corcoran collection, Electric Home and Farm Authority, 1935–38, Folder No. 1, Library of Congress.

[49] My concept of "cooperative capitalism" is in conversation with the recent rich literature on the history of capitalism, which is invested in understanding the "varieties of capitalism" that have emerged in different historical moments as partial economic systems that interact with other forms of economy and social organization in contingent and situated ways. In this paper, I thus use "capitalism" in two ways: specifically, to refer to a particular set of economic relations in a particular place – e.g. "consumer capitalism" or "corporate capitalism" – that situate unique forms of capital accumulation and financial mechanisms in particular places at particular times. I thus place the "cooperative capitalism" of the REA in this camp, as a specific project of naturalizing economic growth, financial mechanisms, and cost-based discrimination within social services. Otherwise, I use "capitalism" to refer to the *ideology* of finance-fueled economic growth and the expansion of markets as an exclusive world-ordering project – e.g., "Western capitalism" or "American capitalism." I interpret these as utopian narratives incapable of full realization. For more discussion on the history of capitalism and its varieties, see: Seth Rockman, "What Makes the History of Capitalism Newsworthy?," *Journal of the Early Republic* 34, no. 3, 2014: 439–466; Sven Beckert, Angus Burgin, Peter James Hudson, Louis Hyman, Naomi Lamoreaux, Scott Marler, Stephen Mihm, Julia Ott, Philip Scranton, and Elizabeth Tandy Shermer, "Interchange: The History of Capitalism," *The Journal of American History* 101, no. 2, 2014: 503–36; Louis Hyman, "Why Write the History of Capitalism?" *Symposium Magazine* (8 July 2013), http://www.symposium-magazine.com/why-write-the-history-of-capitalism-louis-hyman/

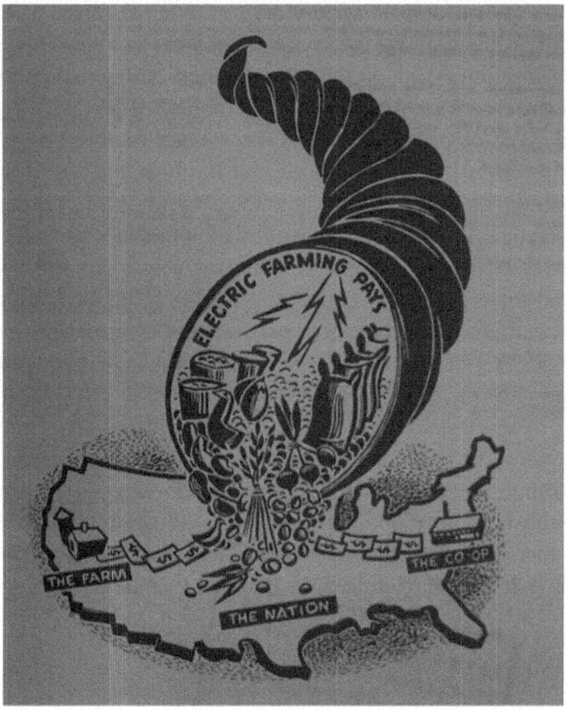

Fig. 8.2 "Electric Farming Pays". (Source: "How Electric Farming Can Help Your Business," One of a Series on the Art of Balanced Management, REA Bulletin 140–1, Information Services REA News Releases, Bulletins, Reports, 1937–53, Box No. 5, REA, NARA)

symbolic nursery of American democracy.[50] REA electric cooperatives were thus anti-urban but pro-industrialization.[51] At a time when cities were growing dense, diverse, and politically radical, they offered a different collective development path for rural America, one that privileged private property, national economic growth, and the conservative influences of small-scale representative governance. In the REA's vision, cooperatives rewarded the individual initiative of the entrepreneurial farmer while tying him[52] to the national economy (see Fig. 8.2).[53] The REA thus crafted a new model of collectivity that allowed its champions to later tout such claims as: "Electric cooperatives are the antithesis of communism."[54]

---

[50]Immerwahr, *Thinking Small,* 30. Also see Westad, *The Global Cold War,* 11.

[51]Spinak, "'Not Quite So Freely as Air,'" 91.

[52]On the REA and gender, see Katherine Jellison, *Entitled to Power: Farm Women and Technology, 1913–1963* (The University of North Carolina Press, 1993), and Leah Glaser, *Electrifying the Rural American West: Stories of Power, People, & Place* (Lincoln: University of Nebraska Press, 2009).

[53]On REA's celebration of individual initiative, see e.g.: *Rural Electrification News* 5, no.10, 1940, 5–6. On the construction of the idea of the national economy and its metrics, see: Timothy Mitchell, *Carbon Democracy: Political Power in the Age of Oil* (Verso 2011) and Michelle Murphy, *The Economization of Life* (Duke University Press, 2017).

[54]Ellis, *A Giant Step,* 220.

Despite the cooperative model being "more or less forced upon the REA" by the constraints of 1930s electricity politics,[55] the idea of democratic electricity nonetheless came to play a large role in the agency's propaganda as it navigated its role in the shifting geopolitics of World War II and postwar American hegemony. Though occasionally electric cooperatives suffered suspicion about being "something communist,"[56] the REA actively promoted them as little incubators of liberal democracy and exemplars of anti-fascism.

The timing demanded it. As the first wave of cooperatives came online at the end of the 1930s, democracy seemed "a system sorely beset." Speaking at the fifth annual REA staff conference on May 8, 1940, Secretary of Agriculture Henry A. Wallace urged REA officials: "we should not be negligent in strengthening our democratic institutions. . . .We can firm our spiritual muscles by cooperative effort in solving the organizational and management problems involved in such expressions of community aspirations and interests as are represented by the REA form of rural electrification." REA co-ops, he declared, have the potential of the "stabilization of our democracy through a more rational and beneficent balance between centralization and decentralization."[57] Declaring cooperatives under siege from the rising fascism in Europe, REA publications during WWII played up the small-town-meeting style democracy that was in theory at the heart of cooperative business management: ". . .Let us be proud of our democratic privileges and responsibilities. Let us exercise them consciously and deliberately. Neglected, they may wither and die. When directors of a rural electric cooperative think, act, vote, and talk at their board meetings, they are living democracy."[58] (See Fig. 8.3.)

Electric co-ops, with their direct ties to Washington, were also ready-made local organizations for coordinating citizens towards the war effort. The REA implored its borrower cooperatives to "take the lead in combing their areas for every available ounce of scrap metal and rubber" to "smash the foes of freedom." The USDA, along with the Office of Civilian Defense and the Office of Price Administration, encouraged co-ops to organize their memberships to host group discussions about "what they can do to help win the war as quickly as possible." "Getting together, talking things over, deciding what needs to be done and how to do it is an old American custom. It is the democratic way. Well-informed citizens thus are better able to deal with their local problems and to make voluntary contributions and sacrifices to the national effort needed to insure victory."[59] The idea that cooperatives were a line of defense against fascism carried over into the proxy wars of the Cold War. In 1951, Claude Wickard, then administrator of the REA, sent a letter to REA co-ops titled, "Rural Electric Co-ops and National Defense," in which he appealed to them: "Rural

[55] Rall, "Cooperative Rural Electrification in the United States," 227.

[56] Curl, *For All the People,* 5; Martin Lowery, oral interview (NRECA, Arlington, VA), 26 April 2010.

[57] *Rural Electrification News* 5, no.10, 1940, 3–4, NRECA.

[58] *Rural Electrification News* 8, no. 9, 1943, NRECA.

[59] *Rural Electrification News* 8, no. 2, October 1942, 3, NRECA.

Fig. 8.3 "Fascism has silenced cooperatives". (Source: Rural Electrification News 9(2), 1943, NRECA Library)

electric co-ops, in my opinion, have a unique opportunity in this present situation to demonstrate to the Nation and to the world what the voluntary cooperation of free men can accomplish."[60]

REA antifascist rhetoric obscured key contradictions inherent in the electric cooperative model in practice, however, which ultimately made their way into USAID electrification efforts:

First, federally-financed cooperative electricity had a complicated record of success at home that the USAID cooperative agenda swept under the rug. Even as REA promotional materials financed by USAID claimed that "electrification makes

[60]Claude R. Wickard, "Co-op Chat: Rural Electric Co-ops and National Defense," July 2, 1951, Information Services REA News Releases, Bulletins, Reports, 1937–53, REA, NARA.

Figure 2. Rural electrification makes farms more prosperous by increasing agricultural production and improving quality.

Fig. 8.4 The Electrified Family Farm. (Source: USDA, The REA Pattern, REA Bulletin 1–8, April 1963, 3)

farms more prosperous by increasing agricultural production and improving quality," REA reports in the early 1960s expressed concern about the "decline" of rural areas, including increasing out-migration and rural poverty.[61] Pictures of fields where electricity lines stretched benevolently over machine-harvested hay bales, a well-tended homestead just at the horizon, painted a picture of family farm modernity that was rapidly being eroded as the twentieth century played out (see Fig. 8.4). In practice, the efficiency gains of electrified farming and other technological advances, in the absence of concomitant labor programs, catalyzed a new rural electric prosperity only possible under increasingly consolidated farms run by fewer farmers, the rapid spread of suburban lifestyles in rural places, and uneven development of the countryside marked by growth of rural tourism in some places and intensified industrial production in others. At odds with early REA rhetoric, which argued that the substitution of electricity for labor would free farmers for leisure and politics, leading to a more vibrant community life that would protect the unique character of rural America, the growth-oriented REA model of rural

[61] See, e.g.: "Report of the Administrator of the Rural Electrification Administration, 1964," USDA, 10–11; *Rural Lines*, 9, no. 12, May 1963, 8, HT; USDA, *The REA Pattern*, REA Bulletin 1–8, April 1963, 3.

electrification created economic incentives that pushed many rural residents out of farming.[62] If, as they intended, USAID programs meant to stem rural-urban migration and decrease economic inequality, REA-style electric co-ops were, by the 1960s, empirically not the right model.

Second, when USAID brought REA-style electrification projects to the global South, they did not encounter the equivalent of the American countryside, but rather a complex and contradictory set of rural needs and desires embedded in distinctly non-American political economies. Patterning the global South on rural America failed to take into account that rural electrification in the States was a top-down national-scale project that launched with the full buy-in of the federal government. USAID-style rural electrification, by contrast, competed against a diversity of national desires for development that often did not have a place for American-style co-op democracy in their ideas of how electricity should support the construction of the nation.[63]

In many ways, USAID became victims of their own government's propaganda. Throughout the 1930s and especially in WWII, the REA explicitly downplayed government involvement in the design and management of electric co-ops, publicly promoting instead an image of grassroots electric democracy made possible first and foremost by local organizing. In this revisionist history, REA had only *invested* in these co-ops, not *organized* them, and it was only the prior lack of capital liquidity that had prevented rural electrification from spontaneously occurring earlier. That was very much not the case, as I have argued elsewhere – the REA had a heavy hand in both designing and managing its borrower cooperatives[64] – but the REA's promoted image of grassroots self-help organizations benignly supported by federal financing gave electric cooperatives an aura of inevitability in American policymaking, and likely delayed realization at USAID that cooperative electricity both relied on and actively constructed, through the heavy hand of REA bureaucrats, a particular political economy that was not always in the interests of other developing states. One can trace this narrative of federal financing opening the floodgates of grassroots rural electrification throughout USAID documents as well as in how they justified (or did not have to justify) their operations to Congress – for example, in the interchange between Frank M. Coffin, AID Deputy Administrator for Operations and Representative John J. Rhodes from Arizona, in which Rhodes opines: "Of course…the genius behind electric cooperatives is cheap money. If you can get 2-percent money in the United States it doesn't take an expert to tell you how to start

[62] Ian D. Wyatt and Daniel E. Heckler, "Occupational changes during the twentieth century," U.S. Bureau of Labor Statistics Monthly Labor Review, March 2006: 54–55; Jayson L. Lusk, "The Evolving Role of the USDA in the Food and Agricultural Economy," Mercatus Research, Mercatus Center at George Mason University, Arlington, VA, June 2016.

[63] See, e.g., Anton Mohsin, "The Indonesian Electric Cooperatives," conference paper, SHOT St. Louis, 2018.

[64] Spinak, "'Not Quite So Freely as Air.'"

a cooperative."[65] It had never actually been that simple in the American context, though: the top-down REA was an agency full of federal "experts" telling rural Americans how to start and run their co-ops.[66]

8.3 Mr. Rural Electrification Refuses to Buy a Baby

USAID did not turn to the REA directly for expertise on global rural electrification, but rather to a related private trade association, the National Rural Electric Cooperative Association (NRECA). The NRECA was an independent advocacy body formed by REA co-ops during WWII, originally to navigate wartime material shortages and provide co-ops insurance, and not with the full blessing of co-ops nationwide.[67] By the 1960s, the NRECA self-identified as "one of Washington's most powerful lobbies" with a large technical and professional staff.[68]

USAID's partnership with NRECA – and in fact its particular focus on electric cooperatives – can be traced largely to the missionary zeal of one man: Clyde T. Ellis, or, as he would come to be called, "Mr. Rural Electrification."[69] A corn-fed farm boy born in Garfield, Arkansas, in 1908, Ellis grew up alongside the electricity industry. Politicized at an early age by the "arrogance" of private power companies' dismissive attitudes towards rural communities, Ellis began running for public office in his twenties and served in the Arkansas state government throughout much of the tumultuous years of the 1930s, during which time he helped pass legislation to make electric cooperatives legal in Arkansas. In 1938, he ran for the U.S. House of Representatives on a pro-hydropower platform and won. Ellis subsequently focused much of his 4 years in federal office on rural energy issues, in support of both local electrical development in Arkansas and the Rural Electrification Administration's national program of electric cooperatives. The ending of Ellis's congressional terms fortuitously coincided with the formation of the NRECA; in 1942, the NRECA board of directors recruited Ellis as the first general manager of the new trade association. He accepted, and, apart from a two-year hiatus

[65] Foreign Operations Appropriations for 1964: hearings before a subcommittee of the Committee on Appropriations, House of Representatives, 88th Congress, first session, July 25, 1963, pg. 1719.

[66] For a good recent summary of the rise of the social category of technical experts alongside the modernist state, see Vincent Lagendijk, "From American South to Global South: The TVA's Experts and Expertise, 1933–98," in Frank Trentmann, Anna Barbara Sum and Manuel Rivera, eds., Work in Progress: Economy and Environment in the Hands of Experts (Oekom Verlag GmbH, 2018), 79–101.

[67] Ellis, *A Giant Step*, 70–71, 75. On cooperatives' skepticism about joining the NRECA, see Orcas Power and Light to Walter Harrison, OPALCO records, and Ellis, ix.

[68] Ellis, *A Giant Step*, vii-viii.

[69] Sheila Yount, "Clyde Taylor Ellis (1908–1980)," *Encyclopedia of Arkansas*, Central Arkansas Library System, last updated: April 29, 2014: https://encyclopediaofarkansas.net/entries/clyde-taylor-ellis-2532/; Also see "Retirement of Clyde Ellis, A Great REA Leader," in the U.S. Congressional Record: Proceedings and Debates of the 90th Congress, Second Session, Volume 114 – Part 1 (January 29, 1968), 1271–2.

to serve in the U.S. Navy in 1943 – during which, as Ellis writes in his autobiographical history of the NRECA, he read widely on the philosophy and practice of cooperatives and managed to visit Rochdale and talk to co-op leaders in England[70] – he held the post until his retirement in 1967.[71]

By the late 1950s, Ellis found himself caught up in international diplomacy in ways that made him critical of the government's foreign aid efforts. Under President Truman's 1949 Point Four program, a variety of technical exchange programs had regularly brought planners, engineers, and policymakers from "underdeveloped" nations to the United States. Electricity being the symbol of modernity that it was, many of these visitors eventually found their way to the NRECA for advice on rural electrification. "Most of them spoke sharply of [American] foreign aid efforts," Ellis recalled, "and they were generally responsible people whose views should be heard and respected. More and more I became convinced that at least part of [foreign] assistance programs should be designed directly to help the people help themselves, and that the people must have a strong sense of personal involvement from the beginning. If we were to demonstrate that democracy was superior to communism, then the people must have some experience with democratic methods and procedures, and they seemed to be getting precious little of this."[72] These encounters with foreign visitors convinced Ellis that the NRECA should expand its role from supporting only American cooperatives to also promoting the electric cooperative model internationally.

The Eisenhower administration proved hostile to Ellis's new vision for the NRECA, but, starting in 1959, Ellis found a receptive audience for the idea of NRECA-led international electrification programs in then-senator John F. Kennedy and, later, his incoming administration. Ellis knew that one of Kennedy's goals was to reorganize American foreign aid, a campaign promise that would be realized in the Foreign Assistance Act of 1961, which folded a number of disparate foreign aid offices into the new umbrella Agency for International Development. During his first summer in office, Kennedy tasked the director of the International Cooperation Administration,[73] Henry R. Labouisse, to convene a Special Advisory Committee on Cooperatives. Labouisse invited Ellis to serve on the committee, where he subsequently contributed to a report, "Cooperatives – Democratic Institutions for Economic and Social Development," that would lead to the 1962 Humphrey amendment to the Foreign Assistance Act.

That fall, Ellis was also part of an American delegation sent to Latin America to study existing cooperatives and assess the potential for funding new cooperatives. Ellis's Latin America tour and subsequent visits to AID recipients bolstered his conviction in the cooperative model as a panacea for development. His history of the NRECA-USAID partnership reads at times like a missionary diary:

---

[70] Ellis, *A Giant Step,* 85.

[71] Yount, "Clyde Taylor Ellis (1908–1980)."

[72] Ellis, *A Giant Step,* 197.

[73] A predecessor agency to USAID which was eliminated with the signing of the Foreign Assistance Act in 1961.

> Along a roadside in a country once rich in natural resources, but whose minerals had been exploited largely by outsiders... Our study team, escorted by AID officials, had stopped to visit farm people planting potatoes. Men were plowing with oxen and wooden plows. Women were sitting by little piles of very small seed potatoes, waiting to begin planting. One was nursing an infant. Several poorly clad children were sitting around motionless. After accepting our tips for letting us take pictures, and then asking for more, one of the men tried to sell us the baby. He was speaking in his language, but the AID people understood him. And the child's mother must have, too, but she showed no emotion. Suddenly I was furious at all the poverty and ignorance and intolerance in the world that could make such a thing possible. I wanted to cry, and I wanted to be sick, and all I could do was turn away and close my eyes for a moment. But I knew I would see the man and the mother and child for the rest of my life. ... .It is no wonder, I thought, that these people revolt – that the siren song of the communists sounds good to them. It was easy to see how they could be duped into believing the line from the Communist Manifesto which says, 'You have nothing to lose but your chains.'[74]

Compare this experience with his writing about Japan:

> In Japan we found striking differences between communities which had electric power and those which were still waiting. Where there was power, we saw mills processing rice and other local products, ceramic plants and factories giving employment and income to rural people who previously had scratched a bare living from the soil. We saw tube-well irrigation pumping making possible an extra crop of rice per year. We saw people painting their houses, and draining and paving the streets and putting in public facilities.[75]

Such experiences bolstered Ellis's commitment to making cooperatives central in the international aid agenda, and he lobbied hard to position the NRECA as the federal "contractor of choice" for rural electrification projects.[76]

In 1962, USAID signed a partnership with the NRECA to provide consultation services for their global electrification programs. Ellis recalls that, as the finalization of the partnership happened to coincide with the Cuban Missile Crisis, President Kennedy requested that the official signing ceremony be moved to the Oval Office, as a sign of American commitment to building democracy in Latin America.[77] (See Fig. 8.5.) Presiding over the signing, Kennedy blessed it with teleological pleasantries that USAID-NRECA documents then quoted ad nauseum: "One of the most significant contributions that we can make to the underdeveloped countries," Kennedy proclaimed, "is to pass on to them the techniques which we in this country have developed and used successfully."[78] Such faith that "what is America today will be the world tomorrow," in historian Odd Arne Westad's words, well describes the early guiding philosophies within the USAID-NRECA partnership.[79]

With the official partnership, NRECA established a central office for overseas rural electrification programs in Washington, D.C., and by 1964, AID approved the

[74] Ellis, *A Giant Step,* 205.

[75] Ibid., 215–16.

[76] Wasserman and Davenport, "Power to the People," A-7-A-10.

[77] Ellis, 210.

[78] *The REA Pattern*, REA Bulletin 1–8, USDA, April 1963.

[79] On the teleological nature of USAID ideology, see Westad, *The Global Cold War*, 9.

Fig. 8.5 President Kennedy presides over the signing of the NRECA-USAID partnership in the Oval Office, November 1, 1962. (Source: AR7570-B, Abbie Rowe, White House Photographs, John F. Kennedy Presidential Library and Museum, Boston)

first international loan for five pilot electric cooperative projects, which aimed to electrify a somewhat modest 16,000 farms, homes, and businesses in Latin America.[80] Over the course of the 1960s, USAID paid NRECA for rural electrification consultation services across Latin America, Southeast Asia, Africa, and the Middle East and funded the translation of REA information guides into multiple languages. Tracing USAID funding disbursements to NRECA only captures part of the American investment in global cooperative electrification, however, as NRECA also arranged for REA co-ops to donate hundreds of thousands of pounds of equipment and materials abroad.[81]

There are several reasons USAID might have partnered with NRECA as opposed to the REA directly. For one thing, the REA's mission had changed since its early days of teaching rural America how to build co-ops. Pressure from the House Appropriations Committee on the REA to "quit spoon-feeding" co-ops and a badly-timed airplane crash, which killed the head of the REA management division,

[80] Wasserman and Davenport, "Power to the People," A-2.

[81] See, e.g., USAID, *Foreign Assistance Program Annual Report to the Congress,* FY 1966–69, HT.

Fig. 8.6 President Kennedy talks with Clyde Ellis, September 23, 1963, in a meeting with NRECA and the American Public Power Association (Ellis is to the right of JFK, holding papers). (Source: AR8137-A, Abbie Rowe, White House Photographs, John F. Kennedy Presidential Library and Museum, Boston)

resulted in a rapid dismantling of REA's business education services for REA borrower co-ops in the early 1950s. Into this new vacuum, Ellis reorganized the NRECA to offer management training programs for REA co-ops, and expanded these services throughout the decade.[82] On the USAID end of the partnership, the NRECA also better fit USAID's foundational mandate that foreign aid should be a partnership with American businesses. The NRECA, as a subsidiary trade association of cooperatives, had strong links to the federal government, but was not a state agency. It was, on the books, a private enterprise. Its consulting services consequently brought USAID funds into the accounts of American private business and encouraged additional private expenditures, thus contributing to the American economy and setting up global market dependencies. This "virtuous" cycle would not have occurred if rural electrification consulting had stayed in house through collaborations with federal REA employees (Fig. 8.6).

[82] Ellis, *A Giant Step,* 91–93.

8.4 The Gospel of Electric Democracy

Spearheaded by the NRECA, global electric co-op development thus became a mission that, as USAID evaluation reports would later clarify, transcended local data collection. Quoting from the "Capital Assistance Papers" in Ecuador and the Philippines, for example, USAID researchers Gary Wasserman and Alice Davenport observed that early USAID programs positioned the development of cooperatives as an "important social objective beyond the delivery of electricity," and justified them as a kind of connective tissue that "laces together all the social and economic levels of the community" and a "'watering place' leading toward further community efforts."[83] This "prevailing 'folklore,'" written by the REA and adopted wholesale by USAID, assumed a set of automatic steps leading from cooperatives to liberal democracy through the pull of electricity.[84]

In their 1978 assessment of USAID electrification efforts in Costa Rica and Colombia, John Saunders et al. succinctly captured these beliefs of natural progression in eleven key linkages:

1. Increasing agricultural productivity resulting from the use of labor-saving electrical equipment;
2. Changing crop patterns resulting from the use of seasonal labor-saving equipment and/or the use of electric pumps to provide irrigation water;
3. Development of rural industries; especially those related to the processing of agricultural commodities in rural areas;
4. Reduction in household labor through the use of electrical appliances;
5. Reduction in rural-urban migration of families and young people as rural living conditions improve and employment opportunities are created through industrial development and expansion;
6. Improvements in household sanitary conditions and family health resulting from the use of electricity for water systems, refrigeration, ventilation and lighting;
7. Relating to rural electric cooperatives; improved sense of community participation and involvement resulting from work on a self-help project; development of community leadership training;
8. Increased levels of living resulting from the availability and use of electricity;
9. Reduction in household energy expenditures with the substitution of electricity for candles, kerosene, gas and wood;
10. Increased satisfaction with life resulting from amenities available following electrification.[85]

[83] Wasserman and Davenport, "Power to the People," 20.

[84] David Pierce and Michael Webb, "Rural electrification in developing countries: A reappraisal," *Energy Policy* (August 1987), 338.

[85] John Saunders, J. Michael Davis, Galen C. Moses, and James E. Ross, *Rural Electrification and Development: Social and Economic Impact in Costa Rica and Colombia* (Boulder, CO: Westview Press, 1978), 3–4.

By the late 1970s, it would be evident to both REA and USAID researchers that these relationships were not automatic and were often contradictory – both in the United States and abroad. As USAID mandates shifted, AID researchers started to question what exactly electrification aid had funded in so many countries. The late 1970s–1980s saw a sudden flurry of USAID-commissioned studies of their earlier rural electrification projects, most of which viewed the USAID electric cooperative ideology of the 1960s critically. Post-"New Directions" USAID evaluation studies began to conclude that electricity did not so much lead to development as electricity *was* development in the performance metrics of early USAID rural electrification programs. USAID researchers criticized these early projects for having the goal just "to electrify," that electrification had become an end for USAID instead of a means.[86]

These studies identified and challenged an ideology that electricity's correlation with development was causative. Data from these studies by contrast often showed that the opposite was the case – that a certain level of "development" – for example, credit, roads, health, and education[87] – was necessary for electricity to have a measurable impact on economic growth and public welfare.[88] In their summary report of USAID electrification in four countries, Wasserman and Davenport note that a "theme emerging clearly from the reports was that the development impact of electricity was a reflection of the existing level of development in the area." They explain:

> ...electricity alone does not lead to spontaneous development activities. The provision of power did not in general release untapped demand for its productive use. Linkages to productive activities had to be planned, and programs involving, say, credit or irrigation systems needed more support than electricity alone provided. Rural electrification can aid development activities, not on its own, but as part of a wider program to develop the productive resources of the area receiving power.[89]

They also noted that the development impacts of USAID electrification projects had reinforced racial inequalities around the world.[90] These later AID studies puzzled over an enduring exceptionalism within electricity programs that continued again and again to take the linkage between electricity and development as gospel without observing and measuring actual electrification efforts on the ground. Unsurprisingly, research studies of USAID electrification projects critiqued the "too-faithful compliance" with the REA model.[91]

86 See, e.g., Wasserman and Davenport, A-1-5.

87 Wasserman and Davenport, vii, 9–10.

88 Kessler et al., 1; Paula O. Goddard, Gustavo Gomez, Polly Harrison, and George Hoover, A.I.D. Project Impact Evaluation Report No. 22, "The Product is Progress: Rural Electrification in Costa Rica" (October 1981), 2–3, 17–18; Edward Butler, Karen M. Poe, and Judith Tendler, "Bolivia: Rural Electrification," A.I.D. Project Impact Evaluation Report No. 16, Agency for International Development (December 1980).

89 Wasserman and Davenport, 12.

90 Wasserman and Davenport, 17.

91 Wasserman and Davenport, A-1.

It has become a historical truism that New Deal era programs contained the seeds of their own destruction.[92] Amy Offner has suggested that growing awareness of New Deal limitations influenced its architects to experiment with corrective ideas across the developing world in the postwar era.[93] One way we might read the USAID-NRECA push for electric co-ops abroad, therefore, is as a partially nostalgic effort to get the New Deal right. Among other ideological prejudices, the Cold War gave unaligned countries an aura of a governance vacuum in American policy circles; they held both the allure and the threat of a tabula rasa for either liberal democracy or communism to insert a developmental state. Electric cooperatives promised to be foot soldiers of democracy in this dividing up of the world, simultaneously rewriting landscapes via powerlines to make legible spaces for industrial development and global trade and rewiring poor rural residents into citizens of a networked bureaucratic modernity. Unsurprisingly, USAID recipient nations rarely complied with these American expectations.

8.5 "A Tiny Piece in a Much Larger Puzzle"[94]

In countries with existing cooperative traditions and strong government support, USAID-NRECA electrification programs succeeded in constructing enduring cooperatives that contributed to – or at least correlated with – periods of economic growth and improving social welfare. Costa Rica was one such a poster child for cooperative electrification aid. In 1963–64, a NRECA-USAID team visited Costa Rica and picked three sites for electric co-ops in diverse areas of the country (Fig. 8.7). Between 1965–1969, USAID loaned $3.3 million to the Banco Nacional de Costa Rica for establishing these co-ops. Power came online for rural residents and businesses by the turn of the decade, and, by the 1980s, all three co-ops were still operational and supplied 23 percent of rural electric consumers in Costa Rica. In a 1981 evaluation report titled, "The Product is Progress," a USAID study team found electrification loans "wisely conceived" and even credited an electric cooperative with the "rebirth" of the canton of San Carlos, after the 1960s "boom-town setting" suffered a brief period of population decline in the early 1970s.[95]

[92] See, for example: Gary Gerstle and Steve Fraser, Introduction, in Steve Fraser and Gary Gerstle (eds.), *The Rise and Fall of the New Deal Order: 1930–1980* (Princeton University Press, 1989); Todd Holmes, "Agriculture and the New Deal," in Aaron D. Purcell (ed.), *The New Deal and the Great Depression* (Kent, OH: Kent State University Press, 2014), 49–64; Jason Scott Smith, *Building New Deal Liberalism: The Political Economy of Public Works, 1933–1956* (New York: Cambridge University Press, 2006); Shane Hamilton, *Trucking Country: The Road to America's Wal-Mart Economy* (Princeton: Princeton University Press, 2008).

[93] Offner, *Sorting Out the Mixed Economy,* Introduction.

[94] Kessler et al., 13.

[95] Paula O. Goddard, Gustavo Gomez, Polly Harrison, and George Hoover, A.I.D. Project Impact Evaluation Report No. 22, "The Product is Progress: Rural Electrification in Costa Rica" (October 1981), ii, 3.

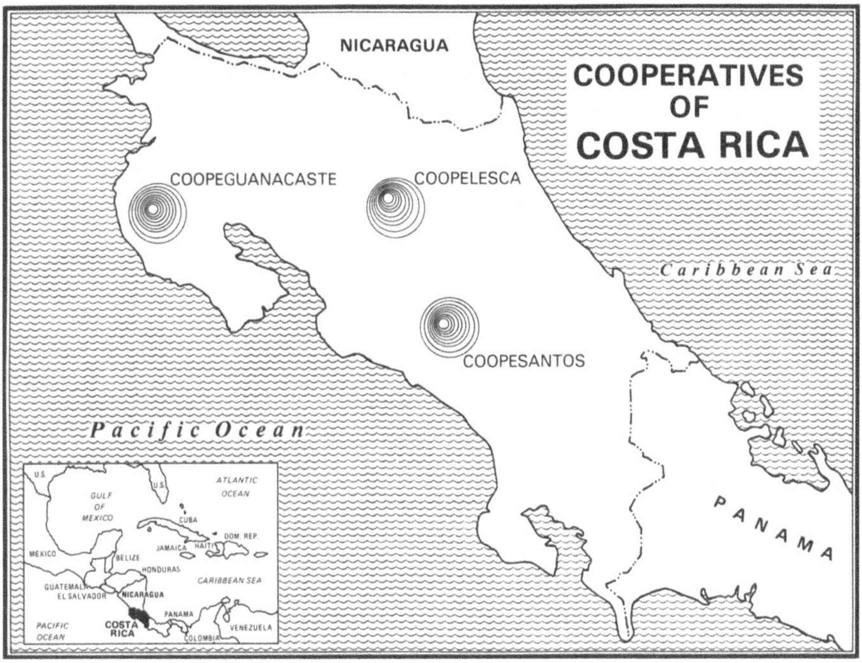

Fig. 8.7 Electric Cooperatives in Costa Rica. (Source: Paula O. Goddard, Gustavo Gomez, Polly Harrison, and George Hoover, A.I.D. Project Impact Evaluation Report No. 22, "The Product is Progress: Rural Electrification in Costa Rica," October 1981)

The study team struggled, however, to disaggregate the influences of cooperative electricity from other development projects. They noted that Costa Rica in the 1960s was more politically stable than many of its neighbors and, of interest to U.S. international development programs, had "a history of democratic institutions" and a "commitment to equitable development unique to the region." Concurrent with USAID-NRECA electrification projects breaking ground, the Costa Rican government had embarked on "unprecedented efforts" towards rural development of health services and education, "which had an implacable rationale and momentum of its own." Additionally, Costa Rica already had a rich history of cooperative organizing. "While one would not want to say that the cooperative was a venerable institution in Costa Rica," the 1981 study noted, "there was certainly adequate precedent by 1960." This cooperative heritage included cooperative housing, buying and agricultural co-ops, savings and loan associations since the 1920s, and more recently the growing success of coffee cooperatives. The canton of San Carlos particularly had "a lively interest in cooperatives, awakened by a local priest," and an umbrella organization, URCOZON (Union Regional de Cooperativas de la Zona Norte). So favorable were conditions in Costa Rica, in fact, that the team questioned whether their observations were "not so anomalous to be irrelevant to other countries."[96]

[96]Goddard et al., "The Product is Progress," 3–5, 17, C-2, D-1-2.

In the absence of such cultural alignment, USAID researchers found that recipient nations were well able to detach electrification aid from its cooperative packaging. In Ecuador, for example, USAID negotiated a variety of development assistance programs through military juntas in both the early 1960s (U.S.-backed) and 1970s. These rapid changes in government and shifting national priorities added risk to foreign aid programs, but amidst fears that Ecuador was "high on the list" of countries heading towards a communist revolution and a commitment to backing the new conservative regime, USAID continued to grant the Ecuadorian government loans throughout the 1960s, a large percentage of which were in support of establishing cooperatives.[97]

Between 1964–1972, USAID allocated $5.8 million to Ecuador for rural electrification projects. In theory, this was supposed to fund the organization of five electric cooperatives across the country, including the expansion of a pre-existing electric cooperative, the Santo Domingo Electric Cooperative, which had been organized with NRECA technical assistance and materials donated from REA co-ops in Kentucky the year before, in 1963[98] (Fig. 8.8). (As part of USAID's "partners of the Alliance" program, Kentucky had "adopted" Ecuador and donated 40,000 lbs. of electrical equipment to the Santo Domingo co-op.) In 1964, Clyde Ellis traveled to Santa Domingo and Daule, where a second cooperative funded by USAID was in the planning phase, and enthusiastically wrote about the trip: "There were many signs of welcome and friendship, but none that said 'Yankee go home' or '*Cuba si, Yanqui no.*' I have seen too many of these latter signs on my trips through Latin America, and I have been caught in the middle of riots, but never in an area where steps have been taken to establish an electric co-op."[99]

Despite Ellis's enthusiasm, retrospective analyses of USAID electric co-op projects in Ecuador found that these projects were met with skepticism from the beginning.[100] After another military coup yet again shifted governance priorities in 1972, the national Ecuadorian Electrification Institute (INECEL) took over the Daule co-op and connected it to the national grid. The final three co-ops planned as a condition of USAID rural electrification aid never made it out of the loan documents – "it is unclear whether [the communities] ever even knew that that they had once been destined to adopt a cooperative form of electrification," Kessler et al.

[97] Kessler et al., 3, Annex 6–2; Jon V. Kofas, "The IMF, the World Bank, and US Foreign Policy in Ecuador, 1956–1966," *Latin American Perspectives,* Issue 120, Vo. 28 No. 5, September 2001: 50–83.

[98] Kessler et al., Annex 5.

[99] Ellis, *A Giant Step,* 212, 214.

[100] Kessler et al., 9–10. They explain: "Since A.I.D.'s zeal in promoting cooperatives as an institutional mechanism for extending rural electrification in Ecuador met with such limited success, it is worthwhile to examine why this concept did not catch hold. At the outset, knowledgeable Ecuadoreans with whom the evaluation team met almost uniformly rejected the overall cooperative model as appropriate for the country. . . .The reasons given for this. . .are varied, but the observation most frequently heard was that Ecuadoreans, in general, are 'uniquely individualistic.'"

Fig. 8.8 The NRECA mascot Willie Wiredhand, an anthropomorphic electric plug, presides over an early meeting to organize a cooperative in Santo Domingo, Ecuador. (Source: Hearings Before the Committee on Foreign Relations, United States Senate, Ninety-first Congress, First Session, on S. 2347, A Bill to Amend the Foreign Assistance Act of 1961, As Amended, and For Other Purposes, Printed on October 8, 1969, 256, HT.)

observed; instead, the funding slated for those co-ops ended up financing the expansion of private company lines to rural towns and agricultural communities.[101]

Taking advantage of volatile U.S.-Ecuador relations in the 1970s, Ecuadorians applied USAID electrification funding through their own institutional framework. As another USAID study concluded about Ecuador: "The fox was in charge of the hen house with predictable results for the cooperatives."[102] In the town of Santa Elena, for example, USAID researchers found little more than lip service to rural electrification as a nod to their USAID financing, as well as rationing for rural clients during peak tourist season (Fig. 8.9).[103] They ultimately conclude: "cooperativism in Ecuador was much more an A.I.D. creation than a response to a local desire. Had we not been so involved in coop promotion in the early 60's as a response to Castro's

[101] Kessler et al., 11.

[102] Wasserman and Davenport, 20.

[103] Kessler et al., 8.

Fig. 8.9 Electricity Lines and TV Antennas in Rural Ecuador. (Source: Judd L. Kessler, Janet Ballantyne, Robert Maushammer, and Nelson Romero Simancas, "Ecuador: Rural Electrification," A.I.D. Project Impact Evaluation Report No. 21, June 1981)

marxist model, A.I.D. might well have discovered more about the political relationship between the municipalities and the central government and designed its electrification activities to strengthen municipal and regional institutions."[104]

There was, unsurprisingly, a wide middle ground of countries that accepted USAID financial and technical aid for cooperatives, but then incorporated the cooperative model into a national governance agenda that did not prioritize local democracy. Thriving electric utilities with limited to no civic engagement proliferated around the world under the guise of cooperatives. In the Philippines, for example, a USAID evaluation team "found no users who had attended the general meeting, attendance at which appeared to vary with the presence or absence of raffles, free food, and transportation," and boards of directors "dominated by a local elite of government officials, businesspeople, professionals, and planters, with no workers or small farmers on the boards." In a corresponding survey, none of the respondents knew the name of anyone serving on their elected board of directors. The only thing maintaining the cooperatives, Wasserman and Davenport found, was the strong interest of the country's National Electrification Administration, an agency that imposed its agenda on the cooperatives it oversaw, sometimes against local interests.[105] Likewise, in Indonesia, the State Electric Company (*Perusahaan Listrik Negara, PLN*) worked with NRECA in the 1970s to establish cooperatives, but then largely used them as an arm of the national government, in coordination with a national resettlement program to shift the population from Java to other islands.[106]

Even in Costa Rica, with its strong tradition of cooperative enterprise, the USAID-NRECA co-op model had few democratic spillover effects. Goddard et al. found that while awareness of membership in USAID-NRECA co-ops was high, attendance at meetings was relatively low, and that there were some criticisms of electric co-ops being "only a business." Their critique of this cooperative implementation strongly evoked REA folklore: "To be sure, the cooperatives have done little in the way of promoting the productive uses of electricity or even educating consumers about the best uses of household electricity...and the sense of 'belonging' even economically, of sharing investment and returns, of exploring possibilities for profitable community uses of electric power, is somewhat lacking."[107] As the REA found in the rural United States, USAID top-down cooperative development required concerted programs of member education for electricity to spark democracy.

USAID researchers further found little evidence of a direct relationship between NRECA-USAID programs and economic growth or poverty alleviation. Despite USAID-NRECA visions of labor-saving technologies revolutionizing rural homes

---

[104] Kessler et al., 15.

[105] Wasserman and Davenport, 19–21.

[106] See Anton Mohsin, "The Indonesian Electric Cooperatives," conference paper, SHOT St. Louis, 2018; Mohan Munasinghe, "Rural Electrification: International Experience and Policy in Indonesia," *Bulletin of Indonesian Economic Studies* 24, no. 2 (August 1988): 87–105.

[107] Goddard et al., D-3.

and farms, few households receiving co-op electricity dove into the all-electric lifestyle with abandon. Surveys of household appliance usage showed uneven adoption of many of the electrical applications rural electrification advocates routinely evoked in the electricity-development dyad. As the REA had found in their early electrification efforts, electric co-op members abroad primarily used electricity for lightbulbs, radios, and irons, and only secondarily and sporadically for larger appliances and expensive renovations such as indoor plumbing.[108] For rural industries, the advantages from central station power were also not so cut and dry. Central station power often replaced diesel equipment or provided an alternative to locally-generated electricity, for example, from a small hydropower plant. Moreover, the "labor savings" from electric farm equipment – e.g. for an electric milking machine on a dairy farm – often decreased employment.[109]

In practice, USAID researchers concluded, the success of USAID-NRECA electric co-ops was not so much a catalyst for as *a sign of* rural development. By the 1990s, retrospective analyses of these programs would conclude that: "Rather than rural electrification bringing about rising living standards, economic dynamism, increased literacy and all the other benefits attributed to it, it is rather the presence of these which creates the conditions under which programmes can be successfully implemented."[110] Cooperative electricity in the end was hardly a panacea for developing nations; it was only "a tiny piece in a much larger puzzle" of bringing American-style development to the world.[111]

8.6 Conclusion: Electricity as Ideology, Electricity as Practice

This chapter, like much of electricity history, is not a neat story of success or failure. As the REA did for rural America, USAID cooperative electrification projects brought quality of life benefits to many, and have even been hailed for helping rebuild social trust in post-conflict states.[112] Still, these programs failed to realize many of their foundational claims, particularly regarding electricity's impact on reducing inequality and increasing civic participation, and it is in these discrepancies that we can learn to more critically understand the history of electricity as a social, political, and diplomatic technology.

Despite the limitations of electrical development in practice, these projects still reshaped the global countryside in ways that continue to reverberate today. Though

---

[108] See, e.g., Saunders et al., 75.

[109] Saunders et al., 88.

[110] Gerald Foley, "Rural electrification in the developing world," *Energy Policy* (February 1992), 146.

[111] Kessler et al., 13.

[112] Ted Weihe, "Cooperatives in Conflict and Failed States," U.S. Overseas Cooperative Development Council (June 2, 2004), 11–12.

many recipient nations effectively co-opted USAID projects for local ends, these new electrical institutions, grids, and products also normalized finance-driven industrialization, the connection of local and global economies, and high-energy lifestyles. In constructing this scaffolding, USAID electrification projects helped to foreclose the possibilities for other visions of development to take hold.[113]

Yet electrification, like most development projects, was rarely a simple case of American imperialism. Though from the privileged vantage point of Western academia in the twenty-first century, we may critique twentieth-century ideas of modernism and development as flawed and neocolonial, we should not discount people's desire to be included in these projects nor the psychological impacts of electricity as a symbol of inclusion in an unequal world – "a sign of their town coming of age or of the attention paid to them by the political leadership" – and a harbinger of a better life to come.[114] The ideology of electricity as modernity clearly took on a life of its own beyond American Cold War machinations. At certain points in the history of development, the social promises of electrical modernity seemed real and possible, and, for some, they evidently were, even as they remained out of reach for others.

USAID documents also suggest that many of the civil servants organizing these programs seemed to authentically believe in American-style electrification as a social good. They sincerely saw electric cooperatives as a tool in building a virtuous cycle between economic growth, liberal democracy, and social justice, and were honestly surprised and disappointed when it didn't turn out that way – which, in Amy Offner's helpful framing, "is not to say that they were benign or that anyone should regard them with nostalgia."[115] Rather, the contradictions raised by these projects show the importance of separating ideology – the stories we tell about electricity – from practice – the social and spatial relations electrification has helped manifest – in electricity history, and to ask what work each has done in shaping the world.

One broader tension this case study raises, for example, is how technological determinism in American policy circles has obscured the state as a relevant actor in technology-based diplomacy. The lack of awareness of the importance of federal intervention in REA co-ops in the U.S. led to USAID programs that ignored the role of central planning in cooperative electrification and created frictions with aid recipient governments at regional and national scales. How could American development ideologies be so anti-state abroad even as they were federal programs dependent on Congressional support at home? How did this normalize ideas of markets and states as separate antagonistic entities?[116]

---

[113] Foley, 146; Kessler et al., 7.

[114] Wasserman and Davenport, 5–6, 8; Goddard et al., C-3.

[115] Offner, *Sorting Out the Mixed Economy*, 16.

[116] Here, I am drawing on critiques of community development as a master narrative in competition with high modernism – see Hodge, "Writing the History of Development (Part 1)," 444–6, and Immerwahr, *Thinking Small*. On other instances of federally-funded programs that challenged sovereignty in practice while being shielded by a rhetoric of neutral technology aid, see: Megan Black, "Prospecting the World: Landsat and the Search for Minerals in Space Age Globalization," *The Journal of American History* 106, no. 1, 2019: 97–120.

A second, related tension that electricity historians may take from this case study is that energy transitions happen differently in different contexts, even when there is a hegemonic source of funding and technical advice attempting to streamline development. This resurgent diversity is not just national, it is, in many cases, very local. Separating ideology from practice in electricity development aid illuminates how electricity has long been harnessed in the service of, but has rarely fully achieved, neocolonial, resource-intensive, inequality-producing world-making. The conceit here is not that Soviet-funded electrification projects would have been any more equitable or inclusive, but that rather we should as historians of electricity seek out the multiplicity of development visions electricity inspired – by Cold War superpowers, by unaligned nations, by rural communities appraising global events, and by many, many others. The outstanding question here is: would any of these visions have built substantially different futures, or was electricity as a global technology already too prescribed to invite that kind of experimentation? A better understanding of where and how such projects created global dependencies and where and how they were able to be tailored to local ends may help us approach our unevenly transitioning present with more humility, creativity, and justice.

Bibliography

Akuno, Kali, and Ajamu Nangwaya, eds. 2017. *Jackson Rising: The Struggle for Economic Democracy and Black Self-Determination in Jackson, Mississippi*. Wakefield, QC: Daraja Press.

Alperovitz, Gar. 2011. America Beyond Capitalism: Reclaiming Our Wealth, Our Liberty, & Our Democracy. Takoma Park, MD/Boston, MA: Democracy Collaborative Press.

———. 2020. Building a Democratic Economy: Sketch of a Pluralist Commonwealth. *Nonprofit Quarterly*. https://nonprofitquarterly.org/building-a-democratic-economy-sketch-of-a-pluralist-commonwealth/. Accessed 14 Apr 2020.

Angelo, Hillary. 2016. From the City Lens Toward Urbanisation as a Way of Seeing: Country/City Binaries on an Urbanising Planet. *Urban Studies* 54: 158–178.

Barnett, Michael. 2011. *Empire of Humanity: A History of Humanitarianism*. Ithaca: Cornell University Press.

Beckert, Sven, Angus Burgin, Peter James Hudson, Louis Hyman, Naomi Lamoreaux, Scott Marler, Stephen Mihm, Julia Ott, Philip Scranton, and Elizabeth Tandy Shermer. 2014. Interchange: The History of Capitalism. *The Journal of American History* 101: 503–536.

Black, Megan. 2014. Interior's Exterior: The State, Mining Companies, and Resource Ideologies in the Point Four Program. *Diplomatic History* 40: 81–110.

———. 2019. Prospecting the World: Landsat and the Search for Minerals in Space Age Globalization. *The Journal of American History* 106: 97–120.

Bronwyn, James. 1995. *The Impacts of Rural Electrification: Exploring the Silences*, EDRC Report Series. Energy & Development Research Centre, University of Cape Town.

Brown, Kate. 2015. *Plutopia: Nuclear Families, Atomic Cities, and the Great Soviet and American Plutonium Disasters*. Oxford University Press.

Christie, Drew. 1984. Recent Calls for Economic Democracy. *Ethics* 95 (1): 112–128.

Cullather, Nick. 2002. Damming Afghanistan: Modernization in a Buffer State. *The Journal of American History* 89 (2)., History and September 11: A Special Issue (Sep., 2002): 512–537.

Cumbers, Andrew. 2019. Economic Democracy. In *Keywords in Radical Geography: Antipode at 50*, ed. Antipode Editorial Collective et al. Wiley Blackwell. https://onlinelibrary.wiley.com/doi/book/10.1002/9781119558071.
Curl, John. 2009. *For All the People: Uncovering the Hidden History of Cooperation, Cooperative Movements, and Communalism in America*. Oakland, CA: PM Press.
Dahl, Robert A. 1986. *Preface to Economic Democracy*. Berkeley: University of California Press.
DeFilippis, James. 2004. *Unmaking Goliath: Community Control in the Face of Global Capital*. New York: Routledge.
Ekbladh, David. 2002. Mr. TVA: Grass-Roots Development, David Lilienthal, and the Rise and Fall of the Tennessee Valley Authority as a Symbol for U.S. Overseas Development, 1933–1973. *Diplomatic History* 26 (3): 335–374.
———. 2011. *The Great American Mission: Modernization and the Construction of an American World Order*. Princeton: Princeton University Press.
Ellis, Clyde T. 1966. *A Giant Step*. New York: Random House.
Essex, Jamey. 2013. *Development, Security, and Aid: Geopolitics and Geoeconomics at the U.S. Agency for International Development*. Athens, GA: University of Georgia Press.
Foley, Gerald. 1992. Rural electrification in the developing world. *Energy Policy* 1992: 145–152.
Gerstle, Gary, and Steve Fraser. Introduction. In *The Rise and Fall of the New Deal Order: 1930–1980, 1989*, ed. Steve Fraser and Gary Gerstle. Princeton University Press.
Glaser, Leah. 2009. *Electrifying the Rural American West: Stories of Power, People, & Place*. Lincoln: University of Nebraska Press.
Granovetter, Mark, and Patrick McGuire. 1998. The Making of an Industry: Electricity in the United States. *The Sociological Review* 46 (1): 147–173.
Hamilton, Shane. 2008. *Trucking Country: The Road to America's Wal-Mart Economy*. Princeton: Princeton University Press.
Haskell, Thomas L. 1985. Capitalism and the Origins of the Humanitarian Sensibility. *The American Historical Review*. Part 1, 90 (2): 339–361 and Part 2, 90 (3): 547–566.
Hecht, Gabrielle. 1998. *The Radiance of France: Nuclear Power and National Identity after World War II*. Cambridge: MIT Press.
Hilson, Mary. 2013. Consumer Co-operation and the Economic Crisis: The 1936 Roosevelt Inquiry on Co-operative Enterprise and the Emergence of the Nordic 'Middle Way'. *Contemporary European History* 22 (2): 181–198.
Hodge, Joseph Morgan. 2015. Writing the History of Development (Part 1: The First Wave). *Humanity: An International Journal of Human Rights, Humanitarianism, and Development* 6 (3): 429–463.
———. 2016. Writing the History of Development (Part 2: Longer, Deeper, Wider). *Humanity: An International Journal of Human Rights, Humanitarianism, and Development* 7 (1): 125–174.
Holmes, Todd. 2014. Agriculture and the New Deal. In *The New Deal and the Great Depression*, ed. Aaron D. Purcell, 49–64. Kent, OH: Kent State University Press.
Hughes, Thomas P. 1983. *Networks of Power: Electrification in Western Society, 1880–1930*. Baltimore/London: The Johns Hopkins University Press.
Hyman, Louis. 2013. Why Write the History of Capitalism? *Symposium Magazine*. July 8. http://www.symposium-magazine.com/why-write-the-history-of-capitalism-louis-hyman/.
Immerwahr, Daniel. 2015. *Thinking Small: The United States and the Lure of Community Development*. Harvard University Press.
Jellison, Katherine. 1993. *Entitled to Power: Farm Women and Technology, 1913–1963*. The University of North Carolina Press.
Jenkins, Kirsten, Darren McCauley, Raphael Heffron, Hannes Stephan, and Robert Rehner. 2016. Energy justice: A conceptual review. *Energy Research & Social Science* 11: 174–182.
Jones, Christopher F. 2010. A Landscape of Energy Abundance: Anthracite Coal Canals and the Roots of American Fossil Fuel Dependence, 1820–1860. *Environmental History* 15 (July 2010): 449–484.

———. 2016. *Routes of Power: Energy and Modern America*. Cambridge: Harvard University Press.

Kaika, Maria. 2017. 'Don't call me resilient again!': the New Urban Agenda as immunology…or…what happens when communities refuse to be vaccinated with 'smart cities' and indicators. *Environment & Urbanization* 29 (1): 89–102.

Keillor, Steven J. 2000. *Cooperative Commonwealth: Co-ops in Rural Minnesota, 1859–1939*. St. Paul, MN: Minnesota Historical Society.

Kelly, Marjorie. 2013. The Economy: Under New Ownership, *yes!*, https://www.yesmagazine.org/issue/issues-how-cooperatives-are-driving-the-new-economy/2013/02/20/the-economy-under-new-ownership/.

Kennedy, David. 2011. *The Dark Sides of Virtue: Reassessing International Humanitarianism*. Princeton: Princeton University Press.

Kline, Ronald R. 2000. *Consumers in the Country: Technology and Social Change in Rural America*. Baltimore/London: The Johns Hopkins University Press.

Kofas, Jon V. 2001. The IMF, the World Bank, and US Foreign Policy in Ecuador, 1956–1966. *Latin American Perspectives*, Issue 120 28 (5): 50–83.

Laakkonen, Simo, Viktor Pál, and Richard Tucker. 2016. The Cold War and Environmental History: Complementary Fields. *Cold War History* 16 (4): 377–394.

Lagendijk, Vincent. 2018. From American South to Global South: The TVA's Experts and Expertise, 1933–1998. In *Work in Progress: Economy and Environment in the Hands of Experts*, ed. Frank Trentmann, Anna Barbara Sum, and Manuel Rivera, 79–101. Oekom Verlag GmbH.

Macekura, Stephen. 2013. The Point Four Program and U.S. International Development. *Political Science Quarterly* 128 (1): 127–160.

———. 2015. Point Four and the Crisis of U.S. Foreign Policy Aid in the 1970s. In *Foreign Aid and the Legacy of Harry S*, ed. Raymond H. Geselbracht. Truman: Truman State University Press.

MacLeod, Greg. 1998. *From Mondragon to America: Experiments in Community Economic Development*. University College of Cape Breton Press.

Malleson, Tom. 2014. *After Occupy: Economic Democracy for the 21st Century*. New York: Oxford University Press.

McGuire, Mark F., and Vernon W. Ruttan. 1990. Lost Directions: U.S. Foreign Assistance Policy Since New Directions. *The Journal of Developing Areas* 24 (2): 127–180.

Mitchell, Timothy. 2011. Carbon Democracy: Political Power in the Age of Oil.. Verso

———. 2002. *Rule of Experts: Egypt, Techno-Politics*. Modernity: University of California Press.

Mohsin, Anton. 2018. *The Indonesian Electric Cooperatives, Conference Paper*, SHOT St. Louis

Munasinghe, Mohan. 1988. Rural Electrification: International Experience and Policy in Indonesia. *Bulletin of Indonesian Economic Studies* 24 (2): 87–105.

Murphy, Michelle. 2017. *The Economization of Life*. Durham: Duke University Press.

Needham, Andrew. 2014. *Power Lines: Phoenix and the Making of the Modern Southwest*. Princeton: Princeton University Press.

Neufeld, John L. 2016. *Selling Power: Economics, Policy, and Electric Utilities Before 1940*. Chicago: University of Chicago Press.

Nye, David E. 1990. *Electrifying America: Social Meanings of a New Technology*, 331–334. Cambridge: MIT Press.

Offner, Amy. 2019. *Sorting Out the Mixed Economy: The Rise and Fall of Welfare and Developmental States in the Americas*. Princeton University Press.

Pierce, David, and Michael Webb. 1987. Rural Electrification in Developing Countries: A Reappraisal. *Energy Policy* 1987: 329–338.

Ranganathan, V. 1993. Rural Electrification Revisited. *Energy Policy* 21 (2): 142–151.

Robertson, Thomas. 2016. Cold War Landscapes: Towards an Environmental History of US Development Programmes in the 1950s and 1960s. *Cold War History* 16 (4): 417–441.

Rockman, Seth. 2014. What Makes the History of Capitalism Newsworthy? *Journal of the Early Republic* 34 (3): 439–466.

Sandwell, Ruth. 2015. Pedagogies of the Unimpressed: Re-Educating Ontario Women for the Modern Energy Regime, 1900–1940. *Ontario History* 107 (1): 36–59.

Scott, James C. 1998. *Seeing Like a State: How Certain Schemes to Improve the Human Condition Have Failed*. Yale University Press.

Smith, Jason Scott. 2006. *Building New Deal Liberalism: The Political Economy of Public Works, 1933–1956*. New York: Cambridge University Press.

Sneddon, Christopher. 2015. *Concrete Revolution: Large Dams, Cold War Geopolitics, and the US Bureau of Reclamation*. Chicago: University of Chicago Press.

Spinak, Abby. 2020. 'Not Quite So Freely as Air': Electrical Statecraft in North America. *Technology and Culture* 61 (1): 71–108.

Swyngedouw, Erik. 2010. Apocalypse Forever? *Theory, Culture & Society* 27 (2): 213–232.

Tarekegne, Bethel. 2020. Just electrification: Imagining the justice dimensions of energy access and addressing energy poverty. *Energy Research & Social Science* 70: 1–6.

Tucker, Richard P. 2010. Containing Communism by Impounding Rivers: American Strategic Interests and the Global Spread of High Dams in the Early Cold War. In *Environmental Histories of the Cold War*, ed. J.R. McNeill and Corinna R. Unger. Cambridge University Press.

Westad, Odd Arne. 2005. *The Global Cold War: Third World Interventions and the Making of Our Times*. Cambridge University Press.

Williams, James H., and Navroz K. Dubash. 2004. Asian Electricity Reform in Historical Perspective. *Pacific Affairs* 77 (3): 416–417.

Yakubovich, Valerie, Mark Granovetter, and Patrick McGuire. 2005. Electric Charges: the Social Construction of Rate Systems. *Theory and Society* 34 (2005): 579–612.

Zeuli, Kimberly. 2002. *The Role of Cooperatives in Community Development*. University of Wisconsin Center for Cooperatives Bulletin.

Abby Spinak is a Lecturer in Landscape Architecture at the Harvard Graduate School of Design, where she teaches environmental history and theory. Her forthcoming book ties the history of rural electrification to the evolution of twentieth-century American capitalism and alternative economic visions. She has published in *Technology and Culture*, and for more public audiences with *Technology's Stories* and the *Harvard Design Magazine*.

Chapter 9
Vehicle-to-Grid, Regulated Deregulation, and the Energy Conversion Imaginary

Matthew N. Eisler

Abstract Automobility and grid electricity are pervasive forms of energy conversion infrastructure that have existed as discrete entities for most of their history. Periodically, however, actors have looked to electric vehicles as a means of solving sociotechnical problems in the business of electricity. In the early days of electricity, systems builders saw electric vehicles as a way of creating demand and storing electricity. Around the turn of the millennium, some analysts believed that electric vehicles mandated by the state of California could be repurposed to supply high-cost ancillary services that the state had unbundled from formerly unified generation, transmission, and distribution services and commodified through market-oriented deregulation. Known as vehicle-to-grid, this imagined massive system of distributed energy storage and power generation was later perceived as an important means of managing the integration of renewable energy conversion systems into the grid. This paper argues that vehicle-to-grid was a product of contradictory public policy impulses that illustrate the challenges neoliberal dirigisme (quasi-planning) faced in renovating legacy energy conversion systems in dynamic environmental and social conditions.

Keywords Electric vehicle · Electrical grid · Vehicle-to-grid · Bidirectional power · Deregulation · Electric utility

9.1 Introduction

In popular and public policy discourse, the electric car figures as a key component of the carbon-neutral green energy revolution. Electrify the automobile, the thinking goes, and society will have made a major stride in the transition to sustainability. Since the early 1990s, policymakers used a mixture of regulations and incentives to encourage automakers to build electric cars, with astonishing results. Since the turn of the millennium, millions of electrics have been manufactured, with millions more

M. N. Eisler (✉)
University of Strathclyde, Glasgow, UK
e-mail: matthew.eisler@strath.ac.uk

W. B. Carlson, E. M. Conway (eds.), *Electrical Conquest*, Archimedes 67,
https://doi.org/10.1007/978-3-031-44591-0_9

projected to enter service in the 2020s and 2030s. Impressive on their face, these figures concealed an energy conundrum. Critics hold that electric cars are only as green as the primary energy sources from which electricity is generated, at least in terms of emissions, yet progress in producing electricity from renewables lagged far behind progress in the industrial technology of the commercial electric car. To be sure, renewable energy production grew by 500 percent from 1950 to comprise more than 10% of aggregate U.S. energy production by the second decade of the twenty-first century. However, renewable energy cannot easily be exploited because it is intermittent and often becomes available at times when the social demand for it is low. For this and other reasons, only a tiny fraction of renewable energy was used for transportation by the 2010s.[1]

However, some planners and policymakers believed that electric cars could be modified in ways that would reconcile the energy conundrum to the benefit of a host of stakeholders. The idea was to reconfigure electrics for bidirectional power flow and repurpose them as power plants on wheels. For utilities, such systems would be a cheap and fast way of storing and utilizing grid electricity, including electricity generated from renewable energy, satisfying rising demand. Ratepayers would have cheaper and more reliable electricity while individual private owners of bidirectional electrics would be empowered as entrepreneurs, selling electricity as a means of paying off their vehicles. In this imagined system, known as vehicle-to-grid (V2G), the electric car as a power plant on wheels represented a potent new public policy rationale for electric automobile technology.

Ideas about vehicle-to-grid and the electric car as an electric utility resource were informed by important assumptions about the relationship between energy and electricity, and about science and technology as public policy more broadly. Unpacking these ideas complicates the premise of the energy transition and the corollary that primary energy resources have determining social power. Indeed, energy poses thorny conceptual problems for historians and sociologists of science and technology. In both policy and academic discourse, energy is often treated as an objective natural force exogenous to technology and society.[2]

Yet human beings access energy through technology, a social entity that as the historian Cyrus Mody notes is importantly shaped by human interactions with and

[1]U.S. EIA, *August 2020*, 3–4, 37. In 2019, renewable energy accounted for .05 percent (1410 of 26,796 trillion BTUs) of energy converted for purposes of transportation in the U.S.; see U.S. EIA, *August 2020*, 44–45.

[2]For examples of historical scholarship that ascribe important causational power to energy, see Kenneth Pomeranz, *The Great Divergence: China, Europe, and the Making of the Modern World Economy* (Princeton, NJ: Princeton University Press, 2000); Christopher F. Jones, *Routes of Power: Energy and Modern America* (Harvard University Press, 2014); and Vaclav Smil, *Energy and Civilization: A History* (Cambridge, MA: MIT Press, 2017). For a critique of the energy transition concept, see Ute Hasenöhrl and Jan-Henrik Meyer, "The Energy Challenge in Historical Perspective," *Technology and Culture* 61, no. 1 (2020): 295–306.

interpretations of natural phenomena.[3] Analyses that privilege energy as the frame of reference often elide "energy" with "electricity," obscuring crucial sociotechnical differences between energy resources and the conversion and carrier technologies that enable useful work to be derived from them. Some forms of primary energy, especially petroleum and its derivatives, serve as stores of energy and, when moved through pipelines and in tankers, as carriers of energy. Electricity is an energy carrier with special properties. It can be generated from all forms of primary energy but can be stored in only a few physical forms at great cost. For these reasons, central station and grid managers tended to seek ways of using electricity as soon as it was generated.[4]

The physical qualities of energy resources and energy carriers, especially as they pertain to storage, do not in themselves determine social relations. However, they do have important implications for when and how people develop energy carrier/conversion regimes. Contemporary energy infrastructure consists of a number of such regimes with varying degrees of sociotechnical compatibility.[5] Automobility and electricity are two such systems. Historically, automobility was constructed as a system of privatized public transportation around a durable consumer good utilizing petroleum-based internal combustion propulsion.[6] Conversely, electricity was

---

[3]Cyrus C.M. Mody, *The Long Arm of Moore's Law: Microelectronics and American Science* (Cambridge, MA: MIT Press, 2016), 9–10. Scholars nominally committed to the constructivist position have long sought to understand how or if the qualities or affordances of technologies make possible or necessary social outcomes; see, for example, Edmund Russell, James Allison, Thomas Finger, John K. Brown, Brian Balogh, and W. Bernard Carlson, "The Nature of Power: Synthesizing The History of Technology and Environmental History," *Technology and Culture* 52, no. 2 (April 2011): 246–59; Christophe Lécuyer and David C. Brock, "The Materiality of Microelectronics," *History and Technology* 22, no. 3 (September 2006): 301–25; Frank N. Laird, "Constructing the Future: Advocating Energy Technologies in the Cold War," *Technology and Culture* 44, no. 1 (2003): 27–49; Langdon Winner, "Do Artifacts Have Politics?" *Daedalus* 109, no. 1 (Winter, 1980): 121–36.

[4]Compare this characteristic of electricity networks with gas networks, which can store gas; see, for example, Leslie Tomory, "Building the First Gas Network, 1812–1820," *Technology and Culture* 52, no. 1 (2011): 75–102, on 82.

[5]Virtually all energy conversion/carrier regimes including electricity are integrated into industrial structures and include industrial/metallurgical coal, industrial gas, utility natural gas, petrochemicals, transportation fuels, and domestic heating oil. In his pioneering studies of the development of early electricity systems, Thomas P. Hughes acknowledged the geophysical qualities of the primary energy resources of coal, oil, and hydro as sociotechnical factors in electricity generation but did not root the social relations of primary energy in the context of energy conversion/carrier regimes; see *Networks of Power: Electrification in Western Society, 1880–1930* (Baltimore: Johns Hopkins University Press, 1983), 262, 367, 406, 418.

[6]Scholars sometimes use the term "automobility" to express the pervasiveness of the modern system of light-duty passenger transportation. The term encompasses technology and physical infrastructure as well values, habits, and folkway. For analyses of the automobile as a social system, see Matthew N. Eisler, *Age of Auto Electric: Environment, Energy, and the Quest for the Sustainable Car* (Cambridge MA: MIT Press, 2022); Peter D. Norton, *Fighting Traffic: The Dawn of the Motor Age in the American City* (Cambridge, MA: MIT Press, 2008); Deborah Clarke, *Driving Women: Fiction and Automobile Culture in Twentieth-Century America* (Baltimore: Johns Hopkins

constructed essentially as a service around a variety of energy conversion devices utilizing coal, oil, falling water, and later natural gas, uranium, wind, and solar radiation. Automobility and electricity emerged in parallel in the late nineteenth century but remained discrete, with largely incompatible technologies and business models.

Technologists and policymakers long imagined ways of bridging the worlds of stored energy and electricity and constructed sociotechnical imaginaries to work out how this might be accomplished.[7] A favored such imaginary was the fuel cell, a family of electrochemical energy conversion devices that in principle can electro-oxidize any hydrogenous fuel. In the 1960s, gas utilities believed fuel cells would enable them to compete with electrical utilities by allowing homeowners to produce their own electricity from utility gas. Other energy analysts envisioned using fuel cells in an even more ambitious scheme that involved repurposing the gas grid for hydrogen as means of storing and carrying energy produced by nuclear and renewable energy.[8]

Vehicle-to-grid is a similar type of energy conversion imaginary. Like the fuel cell, the idea of the electric car as a power plant on wheels was informed by assumptions of the determining powers of science and technology held by neoliberal planners, a subject of much STS analysis.[9] Neoliberal ontology attempts to 'save the phenomenon' of the classical liberal notion of the steady-state self-regulating marketplace by acknowledging the reality of market failure and the need for some form

University Press, 2007); Tom McCarthy, *Auto Mania: Cars, Consumers, and the Environment* (New Haven, CT: Yale University Press, 2007); Matthew Paterson, *Automobile Politics: Ecology and Cultural Political Economy* (Cambridge, UK: Cambridge University Press 2007); John Urry, "The 'System' of Automobility," *Theory, Culture and Society* 21, nos. 4–5 (2004): 25–39; Mimi Sheller and John Urry, "The City and the Car," *International Journal of Urban and Regional Research* 24, no.4 (2000): 737–57; David Gartman, *Auto Opium: A Social History of American Automobile Design* (London: Routledge, 1994); and Virginia Scharff, *Taking the Wheel: Women and the Coming of the Motor Age* (Albuquerque, NM: University of New Mexico Press, 1991).

7 On the idea of the sociotechnical imaginary, see Sheila Jasanoff, "Future Imperfect: Science, Technology, and Imaginations of Modernity," in *Dreamscapes of Modernity: Sociotechnical Imaginaries and the Fabrication of Power,* eds. Sheila Jasanoff and Sang-Hyun Kim (Chicago: University of Chicago Press, 2015), 1–33.

8 See Matthew N. Eisler, *Overpotential: Fuel Cells, Futurism, and the Making of a Power Panacea* (New Brunswick, NJ: Rutgers University Press, 2012); see also Dennis Anderson and Matthew Leach, "Harvesting and Redistributing Renewable Energy: On the Role of Gas and Electricity Grids to Overcome Intermittency Through the Generation and Storage of Hydrogen," *Energy Policy* 32 (2004): 1603–14.

9 Such work is often traced to Philip Mirowski's study of the influence of physics on neoclassical economics; see *More Heat than Light: Economics as Social Physics, Physics as Nature's Economics* (Cambridge: Cambridge University Press, 1989). Mirowski and his collaborators observed that scholars studied the effects of neoliberal ideology on material practice of science and engineering mainly in the context of biomedicine and biotechnology; see Rebecca Lave, Philip Mirowski, and Samuel Randalls, "Introduction: STS and Neoliberal Science," *Social Studies of Science* 40, no. 5 (2010): 659–75, on 663; and Philip Mirowski, "The Future(s) of Open Science," *Social Studies of Science* 48, no. 2 (2018): 171–203, on 178–79.

of state intervention. Such thinking emerged out of the environmental, energy, and socio-economic crises of the 1960s and 1970s. In addressing these crises, policymakers at the state and federal levels devised a range of sometimes contradictory measures that I have elsewhere referred to as quasi-planning.[10] On the one hand, the neoliberal state deregulated capital and deconstructed organized labor.[11] On the other, it tightened environmental regulations and made available vast public resources through what sociologists sometimes refer to as the national developmental state, a decentralized network of institutions set up as a politically acceptable means of producing technology to solve social problems.[12] This state's principal tool was the public-private research and development consortium, a structure that aligned basic science and early-stage technology development with energy and environmental regulations in purposely uncoordinated fashion in order to avoid the appearance of overt winner-picking.[13]

Vehicle-to-grid emerged in a context of intensifying quasi-planned research and development of solar and hydrogen energy conversion systems.[14] The imaginary

---

[10] I first used the expression quasi-planning in "Energy Innovation at Nanoscale: Case Study of an Emergent Industry," *Science Progress*, May 23, 2011; http://scienceprogress.org/2011/05/innovation-case-study-nanotechnology-and-clean-energy/ (https://web.archive.org/web/20120504054443/http://scienceprogress.org/2011/05/innovation-case-study-nanotechnology-and-clean-energy/).

[11] The sociologist Daniel Bell described the landscape wrought by neoliberal policy as "postindustrial," a sociotechnical regime in transition from vertically-integrated enterprises of heavy manufacturing to decentralized specialized enterprises of so-called intellectual technology including digital electronics and software; Daniel Bell, "Foreword 1999," in *The Coming of Post-Industrial Society: A Venture in Social Forecasting* (Basic Books: New York, 1999), xxxiv-xliv.

[12] Fred Block, "Swimming Against the Current: The Rise of a Hidden Developmental State in the United States," *Politics and Society* 36, no. 2 (2008): 169–206; Meredith Woo-Cumings, ed., *The Developmental State* (Ithaca, NY: Cornell University Press, 1999); Onis Ziya, "The Logic of the Developmental State," *Comparative Politics* 24, no. 1 (1991):109–26.

[13] On trends in U.S. science and technology policy in the neoliberal postindustrial era, see Ann Johnson, "The End of Pure Science: Science Policy from Bayh-Dole to the NNI," in *Discovering the Nanoscale*, eds. Davis Baird and Alfred Nordmann (Amsterdam: IOS Press, 2004), 219–22; and David C. Mowery, "Collaborative R&D: How Effective Is It?" *Issues in Science and Technology* 15, no. 1 (1998): 37–44.

[14] See, for example, Frank N. Laird, "Constructing the Future: Advocating Energy Technologies in the Cold War," *Technology and Culture* 44, no. 1 (2003): 27–49; Eisler, *Overpotential*; Lillian Hoddeson and Peter Garrett, *The Man Who Saw Tomorrow: The Life and Inventions of Stanford R. Ovshinsky* (Cambridge, MA: MIT Press, 2018). There is a good deal of advocacy literature around alternative single-form energy or energy carrier regimes with techno-utopian undertones; see, for example, Eduard Justi, *Leitungsmechanismus und Energieumwandlung in Festkörpern* (Göttingen: Vandenhoeck and Ruprecht, 1965); Laurence O. Williams, *Hydrogen Power: An Introduction to Hydrogen Energy and Its Applications* (London: Pergamon Press, 1980); Peter Hoffmann, *The Forever Fuel: The Story of Hydrogen* (Boulder, CO: Westview Press, 1981); Peter Hoffmann, *Tomorrow's Energy: Hydrogen, Fuel Cells, and the Prospects for a Cleaner Planet* (Cambridge, MA: MIT Press, 2001); Jeremy Rifkin, *The Hydrogen Economy: The Creation of the World-Wide Energy Web and the Redistribution of Power on Earth* (New York: Jeremy P. Tarcher/Penguin, 2002); Seth Fletcher, *Bottled Lightning: Superbatteries, Electric Cars, and the New Lithium Economy* (New York: Hill and Wang, 2011).

appeared at the intersection of the near-concurrent restructuring of electricity markets and the transformation of passenger vehicle regulations, primarily via California's zero-emission vehicle mandate, an air quality certification category that forced automakers to produce electric vehicles. For most of the twentieth century, electricity was treated as a "natural" monopoly (hence, paradoxically, requiring regulation) thanks both to systems imperatives calling for a balance of supply and demand and public policy perceptions that reliable power provision was a social necessity.[15] By the end of the century, however, planners and pundits assumed that subjecting electricity to market forces would yield optimal "efficiency," the idealized end of economic systems in which the interests of all groups are said to be reconciled: profits for sellers, low prices for buyers, and, in the green energy era, environmental sustainability for all.

In California, however, deregulation and marketization destabilized electricity, a problem some observers believed could be solved using mandated electric cars modified for bidirectional power flow. At one level, the promoters of vehicle-to-grid shared assumptions derived from precedent and analogy in historical context, making what the sociologist Trevor Pinch referred to as a "similarity judgment," a comparison of superficially-similar classes of artifacts.[16] They believed that most of the requisite technologies already existed in the automobile, electricity, and information technology sectors and had only to be applied in creative new ways consonant with public policy. In the scaled bidirectional electric car, the architects of vehicle-to-grid perceived an efficient and ready-made private sector solution to public policy problems.

This paper uses the case study of the history of the electric car as an electric utility resource as a means of understanding automobility and electricity as co-constructed and co-located yet discrete sociotechnical regimes with distinct material and social affordances. The history of V2G problematizes the idea of the energy transition by highlighting the importance of the relationship between energy carriers and energy conversion technologies. And it illuminates tensions between the theory and practice of localized and centralized systems, between private and public interest in the business of energy conversion, and between "free" and "regulated" markets in this context. This paper joins works in this volume by Julie Cohn, Cyrus Mody, and David Nye in engaging the problems and paradoxes of attempts to renovate legacy energy conversion infrastructure, and by Stathis Arapostathis, Nathan Kapoor, and Abby Spinak in engaging the ways the technologies of electricity have been used to engineer social landscapes.

[15]On this idea, see Adam Plaiss, "From Natural Monopoly to Public Utility: Technological Determinism and the Political Economy of Infrastructure in Progressive-Era America," *Technology and Culture* 57, no. 4 (2016): 806–30.

[16]Trevor Pinch, "'Testing-One, Two, Three. . .Testing!' Toward a Sociology of Testing," *Science, Technology and Human Values* 18, no. 1 (1993): 27–31.

9.2 Infrastructure as Enterprise

At various points in the development of automobility and electricity, entrepreneurs wrestled with the question of whether they were in the business of providing a product or a service, a problem conditioned by a host of assumptions around the social relations of the market in changing historical context. Industrial capitalism is often associated with the production of consumer products, with conventional economic thinking long assuming that such goods were manufactured first in batches as relative luxuries and then at scale as cheap necessities. In practice, advanced industrial economies began to concurrently manufacture luxury and mass-market consumer products both ephemeral and durable from the later nineteenth century. Readers of automobile history are familiar with this dynamic in the contrasting interwar business models of the Ford Motor Company (single standardized low-cost product) and General Motors (variegated tiered-price product lineup).[17]

In contrast, electricity does not readily fit most spatial-temporal and socio-cultural assumptions of what constitutes a consumer product. Producers and consumers alike long struggled to interpret electricity as an economic entity. As a natural phenomenon, electricity is evanescent, variously invisible and visible depending on conditions. As a sociotechnical phenomenon, electricity is usually visually shrouded by the systems that promulgate and maintain its supply and generally may be discerned only indirectly, through its supporting infrastructure and effects mediated by technology. Although electricity was first supplied to customers of means, it lacks the cultural connotations embedded in material "thingness" that helps actors distinguish a luxury from a necessity. Economists sometimes classify electricity as a commodity, a class of substance including agricultural and raw mineral resources whose material qualities are said to complicate differentiation within their class, leading the market to treat each class as equivalent and universal.[18] The perception of electricity as a type of commodity is reinforced by the difficulty of tracing the provenance of the primary energy used to produce it. As one contemporary observer put it, electricity is "not like a fine wine; it is all the same and the consumer does not care whether it comes from the Napa Valley or the Finger Lakes."[19]

While economists interpreted electricity as a commodity, builders of electricity historically defined their business in terms of a service, one whose technological

---

[17]See Thomas P. Hughes, *American Genesis: A Century of Invention and Technological Enthusiasm, 1870–1970* (Chicago: University of Chicago Press, 1989), 212–19; Sally H. Clarke, *Trust and Power: Consumers, the Modern Corporation, and the Making of the United States Automobile Market* (Cambridge: Cambridge University Press, 2007), 104.

[18]There is actually a great deal of qualitative diversity within classes of certain commodities such as wheat, oil, chocolate, and coffee that producers and traders recognize and price accordingly. Diversity pricing has been extended to account for the perceived social and environmental (im)-morality of the production of some commodities; on the political economy of fair trade, see Martin Richardson and Frank Stähler, "Fair Trade," *Economic Record* 291 (2014): 447–61; Mark Hayes, "On the Efficiency of Fair Trade," *Review of Social Economy* 64, no. 4 (2006): 447–68.

[19]See Benjamin Ross, "California's Regulation Debacle," *Dissent* (Spring 2001): 45–7, on 47.

exigencies warranted treating it as a "natural" regulated monopoly, an argument famously advanced by the Edison protégé Samuel Insull. In this framing, expansion and profit were justified as serving systems stability.[20] Unsurprisingly, such arguments did not always resonate with the public. Early consumers of electricity often did not understand what they were paying for. Even producers were divided on the material and commercial identity of their product. Electricity historian Julie Cohn argues that suppliers of electricity long vacillated between treating electricity as an essential service or a commodity well after they had expanded its delivery beyond wealthy elites to the general public.[21] To be sure, arguments for monopoly and competition and the respective cases made (not always explicitly) for efficiency through stability and instability are not unique to electricity. Historically, the U.S. state sought to balance these tensions, at various times promoting competition (and instability) through anti-trust laws and at other times promoting monopolies (and stability) by tolerating and regulating them.[22]

Unlike in the case of electricity, however, public arguments for regulating automobility were rarely explicitly framed in terms of systems stability, even though automobility is subject to similar destabilizing network effects. As the historian Peter Norton notes, automobility went largely unregulated in its early years partly because policymakers (unlike some engineers) did not analogize the road-vehicle system as a utility infrastructure that functioned optimally as a natural monopoly.[23] Many actors tended to regard space occupied by privately-owned automobiles at rest and in motion as private, a perspective that clashed with other private as well as public interests in urban environments. Regulating parking and operation fell to municipalities and states, while the federal government played an increasing role in planning interstate road infrastructure and establishing industry standards for certain aspects of automobile design.[24] From the mid-1950s, these regulatory interventions massively expanded road surface area and gradually improved the safety and efficiency of the individual internal combustion-engined (ICE) automobile and reduced its emissions. As the auto fleet scaled, however, it produced increasingly

[20] Samuel Insull, *Central-Station Electric Service: Its Commercial Development and Economic Significance As Set Forth in the Public Addresses (1897–1914) of Samuel Insull* (Chicago: Privately Printed, 1915), 70–1, 409–10.

[21] Julie Cohn, "Bias in Electric Power Systems. A Technological Fine Point at the Intersection of Commodity and Service," on *Electric Worlds/Mondes Électriques: Creations, Circulations, Tensions, Transitions (19th–twenty-first C.)*, eds. Alain Beltrain, Léonard Laborie, Pierre Lanthier, and Stéphanie Le Gallic (Brussels: Peter Lang, 2016), 271–93.

[22] For a synopsis of these cycles, see David M. Hart, "Antitrust and Technological Innovation in the U.S.: Ideas, Institutions, Decisions, and Impacts, 1890–2000," *Research Policy* 30, no. 6 (2001): 923–36.

[23] Norton, *Fighting Traffic*, 196.

[24] Regulation in the context of automobility is often traced to the work of public interest advocate Ralph Nader in his debut book *Unsafe at Any Speed: The Designed-in Dangers of the American Automobile* (New York: Grossman Publishers, 1965).

large quantities of pollution and traffic throughput became increasingly congested.[25] Until the advent of the zero-emission mandate, regulations of all sorts had relatively minimal effects on the industrial structures and commerce of automobility.

9.3 Cars, Grids, and the Technopolitics of Storing Electrons

The automobile-grid relationship has been importantly conditioned by the materiality of automobility and electricity as energy conversion regimes. The nexus of this relationship is electricity storage, a sociotechnical entity with affinities and antagonisms around the various interests embedded in these regimes. Batteries generally have shorter lifetimes than electric motors, a temporal mismatch that importantly shaped the business of electric automobility: it implied hidden battery replacement costs for users and that much of the profit to be had lay in battery-making. This durability dilemma was an important factor in causing entrepreneurs of conventional automobility to eschew the electric car. At the beginning of the twentieth century, it also led many entrepreneurs of early electric cars to view their business as a service. The service model insulated users from the cost and inconvenience of battery replacement and maintenance and passed those costs onto the service provider. When companies like Ford started selling millions of cheap and reliable gasoline-fueled ICE vehicles, held the historians Gijs Mom and David Kirsch, the leased electric car could not compete.[26]

Managers of early electricity generating stations also viewed electric vehicles within the service paradigm, but to different ends. At different points in the construction of the grid, grid-builders perceived storage as desirable owing to electricity's problematic qualities as an energy carrier. These entrepreneurs initially used large stationary lead-acid rechargeable batteries to provide spare capacity (known as residual load) on early direct-current electricity networks, whose main generators were typically shut down at night.[27] As grid-builders shifted to alternating-current systems and expanded generating capacity, however, they sought new sources of demand for cheap off-peak night-time electricity and reinterpreted the storage function of large rechargeables accordingly. Stationary battery power was expensive, so some central station managers hoped to outsource storage to the

[25] Vaclav Smil noted that the idea that secular improvements in energy conversion efficiency equated with an aggregate decline in energy consumption was first debunked by the English economist W. Stanley Jevons in 1865; see Smil, *Energy in Nature and Society: General Energetics of Complex Systems* (Cambridge, MA: MIT Press, 2008), 271–72.

[26] Kirsch and Mom argued that had the electric car been marketed as a consumer durable, it might have been able to compete with the gasoline-fueled ICE car; see Kirsch and Mom, "Visions of Transportation: The EVC and the Transition from Service- to Product-Based Mobility," *Business History Review* 76, no. 1 (2002): 75–110.

[27] J. N. Baker and A. Collinson, "Electrical Energy Storage at the Turn of the Millennium," *Power Engineering Journal* 13, no. 3 (1999): 107–12, on 107.

owners and users of the battery electric cars then proliferating in northeastern electrified cities.[28] Central station managers and associations established electric vehicle sections and encouraged the formation of centrally-located charging and maintenance pools, emphasizing electric trucks, which employed comparatively larger batteries than electric passenger vehicles and used vastly more energy than the light bulbs and irons that utilities promoted in concert with manufacturers of consumer durables.[29]

Electric trucks did not serve long in this role, a result, argued Mom, of a conflict of interest in the business of supplying electric mobility and the business of supplying electricity. It cost utilities money to support electric vehicles and the immediate beneficiaries were equipment suppliers. Mom argued that the asymmetrical pace of electrification in the U.S. also inhibited proliferation of the electric truck into rural areas where gasoline-fueled ICE propulsion instead became established, leading utility managers in these areas to promote non-propulsion forms of demand, especially electric irons.[30]

Early electric utilities used the battery electric vehicle as a load management tool for a short time before abandoning the technology in this role after the First World War, likely because the storage conundrum disappeared with the massive expansion of the grid after the First World War. With general electrification, a continental energy conversion and distribution system was created, one that was tied into diverse sources of supply and demand across time zones and national borders. Large-scale electrochemical energy storage essentially disappeared from the grid.[31] When utilities considered storage technology at all, they preferred pumped storage, which is capital intensive and limited to hydropower plants. While the capacity of a rechargeable battery declines over the relatively short lifetime of the device, pumped storage infrastructure is as durable as hydropower infrastructure and, with maintenance and ideal environmental conditions, can function near its full capacity for decades. First used in the late 1920s, pumped storage was not widely adopted until the 1970s.[32]

---

[28] Gijs Mom, *The Electric Vehicle: Technology and Expectations in the Automobile Age* (Baltimore: Johns Hopkins University Press, 2004), 204–07.

[29] Mom, *The Electric Vehicle*, 208.

[30] Mom, *The Electric Vehicle*, 233.

[31] Richard H. Schallenberg, *Bottled Energy: Electrical Engineering and the Evolution of Chemical Energy Storage* (Philadelphia: American Philosophical Society, 1982), 391–392.

[32] William F. Pickard, "The History, Present State, and Future Prospects of Underground Pumped Hydro for Massive Energy Storage," *Proceedings of the IEEE* 100, no. 2 (2012): 473–83. One of the most ambitious such schemes was Storm King, planned in the early 1960s by New York's Consolidated Edison as what would have been the nation's largest pumped-storage hydroelectric plant. Con Ed chair Harland C. Forbes analogized this installation, ultimately never built, as an enormous battery that would help the utility reconcile the discontinuities of supply and demand. For the definitive account of the environmental history of this project, see Robert Lifset, *Power on the Hudson: Storm King Mountain and the Emergence of Modern American Environmentalism* (Pittsburgh: Pittsburgh University Press, 2014).

9.4 Regulation, Deregulation, and the Limits of Growth

In the 1920s and 1930s, utilities expanded and became heavily concentrated, precipitating an ideological clash around the question of whether optimal efficiency in this context derived from stability or competition. In the 1920s and early 1930s, 75% of gas and electric utilities were owned by 13 holding companies, many of which went bankrupt in the stock market crash of 1929. In a pivotal episode of New Deal technopolitics, President Franklin Delano Roosevelt singled out the "Power Trust" and Insull in particular as exemplars of predatory monopoly capital and championed the Public Utilities Holding Companies Act as a remedy. Administered by the Securities and Exchange Commission, the PUHCA was a kind of anti-trust measure that abolished the complex holding company structure and restricted ownership of electric utilities to public and private entities that specifically produced electricity, restructuring the industry mainly around intrastate as well as a few interstate holding companies. Interstate wholesale pricing was regulated by state commissions and the Federal Power Commission and its successor Federal Energy Regulatory Commission (FERC) from 1977. As a result of these reforms, electric utilities effectively became regulated regional quasi-monopolies, a system that persisted until the last quarter of the century.[33]

Regulated vertically integrated electricity utilities provided a variety of supply services that together served as much to order and stabilize the system as meet consumer demand. In principle, the most profitable of these services was peaking power, delivered when demand was highest. Regulated utilities spread the high cost of peaking power across the much less lucrative stabilization services, including spinning reserve, or reserve power generation, and frequency regulation, the balance of load and generation that maintained the system at 60 Hertz. They employed the so-called "declining-block" rate principle, where rates were averaged across the integrated high and low-cost bundle of services and declined with increasing consumption. In this system, pricing did not reflect real-time shifts in supply and demand as in other consumer commodity markets, partly because it was not then possible to accurately meter electricity consumption in real time. But regulated electricity was cheap largely because it was based on cheap energy, a consequence of post-Second World War federal policy that made cheap energy a national security imperative.

This sociotechnical system socialized the rhythms of postwar American industrial and middle-class domestic life. It incentivized consumers to use more electricity and utilities to build ever-larger and more efficient generation plant. By the early 1970s, however, higher energy costs, slowing economic growth, inflation, and mounting public opposition to energy megaprojects were eroding the premise of endless growth that underpinned the U.S. energy conversion regime. At Consolidated

---

[33] James W. Moeller, "The Revenge of Wendell Willkie," *Public Utilities Fortnightly* 134, no. 2 (1994): 14; Lori A. Burkhart, "Does PUHCA Inhibit Diversification?" *Public Utilities Fortnightly* 132, no. 16 (1994): 33.

Edison, New York's largest electric utility, for example, executives responded to changing socio-economic conditions by sharply raising rates.[34] It was in this context that planners made a series of efforts to reform regulated electricity as part of a broader public policy shift to seek efficiencies by fostering competition in enterprises of regulated infrastructure.[35] This process was guided by the doctrine of marginal cost pricing, an economic theory that prescribed pricing reflecting the extra cost of an additional unit of output. Its best-known advocate was Cornell University economist and Carter administration advisor Alfred E. Kahn, an influential ideologue of deregulation. As chair of the New York Public Service Commission in the early 1970s, Kahn implemented marginal cost pricing of electricity on grounds that rates reflecting the "real" cost of energy would negate the need for precipitous rate hikes and rebalance the system. In effect, marginal cost-priced electricity incentivized conservation, aligning proponents of this economic theory with environmentalists opposed to the construction of new energy conversion plant.[36]

The corollary of marginal cost electricity was that nonrenewable energy was a finite resource. This is an irrefutable fact, but the assumption that deregulation and competition were the best tools to determine the marginal cost price of electricity did not account for the social reality that cheap energy remained a national security imperative. The energy crisis of the 1970s proved to be a transitory political event around petroleum that did not reflect the real absolute supply of energy. With the resolution of the Arab oil embargo and the exploitation of new sources of petroleum in places like Alaska and the North Sea as well as alternative energy resources, the energy crisis came to an end in the late 1970s and early 1980s.[37] If marginal cost pricing could be adopted at the level of the local electric utility, it could not easily be imposed at the level of the global energy economy.

Kahn gave relatively little consideration to what deregulation and marketization would mean for utility electricity with the return of cheap energy and economic

---

[34]Thomas K. McCraw, *Prophets of Regulation: Charles Francis Adams, Louis D. Brandeis, James M. Landis, Alfred E. Kahn* (Cambridge, MA: Belknap Press of Harvard University Press, 1984), 240–50.

[35]Contemporary analyses often trace these efforts to the Public Utilities Regulatory Policy Act. Passed by Congress in 1978, PURPA was designed to foster competition and energy diversity by requiring utilities to purchase power from independent producers if they could produce it more cheaply than the utilities. The law encouraged the use of conservation technology and renewable energy resources as well as the decentralization of energy production; see, for example, Paul L. Joskow, "Markets for Power in the United States," *The Energy Journal* 27, no. 1 (2006): 1–36, on 1; Ghazal Razeghi, Brendan Shaffer, and Scott Samuelsen, "Impact of Electricity Deregulation in the State of California," *Energy Policy* 103 (2017): 105–15, on 106; Yohanna M.L. Gultom, "Governance Structures and Efficiency in the U.S. Electricity Sector after the Market Restructuring and Deregulation," *Energy Policy* 129 (2019); 1008–19, on 1009.

[36]McCraw, *Prophets of Regulation*, 237, 242, 249.

[37]Smil, *Energy and Civilization*, 365–66.

growth from the later 1980s.[38] The economist's preferred model for demonstrating the efficacy of deregulation and marginal cost pricing was the airline industry, where he argued that the deterioration in the quality of service was more than compensated by the decline in fares.[39] Kahn was not alone in making such assumptions. One contemporary observer imagined that the deregulation of electricity would resemble deregulation in other industries and would entail the unbundling and shedding of "non-essential" services, the eliminating of cross-subsidization, and the commodification of core competencies.[40]

However, restructuring utility electricity in this manner had the potential to destabilize it.[41] In California, planners recognized this and developed a hybrid approach that essentially represented regulated deregulation. In 1998, the state legislature created the Independent System Operator (CAISO), a non-profit organization that reported to the FERC and was designed to manage the emerging market. Generation was deregulated and separated from distribution, which remained regulated, as no group had an interest in duplicating transmission. Forty percent of installed capacity was sold to newly-created independent power producers from which the three major utilities were compelled to purchase power auctioned on a day-to-day, hour-to-hour basis. Retail prices were capped but wholesale prices were not.[42] In California's partially deregulated, disintegrated system, independent producers dominated peaking power as the most profitable service.[43] The less-lucrative

[38] In the list of 21 self and jointly-authored publications Kahn cited in the bibliography of his *Whom the Gods Would Destroy, or How Not To Deregulate* (Washington, D.C.: The AEI Press, 2001), only two focused expressly on utility deregulation.

[39] Alfred E. Kahn, "Surprises of Airline Deregulation," *The American Economic Review* 78, no. 2 (1988): 316–322.

[40] David W. Wise, "The Tides of Deregulation," *Public Utilities Fortnightly* 125, no. 5 (1989): 39–40.

[41] In the early 2000s, Kahn professed doubts about the assumed benefits of deregulated electricity, especially consumer choice; see William Sweet and Elizabeth A. Bretz, "How to Make Deregulation Work: Alfred E. Kahn, the Father of Airline Deregulation, Firmly Defends it in an Interview with IEEE Spectrum but is Less Sanguine about the Effect on Electricity and Communications," *IEEE Spectrum* 39, no. 1 (2002): 50–6, on 51.

[42] Ross, "California's Regulation Debacle." Policy analyses of electricity do not always address the technological implications of unbundling. For example, University of Arizona researchers Elizabeth Baldwin, Valerie Rountree, and Janet Jock acknowledged that electricity grids must balance supply and demand in order to function properly but also suggested that the three core electricity services of generation, transmission, and distribution were unproblematically "separable." This perspective informed their larger thesis that the sociotechnical problems of distributed generation could be solved with "distributed governance," which ostensibly empowered consumers through "demand-side management;" see Baldwin, Rountree, and Jock, "Distributed Resources and Distributed Governance: Stakeholder Participation in Demand Side Management Governance," *Energy Research and Social Science* 39 (2018): 37–45, on 39.

[43] Ross, "California's Regulation Debacle."

grid stabilization functions were spun off as so-called "ancillary" services that CAISO contractually compelled producers to supply.[44]

These interventions produced a highly unstable sociotechnical regime in which the public bore all risk. In 2000, environmental and sociotechnical conditions converged to cause disequilibrium. Hot weather stoked demand, traditionally met with peaking plant, typically gas turbines, built expressly for this purpose and used only a few times a year. It took time and money to add peaking capacity and deregulation and marginal cost pricing disincentivized new construction. Moreover, nearly 20% of the state's generating capacity was idled, ostensibly for maintenance. As a result, producers operating peaking capacity kept it in operation longer to meet rising demand, causing a wholesale price spike of 800% in May, exacerbated by market manipulation. Independent producers profited but Southern California Edison nearly failed while Pacific Gas and Electric went bankrupt. In early 2001, consumers suffered from rolling blackouts.[45]

9.5 A Power Plant on Wheels

It was in the context of California's electricity crisis of 2000–2001 that the electric car was once again perceived as a solution to problems of grid electricity. To be sure, the energy crisis of the 1970s had already caused electric utilities to consider using electric cars in the traditional role of meeting nighttime demand, and they promoted research to this end in the 1980s.[46] In this application, the electric car had a unidirectional power relationship with the grid.

A bidirectional electric, one that could feed power into the grid as well as draw power from it, implied far more complex sociotechnical dynamics. The concept originated in a collaboration between a group of environmental policy analysts and the principals of a small engineering company at the forefront of the electric vehicle revival in the 1990s and early 2000s. The company, known as AC Propulsion (ACP), was founded by the engineers Wally Rippel and Alan Cocconi, a pair of Caltech graduates who as consultants for AeroVironment, a developer of lightweight experimental aircraft, made major contributions to the propulsion system for Impact, an all-battery electric concept car built under contract for GM in the late 1980s. With its integrated solid state power controls, induction motors, and miniaturized on-board charger, the Impact was widely regarded as the most advanced electric

[44]I credit this insight into the origins of the expression "ancillary" services to David Hawkins, an engineer who served as a chief aide to Kellan L. Fluckiger, vice president of operations for the CAISO in the 1990s and early 2000s; Hawkins, interview by the author, November 10, 2020.

[45]Congress of the United States Congressional Budget Office, "Causes and Lessons of the California Electricity Crisis," September 2001; Timothy Egan, "Tapes Show Enron Arranged Plant Shutdown," *New York Times*, February 4, 2005.

[46]Pandit G. Patil, "Prospects for Electric Vehicles," *IEEE AES Systems Magazine* (December 1990): 15–9, on 17.

car of its day and the inspiration for the zero-emission vehicle mandate.[47] According to Cocconi, the vehicle also had bidirectional capability, although it was not fully implemented.[48]

GM's efforts to convert Impact into what would become the EV1 pilot production car, especially the decision to equip the car with an off-board charging station, alienated Rippel and Cocconi. In 1992, they quit the Impact project and founded ACP around the integrated battery electric propulsion concept. The company's second-generation AC-150 power drive had bidirectional capability, intended primarily as a emergency feature to enable electric cars to transfer charge between each other at a time when battery capacity was limited and the prospect of running out of charge and becoming stranded were very real. Through the 1990s and into the early 2000s, ACP earned revenue by selling its integrated propulsion systems to automakers including Honda and Volkswagen for packaging in converted production models.[49]

For the established car companies, this was a quick and relatively inexpensive way to gain experience in the technology and also meet their mandate commitments. But for independent equipment suppliers like ACP, it was an uncertain way to make a living. In 2000, GM ceased production of the EV1, a signal that the automaking establishment was abandoning the all-battery electric format.

With prospects of an EV market fading, ACP looked at other ways of marketing its technology. It saw an opportunity in California's electricity crisis and the ideas of Willett Kempton, an anthropologist and environmental policy analyst at the University of Delaware, and Steven Letendre, a professor of management studies at Green Mountain College. In an academic article published in 1997, Kempton and Letendre argued that bidirectional electric cars could be pressed into service as mobile power plants. They presented their case as a syllogism: because the aggregate generation capacity of the U.S. light duty vehicle fleet was vastly greater than that of stationary generation plant and because the average light vehicle was used only 4% of the time, a fleet of electric cars could constitute an important resource for electric utilities even if it was only a fraction of the size of the conventional fleet.[50]

---

[47] On the effects of Impact on environmental policymaking, see, for example, Michael Shnayerson, *The Car That Could: The Inside Story of GM's Revolutionary Electric Vehicle* (New York: Random House, 1996); John J. Fialka, *Car Wars: The Rise, the Fall, and the Resurgence of the Electric Car* (New York: Thomas Dunne Books, 2015); Gustavo Collantes and Daniel Sperling, "The Origin of California's Zero Emission Vehicle Mandate," *Transportation Research Part A* 42 (2008): 1302–13, on 1306.

[48] Alan Cocconi, interview by the author, November 11, 2020.

[49] Cocconi, interview.

[50] See Willett Kempton and Steven Letendre, "Electric Vehicles as a New Power Source for Electric Utilities," *Transportation Research Part D: Transport and Environment* 2, no. 3 (1997): 157–75, 159–60; Steven Letendre and Willett Kempton, "The V2G Concept: A New Model for Power?" *Public Utilities Fortnightly* 140, no. 4 (2002): 16–26; Steven Letendre, Paul Denholm, and Peter Lilienthal, "New Load, or New Resource?" *Public Utilities Fortnightly* 144, no. 12 (2006): 28–33.

With California plagued by skyrocketing electricity prices and rolling blackouts, ACP president Tom Gage contacted Kempton, who confirmed he had not patented the concept of the electric vehicle as a mobile power plant.[51] The company then launched a project to demonstrate what Gage dubbed vehicle-to-grid.[52] The task was organized by ACP's Alec Brooks, the manager of the production program of the company's tzero sports car. Brooks was a pivotal figure in the electric vehicle revolution, part of a close-knit community of enthusiast-experts that emerged around Caltech. Like Rippel and Cocconi, Brooks was a graduate of the prestigious research university and the three engineers were friends and collaborators. After graduation, Brooks joined AeroVironment and managed Impact as well as its predecessor Sunraycer, recruiting Rippel and Cocconi for both jobs. In the 1990s, Brooks built up a team at AeroVironment that provided support services for GM's electric vehicle programs until he, too, tired of corporate routine and joined ACP in 1999.[53]

Working with Gage and Kempton and consulting CAISO, Brooks set out to understand how marketized electricity was managed and how the electric car might function as a utility resource. Kempton and Letendre had initially envisaged electric cars serving the peaking market.[54] With that market controlled by the independent producers, that left ancillary services. Theorists including Kempton believed frequency regulation was the most attractive of these, constituting about 80% of CAISO expenditures on ancillary services, and it was to this market that ACP looked.[55] With funding from CARB and help from the National Renewable Energy Laboratory, the team staged an experiment. It installed an AC-150 drive in a converted Volkswagen Beetle, where it functioned as an integrated on-board charger that converted AC to DC power for use in the battery and operated the process in reverse. The team demonstrated bidirectional power flow using wireless power dispatch commands simulated from CAISO historical data, not linked in real time, because the power capacity of a single vehicle was too small to be accepted by the operator's energy management system.[56]

The team assumed that all of the basic technologies of vehicle-to-grid were at hand and had only to be applied in novel ways. Nevertheless, the experiment seemed to suggest that vehicle-to-grid did require some technological innovation. A crucial component was the "aggregator," a notional sociotechnical entity that mediated

---

[51] Willett Kempton, email communication with the author, October 13, 2020. According to Kempton, Cocconi developed the bidirectional feature to enable quick discharge of a battery into the grid rather than depleting the battery through automobile use, a process that could take many hours.

[52] Willett Kempton, Jasna Tomić, Steven Letendre, Alec Brooks, and Timothy Lipman, "Vehicle-to-Grid Power: Battery, Hybrid, and Fuel Cell Vehicles as Resources for Distributed Electric Power in California," UC Davis Institute for Transportation Studies, ECD-ITS-RR-01-03, June 2001, xiv.

[53] Alec Brooks, interview by the author, December 6, 2019.

[54] Kempton, email communication.

[55] Kempton, Tomić, Letendre, Brooks, and Lipman, "Vehicle-to-Grid," 56.

[56] Alec N. Brooks, "Final Report: Vehicle-to-Grid Demonstration Project: Grid Regulation Ancillary Service with a Battery Electric Vehicle," Contract number 01–313, October 21, 2002, 11–12.

between the grid operator and connected vehicles, tracking available electrics and representing them as a unified source of controllable capacity. Brooks noted that an aggregator required sophisticated software, technology that was beyond the scope of the experiment.[57]

However, the aggregator also had sweeping institutional-organizational implications. Vehicle-to-grid was predicated on the ambitious project of interconnecting automakers and utilities, industries with dissimilar business models governed by dissimilar regulatory bodies.[58] Brooks held that the chief obstacle to vehicle-to-grid was the lack of an institutional champion. With automakers having no clear interest in the scheme, that champion would have to be a regulatory agency.[59] In order to serve as a solution to problems caused by grid deregulation and marketization, the car-grid energy conversion regime paradoxically required further layers of governmentality.

9.6 The EV Battery Economy

Among the many technical problems raised by the vehicle-to-grid concept was the question of how repurposing the electric car for distributed generation would affect its battery. At one level, this question engages issues at the interface of power electronics and electrochemistry, allied fields of technoscience that had long been alienated from one other.[60] With the advent of mobile computing in the early 1990s, researchers were just beginning to understand how new battery chemistries interacted with portable appliances, whose duty cycle, or proportion of time of operation, had important consequences for the integrity of the chemical structure of the power source.[61]

---

[57] Brooks, "Final Report," 1–3, 8–10.

[58] Kempton, Tomić, Letendre, Brooks, and Lipman, "Vehicle to Grid Power," v.

[59] Brooks, "Final Report," 49.

[60] For a host of reasons, research and development of batteries historically occurred at a great social and intellectual distance from the research and development of consumer devices. While the revolution in electronics and personal computing helped facilitate the revival of power source technoscience and narrow this gap, battery and power source researchers belonged to largely separate research communities; on this question, see, for example, Ralph J. Brodd, *Factors Affecting US Production Decisions: Why Are There No Volume Lithium-Ion Battery Manufacturers in the United States?* (Gaithersburg, MD: National Institute of Standards and Technology, 2005); and Matthew N. Eisler, "Exploding the Black Box: Personal Computing, the Notebook Battery Crisis, and Postindustrial Systems Thinking," *Technology and Culture* 58, no. 2 (2017): 368–91.

[61] The capacity of actors to understand systems integration in this context was further conditioned and complicated by outsourcing and offshoring, a factor in a series of consumer product crises involving the highly combustible lithium ion battery through the 1990s, 2000s, and 2010s. One paper produced for a workshop organized by the National Academy of Engineering in 2006 linked the complex manufacturability of notebook computers to the battery failures and implied a connection with the disintegration of work practices caused by offshoring and outsourcing; see

In the early 2000s, knowledge of the useful lifetime of large electric vehicle rechargeable batteries was just beginning to be developed.[62] Brooks noted that the batteries of electric cars providing grid services would undergo additional cycling in comparison with those of vehicles used solely for mobility, degrading their capacity in the process, although by exactly how much over time was not then known.[63] Understanding and resolving this question was as important to the economic viability of vehicle-to-grid as it was to the commercialization of the electric car.

However, knowledge-making in these contexts was inhibited by the refusal of automakers to build battery electric vehicles at scale, for reasons that had as much to do with the affordances of the technology as with the auto industry's dislike of state meddling. The durability dilemma implied hidden battery replacement costs for users, and that much of the profit to be had in the electric car market therefore lay in battery-making.[64] For automakers, this was anathema. To buy time to neutralize the mandate, they enlisted better battery discourse, a way of talking about the future electric car. Better battery discourse conjured an idealized electric supercar and had origins in narratives of the future around materials science and engineering, an interdiscipline promoted by the national developmental state in the wake of Sputnik to supply advanced compounds for military electronics and aerospace.[65] Informed by military imperatives, materials thinking privileged compounds and devices that enabled high power and energy, often at the cost of durability and cost-effectiveness.[66]

Automakers abstracted this emphasis on performance to construct an argument that existing power sources, especially the lead-acid rechargeable, could not match the cost and energy density of gasoline-fueled ICE propulsion. Industry's mandate

---

Jason Dedrick and Kenneth L. Kraemer, "Impact of Globalization and Offshoring on Engineering Employment in the Personal Computing Industry," in *The Offshoring of Engineering: Facts, Unknowns, and Potential Implications* (Washington, DC: The National Academies Press, 2008), 125–36.

[62] Progress on proving battery lifetime began to be made later in the decade with the advent of the ultrahigh-precision charger, a device that used high-rate cycling to "beat the clock" on problematic side reactions; see J. R. Dahn, J. C. Burns, and D. A. Stevens, "Importance of Coulombic Efficiency Measurements in R&D efforts to Obtain Long-Lived Li-ion Batteries," *Interface* 25, no. 3 (2016): 75–8.

[63] Brooks, "Final Report," 27–8.

[64] One early opponent of the mandate was a member of the air resources board itself and the sole dissenter on the question of battery economics. During the September hearings, Andrew Wortman raised the specter of the durability dilemma, citing a study by the Department of Energy indicating that operators of electric cars equipped with standard lead-acid batteries would have to replace their packs every 15 months at a cost of between $3000 to $4000 for each replacement; see Collantes and Sperling, "The Origin," 1308; see also Jananne Sharpless, interview by the author, September 5, 2019.

[65] Bernadette Bensaude-Vincent, "The Construction of a Discipline: Materials Science in the United States," *Historical Studies in the Physical and Biological Sciences* 31, no. 2 (2001): 223–248.

[66] In power source technoscience, energy is usually defined as the amount of stored energy per unit of volume or mass and power is usually defined as the rate of energy flow per unit of volume.

responsibilities, automakers argued, should therefore be limited to the research and development of advanced power sources including the nickel-metal hydride and lithium ion rechargeable batteries as well as the fuel cell, a technology that offered theoretical advantages over the conventional galvanic storage battery in the electric vehicle application including longer range. Some companies including Honda and Toyota also looked to hybrid battery electric propulsion, a system designed to solve the durability dilemma by narrowing if not eradicating the temporal mismatch of battery and motor.[67]

Through better battery discourse, automakers convinced CARB to allow them to interpret the mandate as an experiment.[68] Car companies fielded handfuls of all-battery electrics in exchange for a promise to develop the fuel cell electric. Around the turn of the millennium, automakers began phasing out their all-battery electric fleets and by the early 2000s, the mandate was in limbo.[69] Industry resistance to the mandate, motivated in good measure by the question of who would bear battery replacement costs, kept battery costs high, fostering a circular argument that demonstrated to regulators the intractability of the durability dilemma and the economic infeasibility of the all-battery electric format.

9.7 Fuel Cell and Hybrid-to-Grid

The technopolitics of green automobility had a decisive practical effect on the vehicle-to-grid concept in this period. Kempton and Letendre asserted that from the perspective of the electric utility, the most desirable electric vehicle battery was the lead-acid rechargeable because it offered the cheapest energy storage.[70] By the

---

[67] See Hideshi Itazaki's semi-official account of the development of Toyota's Prius; Itazaki, *The Prius that Shook the World: How Toyota Developed the World's First Mass-Production Hybrid Vehicle* (Tokyo: Nikkan Kogyo Shimbun, 1999).

[68] Under industry pressure, CARB amended the mandate in late 1995 and early 1996. California eliminated the percentage requirements from 1998 to 2001 and worked out memoranda of agreement with each of the seven largest automakers, committing them to produce a total of 3750 vehicles over calendar years 1998, 1999, and 2000. In addition, regulators gave multiple credits for using advanced power sources, defined as nickel-metal hydride and sodium-nickel chloride batteries. When this was taken into account, the total legal commitment was reduced to just over 1800 vehicles. In exchange, the automakers would have to meet stricter emission standards and agree to continue research and development of zero-emission vehicles. Only the ten percent quota for 2003 remained unchanged; see Brad Heavner, *Pollution Politics 2000: California Political Expenditures of the Automobile and Oil Industries, 1997–2000* (Santa Barbara, CA: California Public Interest Research Group Charitable Trust, 2000), 7–8; Deborah Salon, Daniel Sperling, and David Friedman, *California's Partial ZEV Credits and LEV II Program: UCTC No. 470* (Berkeley, CA: University of California Transportation Center, 2001), 2; California Environmental Protection Agency, *2000 Emission Vehicle Program Biennial Review*, August 7, 2000, 3.

[69] See Matthew N. Eisler, "Public Policy, Industrial Innovation, and the Zero-Emission Vehicle," *Business History Review* 94 (Winter 2020): 779–802.

[70] Letendre and Kempton, "The V2G Concept," 21.

early 2000s, however, car companies were phasing out this power source in electric cars and limiting their participation in research on advanced alternatives as a consequence of their termination of the all-battery electric format.[71]

This left the hybrid and fuel cell vehicles as the auto industry's preferred electric propulsion formats. Kempton, Letendre, and other environmental analysts assumed that all of these formats could be integrated into the grid but the non-all-battery electric systems posed special problems.[72] The fuel cell electric car cannot easily function as a bidirectional power plant because the fuel cell does not store electricity but rather converts it from hydrogenous liquid or gaseous fuels.[73] When such a car fed electrical power into the grid, it consumed fuel, and in the early 2000s it was not clear how the fuel would be replenished. In principle, fuel cells are capable of operating on carbonaceous fuels and automaker had hoped to develop systems capable of operating directly on gasoline. In practice, direct use of carbonaceous fuel in most types of fuel cell cause electrochemical side reactions that degrade the electrolyte and catalyst. By the turn of the millennium, automakers and policymakers were abandoning carbonaceous fuel cell systems in favor of fuel cells using pure hydrogen, a material with attractive operating qualities at the vehicle level but that had significant external costs and complications at the systems level. Planners hoped to resolve these problems with the reformer, a device that converted carbonaceous fuels into hydrogen.[74] Kempton and Letendre envisioned that owners of fuel cell electric vehicles participating in vehicle-to-grid would have access to their own garage-based reformers converting piped utility natural gas into hydrogen, an arrangement that assumed integrating the regulatory and telemetrical systems of the electricity grid with the gas grid.[75]

Such a system was cumbersome but theoretically feasible for the peaking and spinning reserve markets. However, configuring fuel cell electric propulsion for the frequency regulation market would have required an all but impracticable Rube Goldberg-like solution. A reformer-equipped fuel cell electric system could only feed power into the grid, so it could only regulate up. In order to regulate down, the

[71] Heavner, *Pollution Politics*.

[72] Kempton and Letendre, "Electric Vehicles," and Willett Kempton and Toru Kubo, "Electric-Drive Vehicles for Peak Power in Japan," *Energy Policy* 28, no. 1 (2000): 9–18.

[73] For an excellent synoptic history of fuel cells, see Harold D. Wallace, "Fuel Cells: A Challenging History," *Substantia* 3, no. 2 (2019): 83–97.

[74] A 1967 U.S. Army Mobility Command analysis ranked reforming as the most difficult of all the various fuel cell configurations; see James R. Huff and John C. Orth, "The USAMECOM-MERDC Fuel Cell Electric Power Generation Program," in *Fuel Cell Systems II: fifth Biennial Fuel Cell Symposium sponsored by the Division of Fuel Chemistry at the 154th Meeting of the American Chemical Society, Chicago, Illinois, September 12–14, 1967* (Washington, DC: American Chemical Society, 1969), 318.

[75] Letendre and Kempton, "The V2G Concept," 21.

system needed still other auxiliary systems to convert electricity to hydrogen or natural gas.[76]

These questions were largely hypothetical because automakers were no more interested in commercializing the fuel cell electric car than the all-battery electric car, despite much hype. In the mid-2000s, the only type of electric car at commercial scale was the hybrid battery electric and the hybrids of the day could not link to the grid. They were "conventional" hybrids, so-called because their energy conversion units were self-contained and lacked plug-in capability. For proponents of vehicle-to-grid at this time, the last hope for a power plant on wheels lay in the plug-in hybrid, a format increasingly favored by hobbyists and policymakers over the course of the decade.[77]

9.8 Conclusion

The proximate problem for which vehicle-to-grid was conjured as the solution was resolved not by a technological silver bullet but by conventional technopolitical means. In 2001, the state of California intervened robustly to end the electricity crisis, enforcing conservation, purchasing wholesale electricity on behalf of private utilities, raising rates to cover the higher cost of power, and issuing bonds to cover

[76] Letendre and Kempton did note that fuel cell propulsion could not provide regulation down but did not explain why or reconcile this observation with their earlier claims for the utility of the technology in vehicle-to-grid; see "The V2G Concept," 24. Many enthusiasts of all-battery electric vehicles tended to regard fuel cell propulsion as a costly and cynical chimera devised by the car companies to avoid their mandate responsibilities. In his vehicle-to-grid report for CARB, Brooks agreed with the group that fuel cell electric had potential as a grid management tool. In comments submitted to a CARB hearing in March 2003, however, Brooks questioned many of the assumed advantages of fuel cell propulsion over all-battery electric propulsion and criticized the agency for supporting the technology; see Brooks, "Final Report;" and State of California Environmental Protection Agency Air Resources Board, "Supplement to the Final Statement of Reasons for Rulemaking, Including Summary of Comments and Agency Responses," March 27–28 and April 24, 2003, Agenda Item No. 03–02-4, February 20, 2004.

[77] In December 2007, Congress passed the Energy Independence and Security Act, authorizing the Department of Energy to disburse $360 million in grants for cost-shared projects supporting plug-in hybrids from 2008 through 2012. This initiative was part of a much larger initiative called the Advanced Technology Vehicles Manufacturing Incentive Program. This authorized $25 billion in loans to American automakers and their suppliers towards developing and producing cars that met the federal Bin 5 Tier II emission standard and that were 25 percent more fuel efficient than similar vehicles from the 2005 model year; see Public Law 110–140, Energy Independence and Security Act of 2007, Subtitle B-Improved Vehicle Technology, 1510, 1515; see also Sherry Boschert, *Plug-in Hybrids: The Cars that Will Recharge America* (Gabriola Island: British Columbia: New Society Publishers, 2006).

those costs. In effect, the state re-regulated electricity and subsidized "free market" electricity.[78]

Nevertheless, the imaginary of the electric car as power plant on wheels did not die. It persisted and gained impetus with the reversal of fortune of the electric car from the late 2000s, an event due in no small part to ACP, whose prototype propulsion technology informed development of the Roadster, the car that launched Tesla Motors. Tesla was the first automaker to attempt to realize the electric supercar imaginary that the established car companies had invoked as an excuse to evade the mandate, helping set a trend in the manufacture of costly and complex plug-in electrics equipped with ever-larger and more capable and costly batteries. Over time, the bidirectional electric car became the subject of studies, research and development, and start-up enterprises in the United States and Europe.[79] This work revealed that integrating distributed generation and storage technologies into legacy grid systems posed a host of sociotechnical problems.[80] In principle, many of these problems could be ameliorated with new technology but devising appropriate market models remained a challenge. With the resolution of California's electricity crisis, recalled Brooks, the dramatically inflated prices that had made ancillary

[78] Alexander Ritschel and Greg P. Smestad, "Energy Subsidies in California's Electricity Market Deregulation," *Energy Policy* 31 (2003): 1379–91; Ghazal Razeghi, Brendan Shaffer, and Scott Samuelsen, "Impact of Electricity Deregulation in the State of California," *Energy Policy* 103 (2017): 105–15.

[79] For a review of vehicle-to-grid research and development projects, see Adrene Briones, James Francfort, Paul Heitmann, Michael Schey, Steven Schey, and John Smart, *Vehicle-to-Grid Power Flow Regulations and Building Codes Review by the AVTA* (Idaho National Laboratory, U.S. Department of Energy, 2012), 40–5. See also Jonathan Coignard, Samveg Saxena, Jeffery Greenblatt, and Dai Wang, "Clean Vehicles as an Enabler for a Clean Electricity Grid," *Environmental Research Letters* 13 (2018): 054031; Willett Kempton, Keith Decker, and Li Liao, "Vehicle to Grid Demonstration Project DE-FC26-08NT01905: Final Report," May 7, 2011; see also Fermata Energy, "Our Story," accessed October 1, 2020, https://www.fermataenergy.com/our-story; and Enel X, accessed October 1, 2020, https://www.enelx.com/uk/en.

[80] A key problem of distributed generation systems is the phenomenon of "islanding," wherein distributed generation continues to function following grid crash, upsetting the balance between generation, load, voltage, and frequency and creating safety hazards for utility personnel; see Zhang Kai, Liu Kexue, Yao Naipeng, Jia Yuhong, Li Wenjun, and Qin Lihan, "The Impact of Distributed Generation and Its Parallel Operation on Distribution Power Grid," *5th International Conference on Electric Utility Deregulation and Restructuring and Power Technologies*, November 26–29, 2015, Changsha, China, 2041–45; Kari Mäki, Anna Kulmala, Sami Repo, and Pertti Järventausta, "Problems Related to Islanding Protection of Distributed Generation in Distribution Network," *2007 IEEE Lausanne Power Tech*, 467–72. David Hawkins, a CAISO engineer who advised Brooks during the ACP experiment, participated in another experiment in the early 2000s involving the simulated use of electric vehicle storage batteries in the regulation application, revealing an adverse reaction by grid control computers responsible for automatically signaling generators to power up or down. The computers were programmed to control large thermal and hydro plants that gradually responded to signals; when the computers interfaced with storage batteries, they took advantage of the propensity of batteries to respond nearly instantaneously to commands to draw or supply power, leaving the devices completely charged or discharged in a matter of minutes with no spare capacity to regulate the grid; Hawkins, interview.

services an attractive market disappeared.[81] Emerging suppliers of the equipment and services of bidirectional electric vehicle power like Fermata Energy experimented with and ruled out the regulation market in the mid-2010s on grounds that it was small and unscalable.[82]

Where ideas of the electric vehicle as a tool of electricity management were concerned, there was a return to unidirectional demand and storage roles, now with a view to solving problems of intermittent renewable energy.[83] The trend in research on the bidirectional electric vehicle power plant shifted to backup and peaking power for homes and commercial buildings in local or microgrid environments, colloquially expressed as vehicle-to-building (V2B) or, more ambitiously, vehicle-to-everything (V2X). Nissan did much to pioneer these applications. In the wake of the Fukushima nuclear plant disaster in 2011, the automaker developed and marketed an external charger that enabled its Leaf electric car to serve as a source of emergency power for Japanese homes. In 2018, Nissan entered a research and development partnership with Fermata with a view to serving institutional customers capable of fielding vehicle fleets.[84]

This initiative resembled the approach adopted by localized utilities in the early twentieth century. While interest in and demonstrations of bidirectional electric vehicle power seem likely to increase with the scaling of the electric fleet, the prospect of average users becoming power entrepreneurs appears more distant. The notional everyday participant in vehicle-to-grid would face the complex task of attempting to arbitrage the difference between wholesale and retail prices while still paying the retail rate.[85] Advocates argue that advances in software and charging technology in concert with market models based on public subsidies could solve this riddle and enable owners of such vehicles to pay for their batteries.[86] There were grounds to believe that such technology might be more robustly promoted by public policy. After all, the electric vehicle revival that co-produced the vehicle-to-grid imaginary had been stimulated largely by the national developmental state. In 2001, California launched a program to incentivize stationary storage technology in response to the electricity crisis.[87] Indeed, virtually every sector of the contemporary

---

[81] Alec Brooks, interview by the author, August 18, 2020.

[82] Fermata Energy founder David Slutzky, interview by the author, November 12, 2020.

[83] In 2020, much renewable capacity was curtailed or shut down for lack of demand, in turn creating a justification for "demand dispatch," technology that would enable electrics to recharge with renewable energy as it became available; Brooks, interview by the author, August 18, 2020.

[84] Slutzky, interview; and Nissan Motor Corporation, "Nissan Leaf helps to Power Company's North American Facilities with New Charging Technology," November 27, 2018, accessed January 2, 2021, https://usa.nissannews.com/en-US/releases/release-cd8a64d603d241f7b319fc6c1c33397f-nissan-leaf-helps-to-power-company-s-north-american-facilities-with-new-charging-technology.

[85] Hawkins, interview.

[86] Slutzky, interview.

[87] For a review of these subsidies, see California Public Utilities Commission, "Self-Generation Incentive Program," accessed January 2, 2021, https://www.cpuc.ca.gov/sgip/.

U.S. economy depends on federal largesse in one way or another, increasingly so since the Great Recession of the late 2000s and the recession precipitated by the coronavirus pandemic in 2020.

In the era of the quasi-planned neoliberal economy, there is no ideological obstacle to the concept of subsidized commercial bidirectional electric vehicle power. Markets are social constructions, not natural forces, and neoliberal planners demonstrated that they can comfortably accommodate the paradox of regulated deregulation and support technology they believe will accomplish marketization. Still, market modelers of the various applications of the bidirectional electric will probably have to engage the science of battery aging at some point. Owners of electric vehicles would be no further ahead, after all, if they substantially degraded their batteries in the process of paying them off. Understanding how batteries age in electric vehicles would seem to be a prerequisite for meaningfully engaging the sociotechnical problems of repurposing electric vehicle rechargeables for stationary applications.

Historically, however, the political economy of advanced power source research and development disincentivized the knowledge field of battery durability. One of the signal ironies of the electric vehicle revolution is that few of the basic rechargeable battery chemistries originated in research expressly devoted to electric vehicles, let alone stationary generation. Research in nickel-metal hydride and lithium ion rechargeables traced to the energy crisis of the 1970s but these chemistries were first applied in consumer electronics and in that application, technologists privileged energy and power, not durability and cost. Other technologists adapted these chemistries for use in electric vehicles and policymakers subsidized their manufacture in order to help scale production to cost parity with gasoline power. In 2006, M. Stanley Whittingham, one of the pioneers of the lithium rechargeable, suggested that the focus in electric vehicle battery research was likely to shift to durability.[88] Nevertheless, the dominant battery formulas in the electric vehicles of the 2010s were still informed by the bias for energy and power.[89]

The history of the car-grid relationship outlines the limits of the compatibility of electricity and automobility as sociotechnical systems. Where electric utilities periodically experimented with electric vehicles as tools of grid management, most automakers had no clear interest in the bidirectional electric car, even manufacturers who nominally supported the technology.[90] After a century of enculturation of

[88]Kevin Bullis, "Making Electric Vehicles Practical," *MIT Technology Review*, November 29, 2006, https://www.technologyreview.com/2006/11/29/227392/making-ele.

[89]Battery University, "BU-205: Types of Lithium-Ion," accessed January 16, 2021, https://batteryuniversity.com/learn/article/types_of_lithium_ion.

[90]While Tesla co-founder Martin Eberhard supported vehicle-to-grid, his successor Elon Musk was ambivalent about the concept. While Musk supported research on bidirectional automobile power, he expressed skepticism about the use of electric cars as stationary power plants owing to the damage this application caused to the vehicle battery; see Michael T. Burr, "Redefining the Electric Car," *Public Utilities Fortnightly* 144, no. 12 (2006): 34, 36; Fred Lambert, "Tesla Quietly Adds Bidirectional Charging Capability for Game-Changing New Features," *Elektrek*, May 19, 2020,

personal automobility, it was not clear that ordinary motorists did, either. In contemporary society, vehicle-to-grid and vehicle-to-building implied behaviors and economic dynamics contrary to automobile norms. As STS reminds us, however, users are adaptable and sociotechnical conditions are dynamic. There is a precedent in user adaptation of the passenger car as a stationary power plant in the early days of gasoline-fueled ICE automobility in unelectrified rural areas.[91]

The co-production of general electrification alongside general automobilization from the 1920s largely ended this practice. Yet the environmental and economic crises of these systems at the turn of the millennium revived the idea of the power plant on wheels, highlighting the vast disparity in resources invested in mobile versus stationary power generation over the course of a century. For Kempton and Letendre, the salient fact was that the U.S. light duty fleet had around 16 times the capacity of U.S. stationary plant.[92] Their construction of the similarity judgment of the automobile as an underutilized power resource set aside the respective sociotechnical affordances of gasoline ICE and electric propulsion and of automobility and electricity as energy conversion regimes. In effect, the vehicle-to-grid imaginary served as an implicit social critique, much like the techno-utopian energy conversion imaginaries and classical utopias that preceded it. It has problematized basic assumptions of the American lifestyle, including expectations for personal vehicle ownership and reliable grid electricity.

Indeed, vehicle-to-grid is emblematic of the conundrums of the broader energy conversion infrastructure in the era of climate change. Neoliberal public policy's contradictory imperatives of efficiency and growth conceive consumer convenience, corporate profit, and environmental and economic sustainability as complementary, not mutually exclusive objectives. In coming decades, efforts to integrate distributed storage systems with distributed intermittent renewable energy conversion systems will intensify. As this happens, the growth of the electric vehicle fleet and the trend towards more capacious and capable vehicle batteries will likely increase the temptation to try to unlock the utility potential of the personal automobile as an efficient solution to public policy goals.

---

https://electrek.co/2020/05/19/tesla-bidirectional-charging-ready-game-changing-features/; Robert Walton, "Tesla Unveils New EV Battery Design, but Musk Downplays Vehicle-to-Grid Application," *Utility Dive*, September 23, 2020, https://www.utilitydive.com/news/tesla-unveils-new-ev-battery-design-but-musk-downplays-vehicle-to-grid-app/585723/.

[91] Ronald Kline and Trevor Pinch, "Users as Agents of Technological Change: The Social Construction of the Automobile in the Rural United States," *Technology and Culture* 37, no. 4 (1996): 763–95.

[92] Kempton and Letendre, "Electric Vehicles," 159.

Bibliography

Anderson, Dennis, and Matthew Leach. 2004. Harvesting and Redistributing Renewable Energy: On the Role of Gas and Electricity Grids to Overcome Intermittency through the Generation and Storage of Hydrogen. *Energy Policy* 32 (14): 1603–1614.

Baker, J.N., and A. Collinson. 1999. Electrical Energy Storage at the Turn of the Millennium. *Power Engineering Journal* 13 (3): 107–112.

Bell, Daniel. 1999. *The Coming of Post-Industrial Society: A Venture in Social Forecasting*. New York: Basic Books.

Bensaude-Vincent, Bernadette. 2001. The Construction of a Discipline: Materials Science in the United States. *Historical Studies in the Physical and Biological Sciences* 31 (2): 223–248.

Block, Fred. 2008. Swimming Against the Current: The Rise of a Hidden Developmental State in the United States. *Politics and Society* 36 (2): 169–206.

Boschert, Sherry. 2006. *Plug-in Hybrids: The Cars That Will Recharge America*. Gabriola Island/ British Columbia: New Society Publishers.

Briones, Adrene, James Francfort, Paul Heitmann, Michael Schey, Steven Schey, and John Smart. 2012. *Vehicle-to-Grid Power Flow Regulations and Building Codes Review by the AVTA*. Idaho National Laboratory, U.S. Department of Energy.

Brodd, Ralph J. 2005. *Factors Affecting US Production Decisions: Why are There no Volume Lithium-Ion Battery Manufacturers in the United States?* Gaithersburg, MD: National Institute of Standards and Technology.

Brooks, Alec N. 2002. *Final Report: Vehicle-to-Grid Demonstration Project: Grid Regulation Ancillary Service with a Battery Electric Vehicle*. Contract number 01-313, October 21, 2002.

Burkhart, Lori A. 1994. Does PUHCA Inhibit Diversification? *Public Utilities Fortnightly* 132 (16): 33.

California Environmental Protection Agency. 2000. *Emission Vehicle Program Biennial Review*.

Clarke, Deborah. 2007a. *Driving Women: Fiction and Automobile Culture in Twentieth-Century America*. Baltimore: Johns Hopkins University Press.

Clarke, Sally H. 2007b. *Trust and Power: Consumers, the Modern Corporation, and the Making of the United States Automobile Market*. Cambridge: Cambridge University Press.

Cohn, Julie. 2016. Bias in Electric Power Systems. A Technological Fine Point at the Intersection of Commodity and Service. In *Electric Worlds/Mondes Électriques: Creations, Circulations, Tensions, Transitions (19th–21th C.)*, ed. Alain Beltrain, Léonard Laborie, Pierre Lanthier, and Stéphanie Le Gallic, 271–293. Brussels: Peter Lang.

Coignard, Jonathan, Samveg Saxena, Jeffery Greenblatt, and Dai Wang. 2018. Clean Vehicles as an Enabler for a Clean Electricity Grid. *Environmental Research Letters* 13: 054031.

Collantes, Gustavo, and Daniel Sperling. 2008. The Origin of California's Zero Emission Vehicle Mandate. *Transportation Research Part A* 42: 1302–1313.

Dahn, J.R., J.C. Burns, and D.A. Stevens. 2016. Importance of Coulombic Efficiency Measurements in R&D Efforts to Obtain Long-Lived Li-ion Batteries. *Interface* 25 (3): 75–78.

Dedrick, Jason, and Kenneth L. Kraemer. 2008. Impact of Globalization and Offshoring on Engineering Employment in the Personal Computing Industry. In *The Offshoring of Engineering: Facts, Unknowns, and Potential Implications*. Washington, DC: The National Academies Press.

Eisler, Matthew N. 2012. *Overpotential: Fuel Cells, Futurism, and the Making of a Power Panacea*. New Brunswick, NJ: Rutgers University Press.

———. 2017. Exploding the Black Box: Personal Computing, the Notebook Battery Crisis, and Postindustrial Systems Thinking. *Technology and Culture* 58 (2): 368–391.

———. 2020. Public Policy, Industrial Innovation, and the Zero-Emission Vehicle. *Business History Review* 94 (Winter): 779–802.

———. 2022. *Age of Auto Electric: Environment, Energy, and the Quest for the Sustainable Car*. Cambridge MA: MIT Press.

Fialka, John J. 2015. *Car Wars: The Rise, the Fall, and the Resurgence of the Electric Car.* New York: Thomas Dunne Books.
Fletcher, Seth. 2011. *Bottled Lightning: Superbatteries, Electric Cars, and the New Lithium Economy.* New York: Hill and Wang.
Gartman, David. 1994. *Auto Opium: A Social History of American Automobile Design.* London: Routledge.
Gultom, Yohanna M.L. 2019. Governance Structures and Efficiency in the U.S. Electricity Sector After the Market Restructuring and Deregulation. *Energy Policy* 129: 1008–1019.
Hart, David M. 2001. Antitrust and Technological Innovation in the U.S.: Ideas, Institutions, Decisions, and Impacts, 1890–2000. *Research Policy* 30 (6): 923–936.
Hasenöhrl, Ute, and Jan-Henrik Meyer. 2020. The Energy Challenge in Historical Perspective. *Technology and Culture* 61 (1): 295–306.
Hayes, Mark. 2006. On the Efficiency of Fair Trade. *Review of Social Economy* 64 (4): 447–468.
Heavner, Brad. 2000. *Pollution Politics 2000: California Political Expenditures of the Automobile and Oil Industries, 1997–2000.* Santa Barbara, CA: California Public Interest Research Group Charitable Trust.
Hoddeson, Lillian, and Peter Garrett. 2018. *The Man Who Saw Tomorrow: The Life and Inventions of Stanford R. Ovshinsky.* Cambridge, MA: MIT Press.
Hughes, Thomas P. 1983. *Networks of Power: Electrification in Western Society, 1880–1930.* Baltimore: Johns Hopkins University Press.
———. 1989. *American Genesis: A Century of Invention and Technological Enthusiasm, 1870–1970.* Chicago: University of Chicago Press.
Hoffmann, Peter. 1981. *The Forever Fuel: The Story of Hydrogen.* Boulder, CO: Westview Press.
———. 2001. *Tomorrow's Energy: Hydrogen, Fuel Cells, and the Prospects for a Cleaner Planet.* Cambridge, MA: MIT Press.
Huff, James R., and John C. Orth. 1969. The USAMECOM-MERDC fuel cell electric power generation program. In *Fuel Cell Systems II: 5th Biennial Fuel Cell Symposium sponsored by the Division of Fuel Chemistry at the 154th Meeting of the American Chemical Society, Chicago, Illinois, September 12–14, 1967.* Washington, DC: American Chemical Society.
Insull, Samuel. 1915. *Central-Station Electric Service: Its Commercial Development and Economic Significance as Set Forth in the Public Addresses (1897–1914) of Samuel Insull.* Chicago: Privately Printed.
Itazaki, Hideshi. 1999. *The Prius that Shook the World: How Toyota Developed the World's First Mass-Production Hybrid Vehicle.* Tokyo: Nikkan Kogyo Shimbun.
Jasanoff, Sheila. 2015. Future Imperfect: Science, Technology, and Imaginations of Modernity. In *Dreamscapes of Modernity: Sociotechnical Imaginaries and the Fabrication of Power*, ed. Sheila Jasanoff and Sang-Hyun Kim, 1–33. Chicago: University of Chicago Press.
Johnson, Ann. 2004. The End of Pure Science: Science Policy from Bayh-Dole to the NNI. In *Discovering the Nanoscale*, ed. Davis Baird and Alfred Nordmann, 219–222. Amsterdam: IOS Press.
Jones, Christopher F. 2014. *Routes of Power: Energy and Modern America.* Harvard University Press.
Joskow, Paul L. 2006. Markets for Power in the United States. *The Energy Journal* 27 (1): 1–36.
Justi, Eduard. 1965. *Leitungsmechanismus und energieumwandlung in festkörpern.* Göttingen: Vandenhoeck and Ruprecht.
Kahn, Alfred E. 1988. Surprises of Airline Deregulation. *The American Economic Review* 78 (2): 316–322.
———. 2001. *Whom the Gods Would Destroy, or How not to Deregulate.* Washington, D.C.: The AEI Press.
Kai, Zhang, Liu Kexue, Yao Naipeng, Jia Yuhong, Li Wenjun, and Qin Lihan. 2015. The Impact of Distributed Generation and Its Parallel Operation on Distribution Power Grid. *5th International Conference on Electric Utility Deregulation and Restructuring and Power Technologies, November 26–29, 2015, Changsha, China*, 2041–2045.

Kempton, Willett, Keith Decker, and Li Liao. 2011. Vehicle to Grid Demonstration Project DE-FC26-08NT01905: Final Report. May 7, 2011.
Kempton, Willett, and Steven Letendre. 1997. Electric Vehicles as a New Power Source for Electric Utilities. *Transportation Research Part D: Transport and Environment* 2 (3): 157–175.
Kempton, Willett, and Toru Kubo. 2000. Electric-Drive Vehicles for Peak Power in Japan. *Energy Policy* 28 (1): 9–18.
Kempton, Willett, Jasna Tomić, Steven Letendre, Alec Brooks, and Timothy Lipman. 2001. *Vehicle-to-Grid Power: Battery, Hybrid, and Fuel Cell Vehicles as Resources for Distributed Electric Power in California*. UC Davis Institute for Transportation Studies, ECD-ITS-RR-01-03, June 2001.
Kirsch, David, and Gijs Mom. 2002. Visions of Transportation: The EVC and the Transition from Service- to Product-Based Mobility. *Business History Review* 76 (1): 75–110.
Laird, Frank N. 2003. Constructing the Future: Advocating Energy Technologies in the Cold War. *Technology and Culture* 44 (1): 27–49.
Lave, Rebecca, Philip Mirowski, and Samuel Randalls. 2010. Introduction: STS and Neoliberal Science. *Social Studies of Science* 40 (5): 659–675.
Lécuyer, Christophe, and David C. Brock. 2006. The Materiality of Microelectronics. *History and Technology* 22 (3): 301–325.
Letendre, Steven, and Willett Kempton. 2002. The V2G Concept: A new Model for Power? *Public Utilities Fortnightly* 140 (4): 16–26.
Letendre, Steven, Paul Denholm, and Peter Lilienthal. 2006. New Load, or New Resource? *Public Utilities Fortnightly* 144 (12): 28–33.
Lifset, Robert. 2014. *Power on the Hudson: Storm King Mountain and the Emergence of Modern American Environmentalism*. Pittsburgh: Pittsburgh University Press.
Mäki, Kari, Anna Kulmala, Sami Repo, and Pertti Järventausta. 2007. Problems Related to Islanding Protection of Distributed Generation in Distribution Network. In *IEEE Lausanne Power Tech*, 4672. IEEE.
McCarthy, Tom. 2007. *Auto Mania: Cars, Consumers, and the Environment*. New Haven, CT: Yale University Press.
McCraw, Thomas K. 1984. *Prophets of Regulation: Charles Francis Adams, Louis D. Brandeis, James M. Landis, Alfred E. Kahn*. Cambridge, MA: Belknap Press of Harvard University Press.
Mody, Cyrus C.M. 2016. *The Long Arm of Moore's law: Microelectronics and American Science*. Cambridge, MA: MIT Press.
Moeller, James W. 1994. The Revenge of Wendell Willkie. *Public Utilities Fortnightly* 134 (2): 14.
Mom, Gijs. 2004. *The Electric Vehicle: Technology and Expectations in the Automobile Age*. Baltimore: Johns Hopkins University Press.
Mowery, David C. 1998. Collaborative R&D: How Effective Is It? *Issues in Science and Technology* 15 (1): 37–44.
Mirowski, Philip. 2018. The Future(s) of Open Science. *Social Studies of Science* 48 (2): 171–203.
———. 1989. *More Heat Than Light: Economics as Social Physics, Physics as Nature's Economics*. Cambridge: Cambridge University Press.
Nader, Ralph. 1965. *Unsafe at Any Speed: The Designed-in Dangers of the American Automobile*. New York: Grossman Publishers.
Norton, Peter D. 2008. *Fighting Traffic: The Dawn of the Motor Age in the American City*. Cambridge, MA: MIT Press.
Onis, Ziya. 1991. The Logic of the Developmental State. *Comparative Politics* 24 (1): 109–126.
Paterson, Matthew. 2007. *Automobile Politics: Ecology and Cultural Political Economy*. Cambridge, UK: Cambridge University Press.
Patil, Pandit G. 1990. Prospects for electric vehicles. *IEEE AES Systems Magazine*. December: 15–9.

Pickard, William F. 2012. The history, present state, and future prospects of underground pumped hydro for massive energy storage. *Proceedings of the IEEE* 100 (2): 473–483.
Pinch, Trevor. 1993. Testing-One, Two, Three. . .Testing! Toward a Sociology of Testing. *Science, Technology and Human Values* 18 (1): 27–31.
Plaiss, Adam. 2016. From Natural Monopoly to Public Utility: Technological Determinism and the Political Economy of Infrastructure in Progressive-Era America. *Technology and Culture* 57 (4): 806–830.
Pomeranz, Kenneth. 2000. *The Great Divergence: China, Europe, and the Making of the Modern World Economy*. Princeton, NJ: Princeton University Press.
Razeghi, Ghazal, Brendan Shaffer, and Scott Samuelsen. 2017. Impact of Electricity Deregulation in the State of California. *Energy Policy* 103: 105–115.
Richardson, Martin, and Frank Stähler. 2014. Fair Trade. *Economic Record* 291: 447–461.
Rifkin, Jeremy. 2002. *The Hydrogen Economy: The Creation of the World-Wide Energy Web and the Redistribution of Power on Earth*. New York: Jeremy P. Tarcher/Penguin.
Ritschel, Alexander, and Greg P. Smestad. 2003. Energy Subsidies in California's Electricity Market Deregulation. *Energy Policy* 31: 1379–1391.
Ross, Benjamin. 2001. California's Regulation Debacle. *Dissent* (Spring): 45–47.
Russell, Edmund, James Allison, Thomas Finger, John K. Brown, Brian Balogh, and W. Bernard Carlson. 2011. The Nature of Power: Synthesizing the History of Technology and Environmental History. *Technology and Culture* 52 (2): 246–259.
Salon, Deborah, Daniel Sperling, and David Friedman. 2001. *California's Partial ZEV Credits and LEV II Program: UCTC No. 470*. Berkeley, CA: University of California Transportation Center.
Schallenberg, Richard H. 1982. *Bottled Energy: Electrical Engineering and the Evolution of Chemical Energy Storage*. Philadelphia: American Philosophical Society.
Scharff, Virginia. 1991. *Taking the Wheel: Women and the Coming of the Motor Age*. Albuquerque, NM: University of New Mexico Press.
Sheller, Mimi, and John Urry. 2000. The City and the Car. *International Journal of Urban and Regional Research* 24 (4): 737–757.
Shnayerson, Michael. 1996. *The Car That Could: The Inside Story of GM's Revolutionary Electric Vehicle*. New York: Random House.
Smil, Vaclav. 2008. *Energy in Nature and Society: General Energetics of Complex Systems*. Cambridge, MA: MIT Press.
———. 2017. *Energy and Civilization: A History*. Cambridge, MA: MIT Press.
Sweet, William, and Elizabeth A. Bretz. 2002. How to Make Deregulation Work. *IEEE Spectrum* 39 (1): 50–56.
Tomory, Leslie. 2011. Building the First Gas Network, 1812–1820. *Technology and Culture* 52 (1): 75–102.
Urry, John. 2004. The 'System' of Automobility. *Theory, Culture and Society* 21 (4–5): 25–39.
U.S. Congress. 2001. *Causes and Lessons of the California Electricity Crisis*. Congress of the United States Congressional Budget Office, September 2001.
———. 2007. *Public Law 110–140, Energy Independence and Security Act of 2007*, Subtitle B-Improved Vehicle Technology.
U.S. Energy Information Agency. 2020. *Monthly Energy Review*.
Wallace, Harold D. 2019. Fuel Cells: A Challenging History. *Substantia* 3 (2): 83–97.
Williams, Laurence O. 1980. *Hydrogen Power: An Introduction to Hydrogen Energy and its Applications*. London: Pergamon Press.
Winner, Langdon. 1980. Do Artifacts Have Politics? *Daedalus* 109 (1): 121–136.
Wise, David W. 1989. The Tides of Deregulation. *Public Utilities Fortnightly* 125 (5): 39–40.
Woo-Cumings, Meredith, ed. 1999. *The Developmental State*. Ithaca, NY: Cornell University Press.

Matthew N. Eisler is a lecturer of history in the Department of Humanities at the University of Strathclyde, where he researches the relationship between energy and environmental policy and science and engineering practice. Dr. Eisler's latest book is *Age of Auto Electric: Environment, Energy, and the Quest for the Sustainable Car* (MIT Press, 2022).

The manufacturer's authorised representative in the EU is Springer Nature Customer Service Centre GmbH, Europaplatz 3, 69115 Heidelberg, Germany. If you have any concerns regarding our products, please contact ProductSafety@springernature.com

Printed and bound by CPI Group (UK) Ltd, Croydon, CR0 4YY
01/10/2026
02225896-0005